Remote Sensing
of Forest
Resources

REMOTE SENSING APPLICATIONS

Series Editors

E. C. Barrett
Director, Remote Sensing Unit
University of Bristol

the late L. F. Curtis, OBE
Honorary Senior Research Fellow,
University of Bristol

In the same series:

Remote Sensing of Forest Resources

Theory and application

JOHN A. HOWARD

BSc (Wales), MF (Minnesota), PhD (Melbourne),
Dip. For. (Bangor), FICF, FRGS, FLS

CHAPMAN & HALL

London · New York · Tokyo · Melbourne · Madras

Published by Chapman & Hall, 2–6 Boundary Row, London SE1 8HN

Chapman & Hall, 2–6 Boundary Row, London SE1 8HN, UK

Van Nostrand Reinhold Inc., 115 5th Avenue, New York, NY10003, USA

Chapman & Hall Japan, Thomson Publishing Japan, Hirakawacho Nemoto Building, 7F, 1-7-11 Hirakawa-cho, Chiyoda-ku, Tokyo 102, Japan

Chapman & Hall Australia, Thomas Nelson Australia, 102 Dodds Street, South Melbourne, Victoria 3205, Australia

Chapman & Hall India, R. Seshadri, 32 Second Main Road, CIT East, Madras 600 035, India

First edition 1991

© 1991 John A. Howard

Typeset in 10/12pt Plantin by EJS Chemical Composition, Bath, Avon
Printed in Great Britain at the University Press, Cambridge

ISBN 0 412 29930 5   0 442 31347 0 (USA)

A catalogue record for this book is available from the British Library

Library of Congress Cataloging-in-Publication Data available

To Beryl, Shân and Robert

Contents

Preface

When preparing this book, a major problem was not of finding sufficient material for inclusion, but what could be omitted, if a comprehensive text was to be compiled within reasonable limits. Therefore, to comply with the concern of the series editors on the length of the text, several topics have been excluded (e.g. non-imaging sensors, oblique aerial photographic, sensing for urban forestry, sensing for rangeland management). Also at an early stage it was concluded that the presentation of a comprehensive literature review was not feasible, and that the writer would have to make personal decisions on what should be included, while realizing that this would not satisfy all readers and that publications of considerable interest might have to be excluded.

Initially, it was contemplated that more would be written on land use, digital image analysis and photogrammetry. This would, however, have resulted in an imbalance and reduced the content of other topics, essential to the practising forester, such as vertical aerial photography and visual image analysis, or what is needed as a background in decision-making and strategic planning. However, as the book is one of a series being published by Chapman & Hall, certain aspects could be curtailed (e.g. hydrology, atmospheric factors), but overlap remains on technology (e.g. radar), since it cannot be assumed that most foresters will have access to the other texts.

The main thrust of the book is to provide a comprehensive text in a world-wide perspective of remote sensing in forestry; and at the same time it aims to keep a balance between research and forest practice. The writer has drawn on his experience in forestry in Australia, East Africa, the United Kingdom and the United States and his perspective of remote sensing from visits to 109 countries, while serving as Chief of the international Remote Sensing Centre of the Food and Agriculture Organization of the United Nations. The book is not for mathematicians and electrical engineers; but it is hoped that it will prove particularly useful to the many concerned with the management of forest resources, to forestry students and to many researchers commencing studies on the remote sensing of the forest environment. Also, to help the readers of various disciplines the remote sensing terms are usually defined in the text following the *Manual of Remote Sensing* and are then referenced in bold type in the index. Selected terms in English are also cross-referenced in French and Spanish as an appendix; and forest technical terms, following British Commonwealth Forest Terminology (Commonwealth Forestry Association, Oxford), are defined on their introduction in the text or where they are mainly used.

The book has been purposively broken into five parts to provide a step-wise approach to studying the remote sensing of forest resources and also to facilitate the reader in proceeding directly to topics of interest. Following the introduction, the earlier part of the book focuses on the scientific background to remote sensings as related to forestry. This approach facilitates the later presentation of applied remote sensing in forest practice. Emphasis is placed on aerial photography in Part Two and aerial photographic interpretation in Part Five, since, in field forestry, aerial photographs remain the main source of forest data extracted from remotely sensed imagery. This is not to discredit satellite remote sensing, which is world-wide the main field of research and which is increasingly used operationally in forest inventory, forest management and strategic planning.

Finally, mention must be made of the fact that some of the figures and tables in this book have been included through the courtesy and generosity of individual authors, publishers, firms and government and international agencies. If material has been inadvertently adopted or adapted, where permission should have been obtained or acknowledged, it is hoped that the oversight will be excused.

Particular appreciation is extended to colleagues and friends world-wide for providing information; to Z. D. Kalensky (FAO, Rome) for comments on part of the text; to Faber and Faber (London) for being able to draw on the contents of *Aerial Photo-Ecology* (J. A. Howard, 1970a); to the Food and Agriculture Organization of the United Nations (Rome) for being able to include several figures and tables; to the Forestry Section, University of Melbourne (Australia) for permission to use data/figures associated with the author's earlier research; and to NASA and NOAA (USA) and SPOT Image (France) for the supply of satellite imagery.

Acknowledgements are also extended to the following: Figure 2.2 (Director, Science Museum, London); Figure 3.16 (J. W. Salisbury, US Geological Survey, Reston Va.); Figure 5.4 (a), (c) (Markhurd Corporation, St Paul,

Minnesota); Table 6.1 (Environmental Research Institute of Michigan); Table 8.2 (I. Langdale-Brown, University of Edinburgh); Figures 8.7, 9.7 and 9.8 (G. Petrie, University of Glasgow); Figure 10.2 (Optronics International Incorporated); Figures 10.3, 10.4, 10.5 (M. F. Baumgardner, LARS, Purdue University); Figure 11.8 (OMI, Rome); Figure 11.10 (Z. D. Kalensky, Rome); Figure 15.2 (D. Schwaar, Montpellier); Plate 2 (Ushikata Mfg. Co., Tokyo); Plate 10 (R. D. Spencer, Forestry Section, University of Melbourne); Plate 11 (G. Hildebrandt and A. Kadro, University of Freiburg); Plate 12 (c) (W. C. Draeger, Department of Forestry and Conservation, University of California, Berkeley); Figure 15.5 (W. M. Cielsa, US Forest Service, USDA/ G. Hildebrandt, University of Freiburg); Figure 16.7 (G. Hildebrandt and H. Kenneweg, University of Freiburg); Figure 16.8 (W. J. Gittins/University of Melbourne); Table 16.1 (US Forest Service, USDA/T. E. Avery); Plate 13 (J. Schnurr, Universtät Göttingen); Figure 17.2 (H. C. Bodmer, Swiss Federal Institute of Technology, Zurich); Figure 17.3 (M. L. Bryan, Jet Propulsion Laboratory, California Institute of Technology, Pasadena); Plate 14, Figures 17.4, 19.1 (C. J. Tucker, NASA); Figure 18.3, Table 18.1 (V. J. LaBau, US Forest Service, USDA); Plate 15 (Kardono Darmoyuwono, Bakosurtanal, Jakarta, Indonesia).

John A. Howard

Foreword

Dr Howard's new book on remote sensing of forest resources marks the completion of an ambitious and timely task. He has written the book at a time of the most rapid changes in technology and in the role of forestry in our society, in economic development and in our approach to environmental issues. Dr Howard shows how the technology, which he describes, relates to the new demands that society places on the profession of forestry. He provides a good basis for understanding the forestry background that lies behind the methods that have evolved in forest mapping, inventory and monitoring. These methods are put into perspective by discussions of such topics as population growth, the energy crisis, deforestation and concerns about global change and the environment. Each of these issues can be extended into arguments to show the relevance of remote sensing in the management and protection of forest resources at local, regional and global scales.

Dr Howard's book is an international text that is clearly not confined to remote sensing or forestry in any particular region of the earth. The book reflects the author's broad background and experience in international forestry and remote sensing, and especially his many years in a leading position as chief of the international Remote Sensing Centre of the Food and Agriculture Organization of the United Nations. This international perspective adds greatly to the value of the book for students and managers and for those who seek a solid general background on the new technologies of remote sensing. Details on individual methods can, of course, be found in specialized literature.

Dr Howard mentions that one of his problems in preparing this book was not to decide what should be included, but what should be omitted. He has solved this problem well; the result of his effort is a text that covers a wide range of topics and touches on many subjects. It is fortunate that Dr Howard has not omitted the description of traditional approaches such as aerial photo interpretation and conventional mapping. One has to understand these methods even though they will eventually be replaced in part by new technology; they are still widely in use and have to be understood by the practising forester. Also, they often define the niche that will be occupied by the new technology and can help to set the goals for research and development. The state of the art of technology in forestry and remote sensing is always ahead of its use and application. The most appropriate technology is not necessarily the same as the latest invention.

It is inevitable that a book in a rapidly growing field such as remote sensing will soon be followed by new advances and discoveries. In a few years we will be able to look back on experience with radar satellites that are still on the drawing board and we will add a great deal to our knowledge of remote sensing. Airborne electro-optical scanners will make further inroads into aerial photography and there will be rapid progress in the integration of remote sensing and geographic information systems. Dr Howard has indicated these directions and his book will be used as a benchmark for measuring progress. The remote sensing and forestry communities should be grateful for the valuable service that Dr Howard has performed in writing this comprehensive book.

L. Sayn-Wittgenstein
Director General
Canada Centre for Remote Sensing, Ottawa

Part One

Background to Remote Sensing of Forest Resources

1

Introduction

1.1 WORLD POPULATION IMPACT

Many people are aware today through the press, television and radio of our changing natural environment and of the increasing pressure of mankind on its fragile structure. However, few, other than those professionally involved in the practice of forestry, are aware of the dependence of many on the forests and of the rapid deterioration of the forested landscape. The former is illustrated by the facts that an estimated 200 million people live in the forests of the world and that about 2000 million people depend on the forests for domestic energy (Flores-Rodas, 1985).

The forested landscape is not only that part of the earth's surface dedicated to the production of timber, but also extends to areas of woody vegetation sustaining other forest resources and providing protective benefits. This landscape has been threatened or influenced by man's activities for thousands of years and today supports functions ranging from industrial forestry to environmental forestry, rural development forestry and urban forestry. The historic influence of man is seen in the 'fire climax' eucalypt forests of Australia, where the aborigines used fire extensively during 40 000 years of residence. In the tropics, many types of forest cover are now rapidly declining, because of mismanagement and the 'slash-and-burn' practices of the increasing human population. In the cool temperate forests of Europe and North America, external pollution, resulting from man's preoccupation with ever higher standards of living, is seriously affecting the health of the forests and is typified by the damage to the forests of the Federal Republic of Germany, which may now exceed 50%.

It seems very unlikely that the direct and indirect effects of the impact of the increasing human population pressure on the forested landscape will decrease in the foreseeable future. By the end of the century, the world population will surpass 6000 million, is estimated by the year 2010 to reach about 10.5 billion and to peak at about 14 billion by 2130 (Sala, 1981). This may be compared with a world population of only 3 billion in the mid-1960s, and offers no respite from the existing pressures on forested lands. In some countries, the population density will become acute. China already has a population exceeding 1.1 billion and the World Bank forecasts a population exceeding one billion in India by the year 2000. Even on the small island of Java in Indonesia, the population is expected soon to exceed 130 million. In the early 1990s, more than 650 million people in the developing countries can be expected to be seriously undernourished (FAO, 1981); and already in the sahelian countries of West Africa, the human population exceeds two times the carrying capacity of the land (Higgins and Kassam, 1981).

Resulting primarily from this accelerating population explosion, and associated with the indirect and the direct threats of pollution, will be the effect of major changes in the patterns of land use and the increasing threats to mankind by deforestation, which is resulting in soil erosion, declining soil fertility, decreasing productivity and in some regions desertification and probably climatic change. Scientific records already indicate that the earth has warmed by about half a degree centigrade since 1850, and if this trend continues, the world will be warmer within a few decades than at any time in the last 100 000 years. Probably during the last ice age, 10 000 years ago, the mean annual global temperature was about 3°C lower than it is at the present time. The impact of climatic change is nowhere more strongly emphasized than in a letter to *The Times* in July 1988, signed by the presidents of seven British learned societies, in which it is pointed out that because of the so-called greenhouse effect, sea levels are likely to rise by 2–4 metres in the next 100 years. Such a rise would be disastrous for some coastal agricultural lowlands (e.g. the Fens, UK), the mangrove forests of the tropics (e.g. Bangladesh) and some island nations (e.g. Maldives).

At the second World Climatic Conference (November 1990), climatic change as a scientific problem also became a major international political issue. The final report acknowledges that unequivocal evidence of climatic change will not be achieved before a decade or more, but that greenhouse gases will result in additional warming of the earth's surface within the range of 0.2–0.5°C in each decade.

These are the circumstances now being imposed by the burgeoning world population on the forest sector, where output is determined by naturally built-in limits to biological growth; and where decisions taken today on insufficient background data will not significantly affect agricultural and forest production before the first decades of the twenty-first century. Moreover, the rural sector is the core around which in many countries socially complex human cultures continue to evolve and which will critically influence future forest policies. It is therefore essential to maximize the use of cost-effective monitoring methods. These include remote sensing to help provide quickly and objectively the best available information for decision-makers and managers.

1.2   WORLD FOREST PERSPECTIVE RELATED TO REMOTE SENSING

Historically, forestry has been concerned mainly with the assessment of timber resources and the management and utilization of closed forests for the production of wood: wood that was needed as a raw material by industry, or as the source of energy by industry or by the local communities. Attention was only occasionally given to other resources of the land sustaining closed forest. In the nineteenth century, some working plans in continental Europe considered not only timber production, but also wildlife (e.g. in France, Denmark and Germany) and/or protection (e.g. sand-dune control in Jutland, Denmark; Landes, France; Formby, UK). Plantation forestry in Europe, based on conifers, was also rapidly expanded; but in the tropics, sub-tropics and North America, the closed natural forests were increasingly exploited for timber.

In the USA, rangeland management became closely associated with forestry, and in many parts of the British Empire, forest survey and sound forest management was extended to all lands bearing trees (e.g. in India) and to minor forest products (e.g. gum arabic in the Sudan). As early as 1856 a forest department was established in Burma by the British; and the concepts of environmental forestry were being applied in India before the end of the nineteenth century.

As viewed by the Food and Agriculture Organization of the United Nations (FAO), established by the Bretton Wood Conference in 1944, **forests** have become widely recognized as

> all lands bearing a vegetation association dominated by trees of any size, exploited or not, capable of producing wood or other products, of exerting an influence on climate or on the water regime or providing shelter for livestock and wildlife (Loetsch and Haller, 1964).

It was not, however, until the oil crisis of 1973 that world-wide interest was re-focused on the long-standing importance of forests as the major source of energy in many countries. This had been the situation in Great Britain in the earliest days of the industrial revolution, prior to the replacement of fuelwood by cheaper coal. In 1973, even if the low price for crude oil had continued, there was little possibility of substitutes for wood energy in developing countries; and in the industrialized countries, it is often not realized how important is the provision of energy as an end-use of wood. A recent European study (ECE, 1986) suggests that over 40% of the volume of wood and bark removed from European forests is used as a source of energy and that energy remains the single most important use of wood in volume terms.

With the uncertainty of energy prices in the future and of the growth world-wide in the consumption of energy, predicted to exceed 2% a year, the demand for wood energy can be expected to continue increasing. Within a decade, the most urgent need of many local communities in the developing countries will be a massive harnessing of their resources to renew forest supplies through plantations and agro-forestry, and, where practised, the abandonment of the 'slash-and-burn' of natural forests based on short cutting cycles. In response to this situation, remote sensing can be expected to be used increasingly to collect urgently needed data, especially as related to the monitoring of changes in forest cover, assessing land use and forest land degradation, evaluating the productivity of the land and providing information not only for forest inventory but also for direct inputs into forest management and strategic planning.

With continued world-wide economic expansion, the demand for sawnwood and modern sector forest products will also rise in both the developing and industrialized countries. This includes meeting the massive urban needs in Latin America, Asia and Africa. Mexico City, for example, is predicted to have a population of 20 million people within a few years. Within Europe, timber production has been expanding steadily for more than 100 years in area, growing stock, increment and removals, and is expected to continue expanding with particularly strong increases from the planted forests of several western European countries (i.e. France, Ireland, Portugal, Spain and the United Kingdom). Removals in Europe, which were 350 million m^3 in 1980, are expected to exceed 400 million m^3 by the year 2000 (ECE, 1986). The output of sawnwood and modern sector products will also rise markedly in several countries of other regions including Australia, Chile, Brazil and New Zealand. In North America, there is a trend towards increased supply costs associated with the expansion of modern sector forest products, which is resulting in the harvesting of less accessible areas, less attractive species at greater distances and the expansion of plantation forestry.

These circumstances are also stimulating the wider use of remote sensing, which includes airborne thermal sensing for fire protection and control, large-scale aerial photointerpretation, updating and preparing thematic maps using photogrammetric techniques and monitoring forest damage and land-use change using data collected by aircraft and satellites. Particularly in many developing countries, but also absent in some industrialized countries, is a workforce competent in maximizing the use of data obtained from aircraft and earth resources satellites; and for which it will be necessary to provide adequate training at both professional and technical levels. There are also major sectors of forestry in which remote sensing has not contributed, or is not contributing as it could. The former includes the entire range of international trade, to which in agriculture satellite remote sensing is already contributing. The latter includes at the local level forest utilization (e.g. siting of mills for sawnwood and plywood) and world forest resource appraisals.

When international trade and the environment are seen as associated with the national and regional area distribution and condition of forests, satellite remote sensing can be viewed as having an increasingly important role particularly as related to global and regional monitoring. The monitoring of forest stand characteristics has its roots in the periodic national forest timber inventories of Sweden, which have been carried out at regular intervals for fiscal purposes for more than 60 years and more recently with the help of aerial photographs. However, the objective of these surveys was to assess existing forest characteristics and not to compare the periodic inventory results; and it was not until the advent of continuous forest inventory in the 1960s, particularly in Finland, Sweden and the USA, that the importance of **monitoring** (i.e. forest change over time) was fully recognized.

At the international level, since 1948, the Food and Agriculture Organiz- ation of the United Nations (FAO) has carried out, through questionnaires sent to national forest institutes, six world forest resources appraisals, i.e. in 1948, 1953, 1958, 1963, 1976 and 1981. In addition, following the United Nations conference on the environment in Stockholm in 1972, the UN Environment Programme (UNEP) oriented its Global Environmental Monitoring System (GEMS) to co-ordinate disparate international monitoring activities throughout the world (Gwynne, 1982), and is co-operating with FAO in global inventories and in the development of methods for future world-wide monitoring of forest cover.

In the 1981 FAO/UNEP world forest resources appraisal, remotely sensed data was included for the first time; although with the increasing availability of satellite data remote sensing inputs could have been advantageously used more widely and earlier, Landsat imagery was visually interpreted to provide up-to- date information on the forest cover and the forest vegetation types of several African countries, where data was lacking or considered to be unreliable. This was combined with the main sources of information, namely national forest departments, universities, and documents of bilateral and multilateral aid organizations. Estimates were made of forest change during the period 1976– 80 and extrapolated to the end of 1985 for the International Year of the Forest with the addition of data provided by the UN Economic Commission for Europe (ECE). Figure 1.1 shows by countries the national forest cover as a percentage of land area. Table 1.1 shows the total continental/regional forest cover subdivided between coniferous and non-coniferous and further between closed forest and other wooded land. A further appraisal of world forest resources is being undertaken in 1990–91 (Singh, 1990).

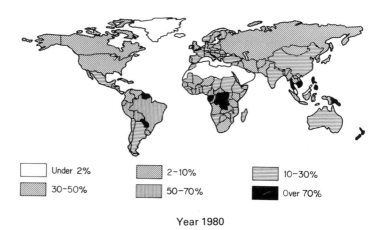

Year 1980

Figure 1.1   World-wide forest cover (1980). By countries, the national forest cover shown as a percentage of land area (FAO, 1985).

1.3   REMOTE SENSING APPLIED TO FORESTRY

In the scenario of a rapidly expanding world population, changes in land use and declining forest cover, remote sensing has the role of an emerging discipline, and provides essential tools of trade to the field forester. In the ensuing text, emphasis is placed on the forest applications of remote sensing; but the technology and its emerging inter-disciplinary role must not be overlooked, and comparisons can be made in this respect with statistics and computer science as applied disciplines. Also, satellite remote sensing has stimulated philosophically a 'one world concept', since man landed on the moon; and it repeatedly draws attention to problems associated with the deteriorating national, zonal and world environments. In its scientific role, remote sensing can often be viewed as a branch of geography with emphasis on viewing landforms and vegetation at a distance; but its importance in refreshing views on biophysical laws and indicating new fields of fundamental study must not be overlooked. The study, for example, of the reflective properties of natural surfaces is not only important to workers in remote sensing applied to forest resources, but also to many other scientists, including plant physiologists and environmentalists concerned with the absorption of solar radiation by plant foliage.

In considering the applications of remote sensing, foresters and geologists have been the main operational users of the collected data for many years.

Table 1.1 Forest area of the world (1980) divided regionally into four cover classes (FAO, 1985)

Forest area (× 100 ha)

Regions	Coniferous — Closed	Coniferous — Other wooded land	Coniferous — Total	Total	Non-coniferous — Closed	Non-coniferous — Other wooded land	Non-coniferous — Total	Forest percentage of land area
World	1 120 805	342 757	1 463 562	4 320 503	1 827 486	1 029 455	2 856 941	32.3
Africa	7 776	8 115	13 891	743 713	228 088	501 734	729 822	24.5
N. C. America	312 830	220 245	533 075	807 092	218 794	55 223	274 017	36.6
South America	19 474	1 000	20 474	215 019	647 498	247 047	894 545	51.4
Asia	56 552	15 688	102 238	468 230	275 270	90 722	385 992	18.5
Europe	86 552	9 576	97 989	158 902	48 592	12 321	60 913	30.5
Oceania	12 060	3 935	15 995	295 947	211 344	71 608	282 952	27.9
USSR	593 700	86 200	679 900	928 600	197 900	50 800	248 700	41.5

Sometimes the remote sensing system provides unique data that cannot be obtained by any other source; but much more often the collected data facilitates fieldwork and enables tasks to be completed at lower cost and much more quickly. It is unlikely nowadays that forest inventory would be undertaken without remotely sensed data. Usually field collected data provide more accurate and precise results, but the collection of the data is slow. In rugged terrain, it may be economically impracticable or confined only to a few ground samples in accessible locations. Ensuing from this is the common forest practice of marrying remotely sensed data and field data and crosschecking continuously the results of image analysis with field samples.

Airborne remote sensing (ARS) in forestry is effectively aerial photography. Operationally, multispectral scanning has been confined to thermal sensing for fire protection. Side-looking airborne radar (SLAR), until very recently, was used only occasionally to provide small-scale imagery for natural resources reconnaissance surveys and for classifying major vegetal cover types of unmapped areas in developing countries. For research the situation is somewhat different, as evidenced by the use of airborne multispectral electronic scanning (MSS) in simulation studies prior to the launching of Landsat TM (Thematic Mapper) and the Satellite Probatoire pour l'Observation de la Terre (SPOT).

As outlined in the next chapter, the role of remote sensing in forestry became established over many years through the use of aerial photointerpretation in forest inventory; and this key role of providing forest information through aerial photography at various photographic scales seems unlikely to change in the near future. The aerial photographs are used directly as a substitute for planimetric and topographic maps, and, with some elementary knowledge of photogrammetry, can be used to derive simple thematic maps incorporating information on the forest stock. Image analysis will provide information on forest cover, forest types, forest condition, some tree and forest stand parameters, information on landforms, land use and land potential, and when related to field samples or aerial volume tables, estimates of the timber volume and forest biomass. This can be extended to obtaining information related to water resources, soil degradation, rangeland management, agro-forestry, etc.

Unlike the development of airborne remote sensing over many years, the application of satellite remote sensing (SRS) in forestry has developed rapidly in association with the digital image analysis of earth resources satellite data; and has led at times to overemphasizing what can be achieved. Care therefore requires to be exercised in distinguishing between sound research and deriving usable information. Sound research has facilitated the development of low-cost computer-assisted analysis systems; but, lack of appreciating the constraints imposed mainly by the pixel size and dependence on spectral differences can lead to overemphasizing the role of satellite remote sensing in forestry, particularly as related to compartments of intensively managed forests.

The advantages provided by the much finer spatial resolution of the second generation satellites (e.g. Landsat TM, SPOT) are now well recognized. In favourable circumstances, thematic maps can be prepared at a scale of 1 : 50 000 and revised at a scale of 1 : 25 000 or possibly larger. Recent studies in Ireland, for example, suggest that some processed image data can be reconciled with stockmaps at a scale of 1 : 10 000 (MacSiurtain *et al.*, 1989). In addition, the finer resolution data of these second generation satellites provides a record of the surface texture of forests, which in classification of the images can be combined with their spectral characteristics. Further, the spectral inclusion of the mid-infrared in Landsat TM sensing is helping to improve the classification of certain categories. These include differentiating several stand size classes and identifying windthrow. Reference to literature suggests that is debatable if the smaller pixel size *per se* helps to improve the spectral classification; but the overall improved spatial and spectral qualities of the sensor and the stereoscopic capability of SPOT can be expected to provide information on land use and land potential comparable to that obtained from aerial photographs at very small scale.

There are also circumstances in which new aerial photography for forest inventory, or management or strategic planning, is not obtainable and the only approach is to combine data obtainable from existing old aerial photographs and current satellite data. In addition, satellite remote sensing can be used to provide quickly information of a more general nature for forest policy at the national level, in providing a permanent visual record of the landscape, and for monitoring forest change over long periods at the continental and sub-continental levels. For such purposes, as will be shown later, satellite remote sensing has a growing role. This is being stimulated by the increasing interest for regional forest studies in the very frequent coverage by the very low-resolution environmental satellites, which in the past have been used primarily to provide meteorological data.

1.4 DEFINITION OF REMOTE SENSING

Finally in this chapter, brief consideration will be given to the meaning of the term remote sensing, since this helps to identify what will be examined in the ensuing text. The term remote sensing was introduced in the USA in the late 1950s to attract funding from the US Office of Naval Research. It was defined by Parker in 1962, at the first Symposium on the Remote Sensing of Environment in Michigan, as covering the collection of data about objects which are not in contact with the collecting device. At the symposium, papers were presented on aerial photointerpretation, aerial photography, airborne radar and thermal sensing. By the early 1970s, comparable terms were in use in French (*télédétection*), Spanish (*telepercepción*) and German (*fernerkundung*).

The *Manual of Remote Sensing* (American Society of Photogrammetry, 1983) continues to define **remote sensing** similarly:

> in the broadest sense, the measurement or acquisition of information of some property of an object or phenomenon, by a recording device that is not in physical or intimate contact with the object or phenomenon under study.

Remote sensing is now generally accepted as not only involving the collection of raw data, but also involving manual and automated raw data processing, imagery analysis and presentation of the derived information. It is usually confined to sensing in the electromagnetic spectrum. That is, using energy which functions in the same manner as light; and covers not only the visible spectrum, but also the ultraviolet, near infrared, mid-infrared, far infrared and radio waves (Figure 6.2). The writer would like to recommend the rider that the energy or signals collected by remote sensing can always, if required, be presented as imagery. In its broadest context, remote sensing is used when reading this paragraph. The eyes, as sensors, are responding to the light differences produced by the reflected light from the dark words on the white page. If, however, the hand is placed on the page to assess its surface temperature, this would not be remote sensing, since the sensor (the hand) is in contact with the object (the page).

It should also be noted that the United Nations Committee on the Peaceful Uses of Outer Space (COPUOS) in several of its publications distinguishes in remote sensing between the collection of data in outer space (i.e. satellite remote sensing) and airborne remote sensing (i.e. inner space). This separation of satellite remote sensing from airborne remote sensing is used in the next chapter in following the historical development of remote sensing. However, as pointed out by Colwell (1979), the development of remote sensing can also be divided conveniently into two general areas, namely prior to about 1960 using only aerial photography and the ensuing remote sensing with the choice of sensors and with the increasing use of satellites as the space platform.

2

Development of remote sensing in forestry

An indication of the stepwise progress of remote sensing, since its beginning in the nineteenth century, is illustrated in Figure 2.1. A year has been set against each major platform development important to forestry. These progress from the first landscape photography in 1838, through airborne photography of forests by balloon in 1887 and by an aircraft in 1919 to continuous multispectral scanning using earth resources satellites in 1972 and the beginning of space shuttles in the 1980s.

2.1 AIRBORNE REMOTE SENSING

Aerial photography

The first known popular interest in the viewing of trees on photographs was in 1838, when Wheatstone demonstrated his reflecting mirror stereoscope (Howard, 1970a). The original stereoscope and stereo-pair of photographs are still available at the Science Museum, London (Figure 2.2). Forty-nine years later, aerial photographs were taken purposively from a balloon near Berlin in order to examine the stand characteristics of beech, spruce and pine woods (Berliner Tageblatt, cited in Spurr, 1960).

If a date is to be given for the beginning of aerial photointerpretation/image analysis in forest inventories, then it is 1920 (Figure 2.3) with the aerial photographic coverage using an H2SL seaplane of half a million acres of boreal forest in Quebec at a cost of 2.6 cents an acre, the interpretation of the photographs was combined with field checking and the preparation of

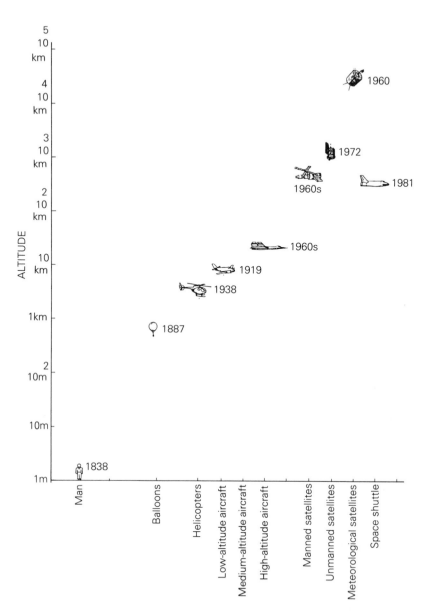

Figure 2.1 Space platforms used in forest resources. The year or period in which the space platform became associated with forestry is shown against the symbols. The forested landscape was viewed stereoscopically as early as 1838 (section 2.1). Altitudes at which the platforms are operated are indicated logarithmically in kilometres.

Figure 2.2 Early days in remote sensing application. The first mirror stereoscope designed by Wheatstone in 1838. This is the forerunner of modern stereoscopes used in forest photointerpretation.

stockmaps (Wilson, October, 1920). At about the same time, aerial photographs were taken for forestry in northern Ontario (Craig, 1920). The taking of the photographs in continuous strips along the flight lines was employed in both surveys in preference to pin-point photography of selected target areas. Strip photography had been introduced successfully during the 1917 Palestine campaign of the First World War.

Between the two world wars, aerial photography came to be widely used on an experimental basis in forestry, particularly in the British Commonwealth and the USA. As early as 1922–23, vertical aerial photography was being planned of the Irrawaddy delta, Burma; and, in 1924, 2500 km^2 of the mangrove forest was covered with black-and-white plate photographs (4 inches × 5 inches format) at a scale of 1 : 20 000 (Blandford, 1924; Howard, 1970a) (Figure 2.3(b)). The endlap between contiguous photographs was sufficient for stereoscopic viewing. Shortly afterwards, experimental aerial photography was undertaken of forested land in Indonesia (formerly Dutch East Indies).

In 1928, aerial photography was applied experimentally in Africa to forest survey in Zambia (Bourne, 1928). Bourne's work in the UK as a forester and photointerpreter (1931) is of much wider interest, since while working from

(a)

(b)

Figure 2.3 (a) Part of one of the 3000 aerial photographs taken of boreal hardwood forest in Quebec in May 1920. The investigator, Ellwood Wilson, commented, 'the actual number of trees can be determined'. Areas of cleared land, burnt areas, forest regrowth and timber stands were delineated on the photographs and their areas measured. (b) In 1924, aerial photographs were used in the tropics (Burma). Mangrove forest of the Irrawaddy delta is shown. Forest types mapped on the photographs at a scale of 1 : 20 000 were *Heritiera minor* (1), *Sonneratia apetala* (2) and *Cynometra ramiflora* (3).

Oxford University in this period, he developed the basis of modern land-unit mapping using aerial photographs (Howard and Mitchell, 1985). Also originating in the 1920s was the serious attempt by Hugershoff at the Tharandt School of Forestry in Germany to introduce the teaching of aerial photointerpretation (API) into forest management (Loetsch and Haller, 1964). In postgraduate studies, tree species were classified, tree and stand parameters estimated using stereoscopic pairs of aerial photographs (Jacobs, 1932) and a photogrammetric plotter (Autocartograph) was used in mapping the volume classes of coniferous stands (Zieger, 1928).

By the advent of the Second World War, black-and-white aerial photography was widely recognized as a valuable aid to forestry and geology, particularly to assist in the planimetric and thematic mapping of unsurveyed lands. Virtually all aerial photographs for forestry were taken with a minus-blue haze filter using panchromatic or orthochromatic films and 152 mm or 205 mm focal length lenses. The combination of panchromatic black-and-white aerial film (although now improved), minus-blue filter and medium focal length lenses is still the most popular for forest applications.

Following the end of the Second World War, there was an enormous surge forward in the taking of aerial photographs for forest inventory and for planimetric and topographic mapping using wartime experience and improved techniques. Particularly in North America, university schools of forestry were quick to introduce specialized courses in aerial photography, photo-interpretation and thematic mapping into both undergraduate and post-graduate curricula. For example, by the mid-1950s, one-semester courses in forest aerial photointerpretation at the School of Forestry, University of Minnesota, were each attended by more than 100 students – 30 years after Hugershoff's effort at Tharandt! Nowadays, world-wide, most university degree courses in forestry include one or more specific courses in remote sensing as separate from studies in forest inventory.

A textbook by Spurr (1948) and parts of textbooks (e.g. Spurr, 1960; American Society of Photogrammetry, 1960; Avery, 1977; Loetsch and Haller, 1964; Howard, 1970a) reviewed in varying detail the application of aerial photographs in forest practice. These have been followed by textbooks on the wider issues of remote sensing, which include a section or chapter on remote sensing technology applied to forest resources (e.g. Johnson, 1969; US National Research Council, 1970; Sabins, 1978; Lillesand and Kiefer, 1979; Heller and Ulliman, 1983, chapter 34).

As a legacy of the Second World War, films with a wider spectral range became available and there was a choice of aircraft for low-level and medium-altitude work. However, as pressurized aircraft were not readily available, the flying height (with oxygen) was limited to about 9000 metres and consequently, the smallest scale black-and-white aerial photography in commercial use was about 1 : 85 000 using an 88 mm lens. In many countries, this continues to be the smallest available scale.

Foresters rapidly appreciated in North America, Europe and Australia that for many purposes the standard black-and-white panchromatic photography, used in photogrammetric mapping, was not always the most suitable for forest studies. Unfortunately, even today, foresters in many parts of the world have to be satisfied with 23 cm format panchromatic black-and-white photography taken at scales between 1 : 25 000 and 1 : 85 000 for planimetric and topographic mapping; and in some countries, the image analyst must rely on satellite imagery, since access to aerial photographs may be militarily restricted.

By the mid-1950s, the value of black-and-white infrared aerial photographs was well recognized for separating stands of conifers (gymnosperms) and broad-leaved species (hardwoods/angiosperms) in the cool temperate natural forests of North America and Europe. However, the technique was found to be unsuitable for separating native conifers from eucalypts in Australia, and later for separating stands of *Eucalyptus globulus* and *Pinus pinaster* in Portugal. Black-and-white infrared aerial photography was widely acknowledged as the most suitable film type for mapping the boundaries of water surfaces and for separating wet and dry land; and in French West Africa under heavy dust haze it was recognized to be superior in resolution to panchromatic photography for small-scale planimetric mapping.

Colour aerial photography

By the 1960s, colour photography was beginning to compete in forestry operationally with black-and-white photography. Natural colour photography received vigorous attention experimentally in forest studies (e.g. Aldrich, Bailey and Heller, 1959; Becking, 1959). The importance of natural colour photography was widely recognized for disease detection and for the identification of some tree species growing in temperate zones. As colour infrared film (camouflage detection film) became more readily available, its versatility in forest practice was recognized (e.g. Fritz, 1967; Hildebrandt and Kenneweg, 1970; Howard, 1970b; Lauer and Benson, 1973; Hudson, Amsterburg and Meyers, 1976).

During the Vietnam War and in preparation for the launch of Landsat-1 in 1972 in the USA, extensive use was made of high-altitude colour infrared (CIR) photographs at very small scales (e.g. 1 : 120 000), which were taken from pressurized aircraft. The latter clearly showed the advantage of the technique for land-use mapping at small scale; and in 1975 high-altitude CIR photography quickly provided national coverage of Sierra Leone (Howard and Schwaar, 1978).

The versatility provided by the widening range of film–filter combinations, film types, improved photographic equipment and flying heights from near ground level to high altitudes has contributed to the growing importance of colour aerial photography in forestry in the satellite era of the 1970s and 1980s. Colour infrared photography has now generally replaced black-and-white infrared; and colour photography at large, medium and small scales is now increasingly used in periodic national inventories and combined with satellite-derived data. The US Forest Service, for example, routinely acquires colour photographs at scales of 1 : 15 840, 1 : 20 000 and 1 : 24 000 over all national forests. Aerial photography at very large scale (e.g. 1 : 1200) has been demonstrated to be suited to the reliable identification of trees of individual species in cool temperate forests; and is used operationally in forest inventories in Canada (e.g. Nielsen, Aldred and Macleod, 1979). Also, aerial photographs taken with small-format cameras (i.e. 35, 70 mm) are now used for many purposes in forestry, and are often used to take photographs at large scale in preference to using large-format aerial cameras.

Image analysis

For the interpretation of aerial photographs, instrumentation has remained relatively simple and is usually analogue, as distinct from the digital image-analysis systems favoured in processing satellite data. The mirror stereoscope, parallax bar and pocket stereoscope, used by foresters for over 60 years, remain very popular; and for fieldwork the interpretation aids, which became commercially available in the late 1940s onwards, have not been replaced (Chapter 8). The high-quality photographic products produced by staff skilled in processing films and prints are needed as much as ever. The black-and-white photographic laboratory, which was popular in the 1950s and 1960s, has been expanded to cover colour processing, and when large, may include automated equipment (e.g. Kodak Versamat). Additive viewers and imaging analogue density slicers were available to the imagery analyst from the late 1960s onwards; but in recent years, the density slicer and, increasingly, the additive viewer have been replaced by digital image analysis equipment.

Other sensors

Although radar was used by Appleton in the UK in 1925 for measuring the height of the ionosphere, and the acronym radar was coined (radio detection and ranging), the military importance of ground-based radar was not demonstrated until the Battle of Britain (1940). By the end of the Second World War, polar imaging radar (plan position indicator radar – PPI) was used at sea and in the air by both the UK and the USA. However, side-looking airborne radar (SLAR), as currently used, was not available for civilian application until the 1960s and parallels the availability of multispectral electronic scanning (MSS). Crude photo-multiplier imaging sensors were first tested for civilian purposes in 1958; and within a few years, with the development of lead sulphide detectors, multispectral scanning was operational for use at wavelengths outside the aerial photographic spectrum.

Both electro-optical systems, which include thermal sensors and side-looking airborne radar (SLAR), provide imagery with a much lower ground resolution for the same flying height than aerial photography. This has been a major constraining factor on their wider use; and in addition most developing countries do not have ready access to this type of equipment. Although electro-optical line scanners, with their ability to sense outside the photographic spectrum ($c.$ 0.38–0.95 μm), were demonstrated in the mid-1960s to assist in tree species identification in cool temperate North America, their operational success has been in the field of forest fire monitoring. A prototype thermal IR system was developed to assist in forest fire-fighting in the USA in the mid-1960s (Hirsch, 1968); and by the early 1970s, the US Forest Service had an operational thermal system for fire detection and its mapping at flying altitudes of up to 5000 m.

In 1964 the Royal Air Force demonstrated the usefulness for civilian purposes of polar radar by mapping the frozen St Lawrence river (Canada); and a year later SLAR was sufficiently declassified militarily in the USA to demonstrate its usefulness for small-scale reconnaissance mapping. Selected sites in several parts of the USA were covered. In the tropics, the first successful coverage of continuously cloud-covered rainforest was completed in Panama in 1968. This was followed by the massive Radam project in Brazil which was primarily for mineral exploration of the Amazon, but which also proved useful for small-scale forest-type mapping by relying mainly on tonal differences to provide the basis to discriminate between broad forest types.

Only in the late 1980s has much higher resolution, synthetic-aperture radar become available commercially. This permits the identification on the imagery of individual larger crowned trees, and allows greater use to be made of textural differences in the forest canopy to identify more forest types, some density classes and more detailed land-use classes.

2.2 SATELLITE REMOTE SENSING

Ensuing from the enormous progress made in recent years in the development of satellite technology, satellite data is being applied more and more in forestry, particularly in **multitemporal** and **multistage** surveys of forest resources. The former refers timewise to the collection of data about a site on more than one occasion, while the latter infers that the data is collected simultaneously for the same site from more than one altitude.

Satellite remote sensing in forestry has been developed over a period of about 25 years, whereas the developmental period of aerial photography exceeds 100 years. Satellite sensing in forestry started effectively with the launching in 1972 of the US Earth Resources Technological Satellite (ERTS-1), which was later renamed Landsat. This, however, had been preceded by the launching of man into outer space in 1961 (i.e. USSR Vostock-1) and the first photographs from outer space by the US Explorer-6 in 1959. The systematic orbital observations of the earth from outer space, from 1960 onwards by the US Television Infrared Observation Satellite (TIROS-1), provided very low-resolution imagery for meteorology; while the manned polar orbiting US satellites of the 1960s (Mercury, Gemini, Apollo) provided high-quality low-resolution photographs of strips of the earth's surface, including information at the regional level on geology and vegetation.

The Skylab project (1973), representing the first US station in outer space, provided 35 000 photographs. These, however, have been little used, probably because of the 'patchiness' of the coverage; but through the thirteen channel scanner system of the sensor package, attention was drawn to the importance of satellite sensing outside the spectral range of Landsat (i.e. $0.5–1.0\,\mu m$). The aborted Seasat radar project (106 days, 1978) indicated that synthetic-aperture radar imagery from the satellite platform was often comparable in usable land information content to the existing side-looking airborne radar (SLAR).

This was followed in the 1980s by the US Space Shuttle, which is reusable, and has been used to test several sensors with earth resources application. The Shuttle's imaging radar experiment (SIR-A) has provided additional information on radar from outer space. The German modular opto-electronic sensor (MOMS) was the first imaging charge-couple detector to be operated in outer space (1983–84). The NASA/European Space Agency Spacelab experiment, carrying a high-quality Germany metric camera, provided high-resolution photographs and valuable experimental results on planimetric and topographic mapping from outer space, including comparisons with thematic mapping using Landsat and later SPOT. Because of the much improved resolution of ground objects provided by SPOT and the Thematic Mapper (Landsat), satellites of this type are often referred to as the **second generation of earth resources satellites**, and will be considered further in Chapter 7.

Comment must also be made on the satellites with very low-resolution sensors (e.g. 1 km, 2.5 km, 5 km), which are primarily used for collecting meteorological data. As a group they are frequently termed 'environmental satellites', and include the US ATS-1 placed in geostationary orbit in 1966, and the USSR Kosmos series (1966–68) and the USSR Meteor series. The US polar-orbiting TIROS-1 satellite was followed by an Improved TIROS series (ITOS) in the 1970s and a third generation TIROS-N, with much improved spatial resolution in 1978 (i.e. NOAA-6). In addition, the US Heat Capacity Mapping Mission (1978–80), with a spatial resolution of 0.6 km, provided data for experimental studies related to geology, hydrology, pedology, etc. The success with the earlier environmental satellites, sensing vast areas of the earth's surface at very frequent intervals, stimulated the development of the earth-synchronous US GOES series (1974 onwards), the European Meteosat (1977), the Japanese Himawari (1977) and the Indian INSAT.

In the USSR, parallel to the developments in the USA, satellite data has been obtained from (i) manned spacecraft and employing cameras and visual observations; (ii) manned space stations (Salyut, Soyuz-Salyut); (iii) unmanned satellites of the Meteor series used primarily for collecting meteorological data and low-resolution environmental data and (iv) unmanned Cosmos series used as a quick-look system with electro-optical scanners and direct data transmission to the ground (cf. Japanese MOS-1). Currently, USSR satellite photography using spectral zonal film records ground objects as small as 5 m in diameter. Earlier (1976 onwards), the multispectral MKF-6 camera, manufactured in the German Democratic Republic, was employed on board Soyuz to obtain several thousand photographs in six spectral bands and with a spatial resolution of 20–30 m. Thematic maps were prepared; environmental problems studied included the rapidly declining area of the Aral Sea in central Asia, and comparisons were made with aerial photographs for agriculture, forestry and the earth sciences.

For the survey and monitoring of forest resources in the late 1980s using satellite data, reliance has been placed mainly on the products of the US Landsats and the French SPOT. Demand for the data products of these satellites continues to increase, particularly for digital tapes; but the increasing purchase price of all types of data is probably now a constraining force on their maximum use. Landsat-5, like Landsat-4, had on board, in addition to the 4-channel multispectral scanner (MSS) of the type used on Landsat-1, a much higher spatial resolution 6-channel scanner. This, the Thematic Mapper, has a spatial resolution of 30 m versus the 80 m of the older Landsats. The higher resolution combined with the extended spectral range is helpful in the thematic mapping of forest resources at scales of 1 : 100 000 and map revision at 1 : 50 000. SPOT provides data with a further increase in spatial resolution (i.e. 10 m, 20 m), which is comparable to the USSR–GDR MKF-6 camera, and with its capability of recording in a sidewards direction enables the same forested target area to be viewed obliquely every 2–5 days (cf. Landsat 16 days). Much information on the results of studies in forestry using Landsat-1 (1972–78), Landsat-2 (1975–79, reactivated 1980) and Landsat-3 (1978–80) will be found in the proceedings of the Environmental Research Institute of Michigan (ERIM) symposia, NASA Scientific and Technical reports, *Manual of Remote Sensing* (American Society of Photogrammetry), the several technical journals (e.g. *Photogrammetric Engineering and Remote Sensing, Remote Sensing of Environment*) and published FAO documents (e.g. FAO Remote Sensing Centre series 1–40).

From the launching of Landsat-1 in 1972, digital imagery analysis has played an important and increasing role in the processing of satellite-derived data and was preceded by computer-assisted technology applied to the handling of data collected by multispectral scanners and radar on aircraft. The earliest software programmes applied in forest studies (1972 onwards) used cluster analysis to define the training classes and Gaussian maximum-likelihood algorithms to classify the data. The rapid improvements in small computers in recent years and the reduction in the purchase price of equipment has greatly benefited the development of digital image analysis systems. A PC-computer-supported image analysis system can now undertake with its specialized software most of the functions of a mainframe system of 20 years ago and at a purchase price of about one-tenth to one-twentieth.

3

Reflective properties
of tree foliage

An understanding of the biophysical factors constraining the operation of passive remote sensing systems and influencing image analysis is basic to the successful use of remote sensing in forestry. The principal atmospheric factors affecting the use of remote sensing systems have been incorporated, when appropriate, in Part Two; but as the reflective properties of foliage are so much a part of the forest environment, they are examined here as a separate chapter. The reflective properties will be considered under three main headings, namely leaf structure as influencing reflectance, the radiant geometry of leaves as influencing the quantity of energy directed towards the sensor and the spectral reflectance of leaves.

Most remote sensing systems are **passive** in that they rely on capturing at shorter wavelengths directly reflected electromagnetic energy originating from the sun, or occasionally on capturing indirectly solar energy which has been absorbed by the earth's surface and then emitted at longer wavelengths (i.e. as thermal energy). If the image-forming electromagnetic energy is generated artificially by the man-made remote sensor, as in the case of radar, the term **active sensing** is applied. A simple example of active sensing is flashlight 35 mm amateur photography; and a simple example of passive sensing is human vision.

The most familiar source of passive energy within the electromagnetic spectrum is sunlight. This provides the energy for human vision and contains the well-known primary colours of blue, green and red or the artists' colour components of violet, blue, green, yellow, orange and red. Aerial photographs are taken in the visible spectrum and the near infrared (IR) which are relatively narrow electromagnetic bands; electro-optical scanning operates in these bands, the ultraviolet (UV), the mid-IR and the thermal IR; and radar operates as an active system with radio waves at much longer wavelengths. Most multispectral sensing by earth resources satellites has been of reflected solar energy in the visible and near IR, but environmental satellites usually also record the thermal energy emitted by the earth.

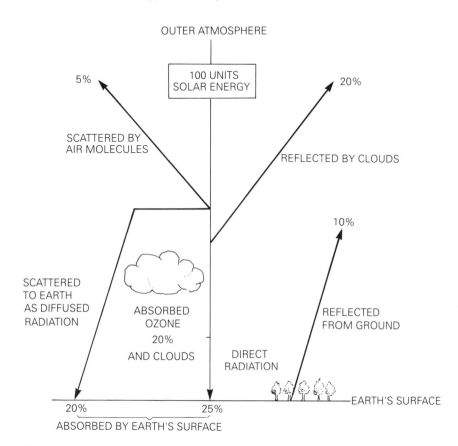

Figure 3.1   Solar flux. One hundred units of solar energy is assumed to enter the outer atmosphere at a rate of 2 cal cm^{-2} min^{-1}. About 25% may be reflected back directly to 'outer space', about 20% is absorbed by clouds etc., some 20% reaches the earth as diffuse radiation and some 25% reaches earth as direct solar radiation. Some 8–15% (shown as 10%) will be reflected back by the forest towards the remote sensor and provides the image forming energy. The remainder is absorbed by the earth's surface, which is emitted later as long-wave radiation (earth glow). Most of the unwanted scattered and reflected energy in the direction of the sensor is eliminated by sensing in those parts of the solar spectrum in which it is minimal and by careful selection of filters.

On average, slightly less than half of the solar radiation entering the outer limit of the atmosphere reaches the earth's surface (Figure 3.1); and of this, about half is direct solar radiation, so important in remote sensing. That is, an average of about $0.228\,\text{cal cm}^{-2}\,\text{min}^{-1}$. Probably under 10% of this will be reflected back directly towards the remote sensor – possibly 15% from snow, 8% from a wet soil, 10% from closed forest and under 1.5% from water surfaces. At the same time the earth's surface is heated by the absorption of some of the solar radiation; and it becomes a source of long-wave radiation, which is mostly emitted in the far IR spectrum and used as the image-forming source of thermal remote sensing.

3.1 TERMINOLOGY

Before proceeding to examine the reflective properties of vegetation, it will be helpful to define some of the terms to be used. Following the International Commission on Illumination (1957), **reflectance** (US) or reflection factor (UK) refers to the ratio of the **total** (hemispherical) radiation reflected by a body to the incident radiation on the body. If the reflectance is restricted to a fixed angle of viewing, the term **directional reflectance** must be used – or **bi-directional reflectance** since the observation is for fixed angles of incidence and reflection. The term **reflectivity** implies that any increase in the thickness of the body does not result in an increase in the reflectance. That is, reflectivity equates with 'maximum reflectance'. Thus, in studying the phenomenon of reflection in the laboratory, spectroradiometer readings for a single leaf of a broad-leaved species (i.e. angiosperm) record reflectance, but readings for a bunch of needles of a conifer are likely to be recording reflectivity; and hence it would be incorrect to make direct comparisons between the readings. When the adjective **spectral** is added to the terms reflectance and reflectivity, this infers that the measurements are taken at a specific wavelength or in a band over a sequence of wavelengths (e.g. visible spectrum, $0.38–0.70\,\mu\text{m}$). **Albedo** refers to the ratio of the solar radiation by a body to the total solar radiation incident on the body.

Transmittance and transmission factor, **absorptance** and absorption factor are the comparable terms to reflectance; but transmissivity and absorptivity imply the transmission and absorption of energy by a body of unit thickness. A theoretical body which absorbs all incident radiation irrespective of wavelength is termed a **black body** and a cavity with these properties is termed a **black cavity**. In practice, as this ideal of 100% does not occur, the term grey body is sometimes used. The relationship of absorptance ($A\%$), transmittance ($T\%$) and reflectance ($R\%$) can be expressed as $A\% + T\% + R\% = 100\%$ (i.e. where 100% is the incident energy). If $T = R$, then $A\% = 100 - 2T\%$, which is used in spectroradiometry for calculating absorptance or absorptivity, but is not applicable to the study of leaves, since reflectance and transmittance usually differ.

For reflection studies, in the laboratory, the most favoured standard reference is freshly smoked magnesium oxide, which has a very high and relatively uniform spectral reflectivity and compares favourably for normally incident radiation with a perfect diffusing surface. A uniform or perfect diffusing surface has equal brightness in all directions and is termed a **Lambertian surface**; and in accordance with Lambert's **cosine law**, its reflectance will decline proportionally to the cosine of the angle of the incident energy. As smoked magnesium oxide is fragile other reference materials are often used in field studies (e.g. barium sulphate, Fiberfax; Watson, 1971). White blotting paper, and similar materials, are unsuitable as references, although occasionally recommended, due to their spectral characteristics.

In addition to uniform diffuse reflection, other types of reflection are of interest as being associated with natural surfaces. These are specular reflection, reflex reflection and spread reflection. **Specular reflection**, as illustrated in Figure 3.2(a), implies that in accordance with the **sine law** the angle of reflected light equals the angle of incident light (i.e. $(\sin i)/(\sin r) = 1$). This mirror-type reflection, although the classic example of reflection in many school textbooks, occurs rarely in nature and on passively sensed imagery is considered as confined to relatively calm water surfaces, some ice sheets and very occasionally other surfaces (e.g. some salt-pans). When present on aerial photographs taken in the visible spectrum and the near IR, specular reflection will be observed as an area of high over-exposure.

The **reflex reflection** zone (or point) on an aerial photograph occurs at the intersection of the line of the sun's rays through the camera lens to the earth's

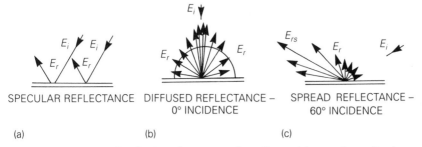

SPECULAR REFLECTANCE   DIFFUSED REFLECTANCE –   SPREAD REFLECTANCE –
                 0° INCIDENCE   60° INCIDENCE

(a)   (b)   (c)

Figure 3.2 Types of reflection from natural surfaces: (a) specular reflection as occurring from a calm water surface; (b) uniform diffuse reflection with normally incident energy (E_i) as associated with foliage and exposed soil – the scribed solid line represents the hemisphere of a perfectly diffusing (or Lambertian) surface; (c) typical spread diffuse reflection from a sunleaf with oblique incident energy (E_i) – note the spread reflection about the vector E_{rs}.

surface. The reflex reflection zone, when present on remote sensing imagery as a small spot, is seen only over land surfaces but can be a constraint to forest photointerpretation and photogrammetric mapping. Over forest it is associated with minimum shadow, and results from the increased backwards reflection of the reflecting surface.

In contrast, **spread reflection**, together with **uniform diffuse reflection**, are the two major components of reflection associated with the image-forming process. Spread reflection is illustrated in Figure 3.2(c) and uniform diffuse reflection in Figure 3.2(b). Sometimes spread reflection is termed **mixed reflection**, since fundamentally the reflection comprises closely integrated specular and diffuse components, which are not readily separated. In order to obtain a two-dimensional visual impression of the directional brightness of a surface, its measured luminance factors at defined angles of observation are plotted on polar graph paper; and the derived envelope of luminance factors is termed a **luminance indicatrix** (Figure 3.6). A **luminance factor** refers to the directional brightness of the surface divided by the directional incident luminous energy.

Finally, comment needs to be made on the physical limits of the **visible spectrum** and the infrared spectrum. The exact limits of the visible spectrum cannot be defined precisely, since it depends on the spectral sensitivity of the human eye and thereby coincides with that part of the electromagnetic spectrum termed 'sunlight'. The boundary of the visible spectrum has been variously placed between $0.38-0.40 \, \mu m$ and $0.70-0.80 \, \mu m$ and even at $0.835 \, \mu m$. In this book, the visible spectrum is accepted as extending from $0.38 \, \mu m$ through the basic primary colours of blue ($0.40-0.50 \, \mu m$), green ($0.50-0.60 \, \mu m$) and red ($0.60-0.70 \, \mu m$) to the commencement of the near IR at $0.70 \, \mu m$. The mid-IR is arbitrarily considered as commencing at $1.5 \, \mu m$ (although sometimes placed at $1.3 \, \mu m$) and the far IR as commencing at $3.0 \, \mu m$ and extending to the range of microwaves commencing at $1 \, mm$ ($3000 \, \mu m$ or $0.3 kHz$). The ultraviolet of the solar spectrum occurs at wavelengths below about $0.38 \, \mu m$.

3.2 LEAF STRUCTURE

Before considering the geometric and spectral properties of foliage, it is important to have an appreciation of the primary type of surface with which we are most concerned. The leaf (or needle) is a complex structure with its internal tissues and external morphology varying between genera and species. Even for the same species there may be differences in leaf structure as influenced by environmental conditions; and on the same tree differences will exist between leaves fully exposed to sunlight (i.e. **sunleaves**) and shade-bearing leaves (i.e. shade leaves). However, the shade-bearing leaves are unlikely to contribute to the energy recorded by the remote sensor. Sunleaves (needles) will usually provide all or most of the reflected energy recorded by the sensor unless the trees are leafless or heavily defoliated. The greatest contrast in reflectance within the same tree crown will be provided by the leaf ontogeny. This often results in crown patterns seasonally conspicuous in large-scale images and helps in the identification of some tree species.

It can be demonstrated that even when there are conspicuous differences between the tissues of mature healthy green sunleaves, their spectral reflectances may be similar (Figure 3.3); and that the presence of spongy parenchyma is not required to provide the spectral reflectance, as sometimes stated in publications (Howard, 1967). As pointed out by Knipling (1969), the importance of air cavities within the leaf has been overemphasized, and it is doubtful if the spongy mesophyll contributes more than the palisade cells to reflectance. It is probable that all plant cells of the mesophyll reflect irrespective of their shape, size and arrangement. It is likely that variation in spectral reflectance is dominated by differences in the refractive indices between the cytoplasm, cell wall, middle lamella, etc. and as modified by the two-way absorption by leaf pigments (Howard, 1971; Gausman, 1974).

In conclusion, the following hypothesis is formulated for mature healthy sunleaves (Figure 3.4). The incident **sunlight** (i.e. the visible part of the solar spectrum) passes through the cuticle and epidermal layer of cells and is partly absorbed by the chloroplasts in the blue and red bands and by carotenes peaking at about $0.50\,\mu$m. The unabsorbed light is then scattered/diffused and multiply reflected by the cell walls of the mesophyll tissue before being further partly absorbed by the leaf pigments. It may need emphasizing that leaf pigments do not contribute directly to the leaf reflectance, as sometimes stated in publications.

In the near-IR part of the solar spectrum, the reflection provided by the cell walls is similar to reflection by cell walls in the visible spectrum; but there is no absorption by the leaf pigments. Both the reflectance and transmittance of solar radiation is much higher, with only a small percentage being absorbed by the leaf tissue (e.g. 10%). The cuticle is seen to serve as a protective layer and transmitting window.

If hairs or abundant waxes are present on the leaf surface, then the spectral reflectance in the visible spectrum will be increased considerably. Schulte (1951), for example, observed for white poplar in Minnesota that the top surface had a reflectance of 12%, while the hairy white ventral surface had a reflectance of 50%. If abundant waxes are present as on some young leaves, these will increase considerably the spectral reflectance in the visible and near-IR spectra; but on old mature leaves the waxes may be lost and dust particles contribute more to the spectral reflectance (e.g. 2%, Howard, 1971). The lignified tissue of sunleaves, which may be exposed during insect attack, has a low spectral reflectance in the visible spectrum up to about $0.65\,\mu$m and then rapidly increases in the near IR (Howard, 1969).

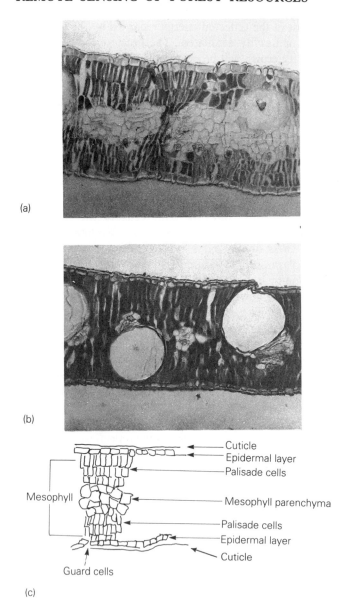

(a)

(b)

(c)

Figure 3.3 Sunleaves with conspicuously different internal structures may have similar reflectance curves. Cross-sections of (a) *Eucalyptus radiata* leaf and (b) *E. obliqua* (250 × magnification); (c) diagram identifying the tissues. Note the abundant mesophyll parenchyma present in *E. obliqua* and the large oil ducts in *E. radiata*; (d) spectral reflectance curves of the leaves of the two species in the vicinity of the cross-sections. Note their similarity.

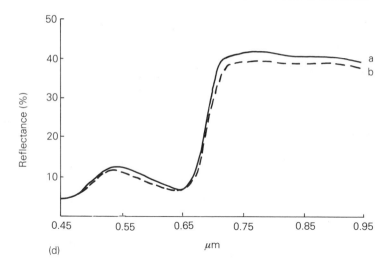

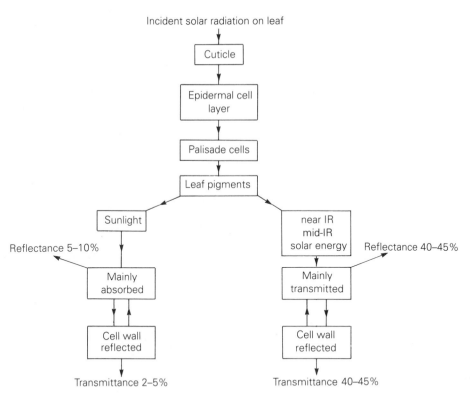

Figure 3.4 Flow chart illustrating the influence of healthy sunleaf tissue/leaf pigments on the reflection of incident solar radiation (see text).

During leaf senescence, leaf dehydration and fungal attack, etc., the leaf will undergo changes, which may affect its spectral reflectance; and hence may result in detection using remote sensing techniques. Each adverse factor requires to be carefully considered in respect of how it could affect the reflectance. Thus in the case of *Puccinia* on wheat, the fungal hyphae penetrate into the mesophyll and will reduce or will stop spectral reflectance in the visible spectrum and in the infrared. With leaf senescence, changes in the leaf physiology may cause an increase in phytocyanins and a decline in the chloroplast content of the leaf. This changes its spectral reflective characteristics in favour of the red of the visible spectrum but may have little or no effect on reflectance in the infrared. Advanced senescence, or irreversible severe dehydration beyond the wilting stage, may cause not only a decline in the moisture content of the leaf but also rupturing of the cell tissue of the mesophyll. This results in increased spectral reflectance, particularly in the near IR. However, minor to moderate wilting, as distinct from severe wilting, appears to have little effect on the reflectance useful to remote sensing in the visible and near IR (cf. Knipling, 1969). Reflectance in the mid-IR is more variable, indicating the potential of this band in studying moisture stress etc.

Leaf pigments

Extensive literature is available on the absorption of light by leaves, and, from this, there is no doubt that the two dominant troughs in the spectral reflectance curves of the visible spectrum are caused by chlorophyll absorption. The minimal reflectances occur in the red band of the visible spectrum at about $0.67 \ \mu$m, which is important to remote sensing, and in the violet/blue band between about $0.36 \ \mu$m and $0.40 \ \mu$m, which is out of the range of aerial photographic sensing.

The primary leaf pigments of higher plants are chlorophyll-a with maximum absorption at about $0.43 \ \mu$m and $0.66 \ \mu$m, chlorophyll-b, with absorption peaks at about $0.45 \ \mu$m and $0.65 \ \mu$m, and the carotenoid pigments (carotene B, xanthophyll). Phytocyanins have a high absorption in the UV, attain a maximum at about $0.50 \ \mu$m and absorb strongly again in the mid-IR. Phytocyanins, when present as red pigment, will contribute considerably to the absorptive and reflective properties of the leaves. With many deciduous forest trees, as leaf senescence approaches in autumn, there is a rapid change in pigments from chlorophyll to anthocyanins.

A logical explanation of the characteristic shape of the spectral reflectance curves of recently mature sunleaves is provided by examining the spectral transmittance curves of suspended leaf pigments in solution. If it assumed that the solar radiation reflected internally by the leaf tissue passes twice through the leaf pigments, the derived curve will approximate to the spectral reflectance curve of an intact leaf (Howard, 1969). Experimentally, it is also

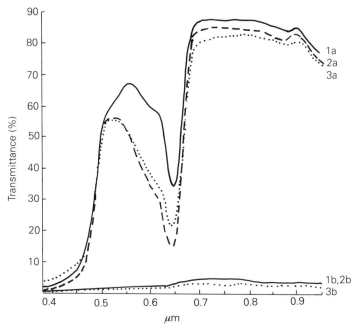

Figure 3.5 Spectral transmittance curves (a) and spectral reflectance curves (b) of the leaf pigments of three tree species: 1. *E. goniocalyx*; 2. *E. regnans*; 3. *E. obliqua*. The spectral reflectance curve of the sample holder (i.e. without leaf pigments) coincided approximately with curve 3b.

readily shown using a spectroradiometer that the spectral reflectance of green leaf pigments suspended in solution is negligible (Howard, 1971); and that the transmitted energy by the leaf pigments provides the characteristic curve shape to be observed for the reflectance of intact leaves (cf. Figure 3.5).

The abrupt change in reflectance and transmittance between about 0.65 μm and 0.75 μm coincides with the change in energy absorption for electron excitation to molecular vibration which requires considerably less energy. A shift to the left or right in this part of the spectral curve is commonly termed **red shift**, and is receiving increasing attention in research related to remote sensing of plant condition.

3.3 RADIANT GEOMETRY OF LEAVES

This refers to the pattern of the electromagnetic energy reflected by the system under examination, and for studies in the visible spectrum is termed **geometric optics**. The radiant geometry of vegetation has been studied successfully in the laboratory using intact leaves of tree species (e.g. Backstrom and Welander, 1953; Tageeva and Brandt, 1961; Howard, 1971), and recently for seedling canopies (e.g. Fern, Zara and Barber, 1984); in the field, mainly the canopies of cereals and pasture, where the earliest detailed study was undertaken in the USSR by Krinov (1947); and occasionally from aircraft (e.g. Brennan, USA, 1969) and using satellite collected data (e.g. Smith *et al.*, USA, 1980). Obviously, the geometric pattern of reflected radiation by the foliage of tall forest trees, other than the radiation directed towards the ground, is extremely difficult to observe. Consequently, laboratory studies of single leaves (or groups of conifer needles) assume much greater importance in forestry than in agriculture and remain the major information source.

Single leaves

Of interest is the measurement of the total reflected energy by the leaf (i.e. reflectance) with varying angles of incident energy and the directional magnitude of the reflected energy at different angles of observation. For the same sun angle, trees with leaves that hang more or less vertically (i.e. **erectophilic**), which includes many eucalypt species, can be expected to reflect differently from tree species with foliage oriented more or less

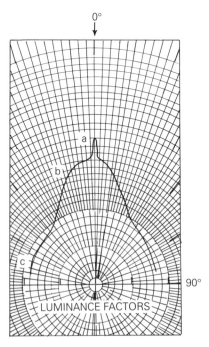

Figure 3.6   Directional reflection recorded for a typical mature sunleaf with normally incident energy (↓): (a) cusp of reflex reflection, (b) spread diffuse reflection and (c) uniform diffuse reflection.

horizontally (i.e. **planophilic**), as is common for broad-leaved temperate deciduous species and many tropical species. Failure to appreciate this fact can result in prescribing the incorrect film–filter combination for tree species identification in forest inventory. However, as indicated by laboratory studies using leaves of several broad-leaved species (Howard, 1969), the leaf orientation in relation to the incident energy has to be considerable before there is a marked change in the reflectance.

As shown in Figure 3.6 for normally incident radiation on the surface of a sunleaf, there is a small cusp of reflex reflection (i.e. about $\pm 2°$) in the direction of the light source. This corresponds to the subtense angle of 3–4° for the reflex reflection to be observed on aerial photographs (Howard, 1971). Most of the reflection, however, can be described as uniform diffuse at angles below about 45°, and as spread reflection between this and the occurrence of reflex reflection. When the angle of incident energy increases, the three types of reflection can still be identified; but against a minor decline in the uniform diffuse component, there is an increasing spread component. Leaves also reflect the incident radiation in the near IR anisotropically, as reported for several cereals (e.g. Dauzet, Methy and Salager, 1984; Kimes, 1983). The occurrence in nature of spread reflection is readily demonstrated by viewing obliquely at arm's length large glossy leaves with the sun's rays towards the observer. As the angle of the incident sunlight becomes increasingly oblique, the brightness of the leaf increases markedly.

Assemblages of leaves (or needles)

On first thoughts, an assemblage of leaves might be thought to reflect similarly to the individual leaf, but this would be ignoring the effect of shadow, transmittance and the spatial arrangement of the leaves in their several layers (i.e. the leaf mosaic). An appreciation of the importance of the shading factor for a group of leaves can be obtained using a photometer and by studying transmittance and reflectance using a spectroradiometer (section 3.4). The influence of the spatial arrangement can be studied most easily using small groups of metal reflectors or by mounting groups of differently arranged pine needles on the plate of a spectroradiometer. Groups of leaves are cumbersome to manipulate.

Within a short distance (e.g. 7 mm) of moving from full illumination of the leaf to deep shade, the luminance will decline rapidly to zero and even in the penumbra at the mid-point, the luminance is low – possibly 15% of full illumination. This suggests that for many purposes the deep shadows occurring between spatially separated overlapping leaves can be considered as 'grey' or black cavities, which have no influence on increasing the reflectance. Foliage can be viewed as a mosaic of reflectors (i.e. leaves) and black cavities. The spatial arrangement or density of the black cavities within the canopy will

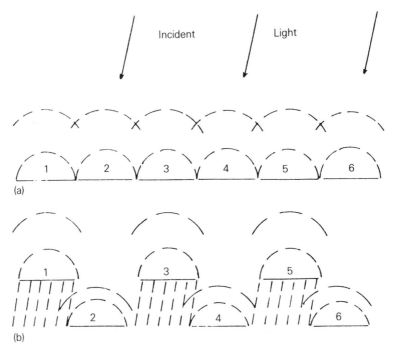

Figure 3.7 Diagram illustrating the effect of the spatial arrangement of leaves on the reflectance and hence the 'containment' of incident solar radiation. Note that the leaves are diffuse reflectors, and the leaf area index remains the same (LAI 1) for spatial arrangements of (a) first order surfaces and (b) first and second order surfaces. In (b), the reflectance is less, as indicated by the wave-fronts and the shading (hatched).

not change the characteristic shape of the plotted spectral reflectance curve, but will change the magnitude of the reflectance. The spectral reflectance in the near IR of sunleaves continues to increase markedly up to a leaf area index of at least four and reaches a maximum at a leaf area index (LAI) of 7–8; but in the visible spectrum, this is usually attained at a LAI of 1–2.

The importance of varying the spatial arrangement of needles (or leaves) on reflectance can be demonstrated using samples of fresh green pine needles mounted on the sample holder of a spectroradiometer. In Figure 3.7, two spatial arrangements of leaves are shown to illustrate this. Reflectance is reduced by arrangement (b), although the leaf area index is the same. In Figure 3.8, the spectral reflectance curves of pine needles are shown for several needle arrangements in the range 0.45–$0.90\,\mu m$. The distances between the layers was approximately $1.5\,mm$. It will be observed that for the needles, which were all mounted in the sample frame of the integrating sphere of Beckman spectroradiometer, the highest spectral reflectance occurs when the needles form a single surface (curve 2) or are compacted (curve 1). When the leaf area

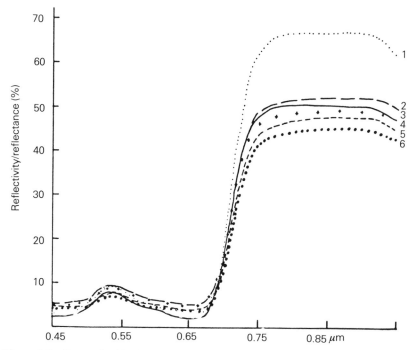

Figure 3.8 Spectral reflective curves showing the influence of the spatial arrangement of pine needles (*Pinus radiata*), as measured using a spectroradiometer: curve 3 (solid line): needles at their natural spacing; curve 1 (dotted line): needles of curve 3 compacted by elastic band to provide maximum reflectance; curve 2 (broken line): 95% of sample frame covered with needles; curve 4 (crosses): frame containing six needles placed between needles of curve 2 and port of integrating sphere; curve 5 (dashes): two layers of six needles between needles of curve 2 and port; curve 6 (large dots): three layers of six needles between needles of curve 2 and port; note that, although the density of needles is increasing, the spectral reflection is decreasing.

index of curve 2 is increased by superimposing second, third and fourth order surfaces (curves 4, 5 and 6), the reflectances decrease (Howard, 1971). More of the incident energy is 'contained' within the foliage due primarily to the diffuse nature of reflection, and has resulted in a decrease in the spectral reflectance of about 2% in the visible spectrum and 10% in the near IR.

Canopies

For information, reference must be made to publications on agricultural crops (e.g. Krinov, 1947; Colwell, 1974). Rigorous studies under field conditions have been undertaken for several cereals (e.g. wheat, barley, maize), several non-cereals (e.g. soyabean, cotton) and grassland. Mostly, emphasis has been placed on absorptance and photosynthesis and the testing of mathematical models designed to cover the interception, scattering and absorption of radiation of non-woody communities. Since the pioneering studies of Monsi and Saeki (1953), many models have aimed to simulate the radiative transfer of energy by entire plant communities (e.g. De Wit, 1965; Idso and De Witt, 1970). As pointed out by Dauzet, Methy and Salager (1984), the albedo and (hemispherical) spectral reflectance were usually derived, and not observed, quantities.

Only recently has remote sensing influenced the development of radiative transfer models and focused interest on models incorporating the optical properties and spectral reflectance of vegetation surfaces (e.g. Kimes, 1983; Sellers, 1985). Usually in the models, the leaves have been assumed to slope equally in all directions irrespective of the sun's azimuth and to behave as simple diffuse reflectors with Lambertian characteristics; but work by the writer with sunleaves suggests that reflectance does not conform to Lambert's law. Also, the reflex reflection of the canopy is usually ignored or cannot be measured with the type of equipment employed.

One consequence is that the model indicatrix based on field observations of the crop is more isotropic than the indicatrix of leaf observations in the laboratory. Often in modelling the optical properties of leaves, the leaf is considered to be an inclined plane with isotropic forward and backscattering, and further, no distinction is made under field conditions between the incident direct solar radiation and the considerable incident diffuse radiation from haze and clouds.

Directional reflectance

This, more correctly termed **bi-directional reflectance**, refers to the angle of incidence of the incoming radiation (i.e. from the sun) and the angle at which the reflected energy is observed (i.e. by the sensor). In aerial photography, the wide angular field of the camera lens results in the recording of images simultaneously on film over a wide range of look angles and consequently diffuse radiation at very differing angles of reflection. In comparison, each signal recorded by an electronic scanner is at a very narrow look angle.

Since the recording of reflex reflection can usually be avoided by careful planning of the flight times (Chapter 6), the imagery of the landscape will be formed usually by uniform diffuse and spread reflection. The common practice in flight planning of avoiding zenith and very low sun angles with its accompanying long shadows provides look angles in which reflex reflection and the extreme angular components of spread reflection are avoided. For complete agricultural crop canopies, several workers (e.g. Kimes, 1983) have commented on the increasing directional reflectance with the increasing off-nadir viewing angle and the contrast between bare soil and vegetation at larger solar zenith angles due to the increasing shadows.

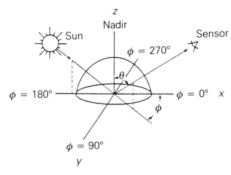

Figure 3.9 Hemispherical geometry and co-ordinate system used for bi-directional reflectance studies in the field. The solar azimuth is set horizontally at 180° with the sensor azimuth angle shown as ϕ. A sensor with 0° azimuth (x) would face directly into the sun and have a zenith angle θ. The y-ordinate is at 90° to the 0°/180° line.

Figure 3.9 (after Kimes, 1983) illustrates the hemispherical geometry involved in making bi-directional measurements in the field using a hand-held radiometer or a radiometer mounted on an observation tower, overhead cableway or 'cherry picker'. The same geometric relationships exist in satellite and airborne remote sensing. The sun's declination and sensor observation angle are defined in degrees off the nadir and zenith angles (i.e. in z-direction). The co-ordinate system of each location requires plotting in the horizontal plane using the azimuth or geographic bearing as commencing with the line in the direction of the sun.

Soil component

It is important before finalizing this section to comment on the geometric radiant properties of soils, since exposed soil often contributes as a background to the reflection of the vegetation. Soils, like vegetation, provide diffuse reflecting surfaces of the types already mentioned (i.e. uniform, spread, reflex), and only in the presence on the soil of surface water, or ice, or a salt-pan is specular reflection to be observed on remotely sensed imagery.

Soil is more anisotropic in its directional reflectance than vegetation; but due to its high opacity the transmitting properties of solar radiation do not require consideration in modelling. Unlike smoked magnesium oxide and other fine powder surfaces, soil (like vegetation) cannot be assumed in models to be a Lambertian reflector.

Under a dense forest canopy, the ground will be completely covered by tree crowns; and it is therefore unlikely that the soil (or the soil plus humus layer) will contribute to the radiant energy recorded by the remote sensor. With a sparse or clumpy tree canopy, the soil can be expected to contribute significantly to the pixel values recorded by the satellite sensor; but this contribution to the image-forming process will be minimal in the presence of shadows. The reflection from the soil is diffuse, and primarily from the first order surface of reflecting soil particles. Their interfaces away from the sun's rays are cast in deep shadow. These provide non-reflecting or very low-reflecting grey cavities which, due to the opacity of the soil particles, will be stronger in shadow effect than foliage. As pointed out by Kimes (1983), the maximum directional reflectance occurs in the general direction of the sun's rays and decreases rapidly in the anti-solar direction. As observed for sand particles in the laboratory, a strong peak occurs in the direction of reflex reflection ($\pm 2°$) (Howard, 1972, unpublished data). An airborne narrow angle sensor pointing at the exposed soil in the direction of the sun's rays, as compared with the sensor pointing in the opposite direction, will usually record the highest pixel values.

3.4   SPECTRAL REFLECTIVE CHARACTERISTICS

Most studies of spectral reflection have been confined, until recently, to measurements of leaf samples in the laboratory. These are now increasingly supplemented for shrubs, herbaceous vegetation and forest seedlings by field measurements of directional spectral reflectance. Publications on airborne measurements of forest stands are few, due mainly to the high cost of measurement and often to the lack of suitable equipment to scan rapidly and spectrally in narrow bands. Laboratory spectroradiometers incorporating an integrating sphere quickly provide readings of the total hemispherical reflection by the sample (i.e. its integrated spectral reflectance); while field instruments quickly provide measurements of directional reflectance, but often in broad spectral bands. To obtain measurements of total hemispherical reflectance in the field would be cumbersome and impractical for forest canopies.

Spectroradiometers designed for field use require to be much more robust than laboratory radiometers, are slower in data collection and do not operate with an integrating sphere. However, they are well suited to broad-band bi-directional studies and band ratioing; but, if used to measure the directional reflectance of tree crowns, the directional component will be different from the directional component recorded by the satellite or aircraft sensor. Field spectroradiometers may be hand-held at ground level (Figure 3.10), or mounted on a tower or 'cherry picker', or operated from a helicopter. Commercially available spectroradiometers operating in narrow wavebands and narrow angular fields are generally not suited for use on fixed-wing aircraft, due to the weak signal strength, signal-to-noise ratio and slow recording rate related to ground speed; but as indicated by Fern, Zara and Barber (1984), a full scan in resolution steps of 2–3 μm for a range of 460–2400 μm can now be achieved with a laboratory instrument in less than a

Figure 3.10 Hand-held spectroradiometer being used to measure the spectral directional radiance of ground vegetation on malachite (Colorado). After each set of readings, readings are also recorded for a diffusely reflecting reference surface, so that directional spectral reflectance values can be calculated.

minute. The derived curves from field measurement of directional spectral reflectance retain the overall characteristic shape of the spectral reflectance curves for healthy vegetation as given later in this section; but the magnitude of the reflectance at all wavelengths may be depressed (Howard, Watson and Hessin, 1971).

Spectral reflectance in the solar spectrum

The solar spectrum for practical purposes can be considered as occurring between about $0.30\,\mu$m and $3.0\,\mu$m; but for passive reflective remote sensing the spectrum can be further restricted. For aerial photography, a lower limit is fixed at about $0.38\,\mu$m by the low spectral reflectance of natural surfaces in this waveband, the transmission properties of the atmosphere and the transmission properties of glass lenses used in aerial cameras. This precludes taking advantage of variations in chlorophyll absorption in the UV/blue band of the solar spectrum. In the mid-IR, the low intensity of the solar irradiation at the earth's surface, combined with the declining spectral reflectance of vegetation and the occurrence of a strong water-absorbing band, favours regarding about 2.3–$2.4\,\mu$m as the upper limit for passive remote sensing in the solar spectrum of forests.

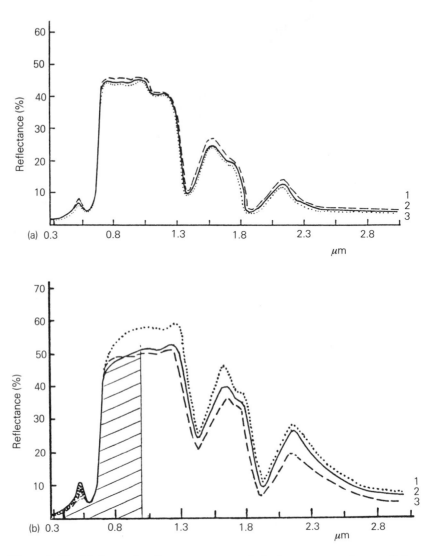

Figure 3.11 (a) Spectral reflectance curves of single mature sunleaves of three evergreen species: *Eucalyptus regnans* (1, broken line), *E. obliqua* (2, solid line) and *E. goniocalyx* (3, dotted line). Each green sunleaf was mature in size. (b) For comparison with (a) spectral reflectance curves of single green mature leaves of deciduous broad-leaved species, *Fraxinus americana* (1, dotted line), *Quercus robur* (solid line) and *Salix sp.* (broken line). The hatched area below the curve indicates the part of the solar electromagnetic spectrum most commonly used in remote sensing.

Figure 3.11 provides spectral reflectance curves between 0.3 μm and 3.0 μm, using a spectroradiometer with an integrating sphere for mature sunleaves of three evergreen species of the southern hemisphere and three deciduous broad-leaved species of the northern hemisphere (Howard, 1971). As indicated by the curves, the shapes are similar. For all species there are four peaks or plateaux of reflectance and troughs of minimal reflectance which occur at about 0.67, 1.4 and 1.95 μm. The trough at 0.67 μm results from leaf pigment absorptance (section 3.2), while the troughs at 1.4 and 1.9 μm are associated with molecular hydroxyl absorptance (water absorptance).

Spectral reflectance and spectral reflectivity in the visible and near-infrared spectra

In biological studies it is important to distinguish between spectral reflectance and spectral reflectivity to avoid misinterpretation of data (Howard, 1966). Soils, rocks, etc. provide readings of spectral reflectivity, since the measurements remain constant irrespective of the thickness of the samples. In contrast, leaves/forest canopies provide measurements which vary with the leaf area index and other factors related to the overlap of the vegetation.

Figure 3.12 provides a comparison between the typical spectral reflectivity curves of several natural surfaces in the range of 0.50–1.0 μm, which is the spectral range covered by Landsat MSS imagery. Spectral curve 1 is typical for overlapping healthy green sunleaves with a leaf area index of 8 (LAI 8). For comparison, the spectral reflectance curve of the single top leaf (curve 2) has been included in the figure (i.e. LAI 1). Spectral reflectivity curves 3 and 4 can be considered typical for dry and moist sandy loams. The magnitude of the spectral curve of the moist soil, although not changing its characteristic shape, will vary greatly with its moisture content and there would be a further decrease in its spectral reflectivity as the soil moisture saturation point is reached. Since only the spectral reflectivity of the soil surface can be measured, no information is available on the sub-surface moisture conditions. Curve 5 for water indicates a low uniform reflection throughout the visible and near-IR spectra; and at least part of its reflection is due to the difference in the refractive indices of the water and its background. Curve 6 is for a dead dry leaf (LAI 1) and draws attention to the fact that when foliage is dead and lacks chlorophyll activity, there is a considerable increase in its spectral reflectivity in the red band of the visible spectrum (0.6–0.7 μm) and beyond about 0.75 μm in the near IR. Spectral reflectivity is attained for dead leaves with a LAI 1. An awareness of such characteristic spectral differences can be helpful not only in planning a remote sensing mission but also in selecting bands and classes for digital image analysis.

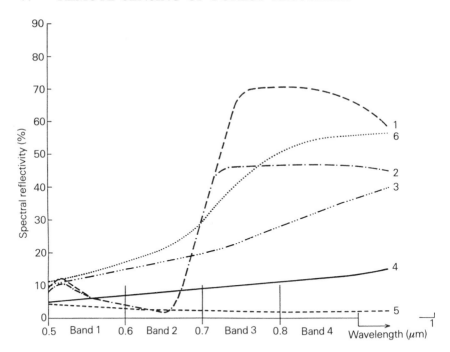

Figure 3.12 Characteristic spectral reflectivity curves of several surface types. 1: healthy green foliage with a leaf area index of 8; 2: for comparison, the spectral reflectance for a LAI of 1; 3: dry soil; 4: moist soil; 5: water; 6: dead brownish foliage for LAI of 1 and 8. The Landsat MSS bands 1–4 are also shown, which correspond to the green and red of the visible spectrum and the near infrared between 0.7 and 1 μm.

In Figure 3.13 typical curves of spectral reflectance at different leaf area indices are shown for mature sunleaves of evergreen broad-leaved tree species and deciduous broad-leaved species. The curves cover the aerial photographic spectrum (0.38–0.95 μm). There is a conspicuous peak in the curve at about 0.540 μm in the green band of the visible spectrum, a trough due to chlorophyll absorption in the red band (0.6–0.7 μm) and a plateau in the near infrared beyond about 0.75 μm. For more than 40 Australian eucalypt species, the writer has observed, using four different spectroradiometers that the spectral reflectance peak in the green band is more precisely located at 0.537 μm ± 2 μm. The shoulder and the gamma of the curve occurring between the visible and near IR of the curve can vary not only between species but also within species, according to the leaf condition. The typical shape of the spectral reflectance curves observed for the leaves of tree species in Australia, North America and western Europe also characterizes the leaves of the trees of the USSR (cf. Kharin, 1975).

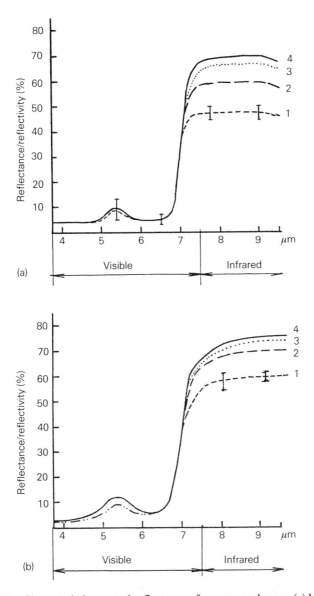

Figure 3.13 Characteristic spectral reflectance of mature sunleaves: (a) broad-leaved evergreen species, *Eucalyptus regnans*. For a mat of 2 leaves in the visible spectrum and a mat of 4 leaves in the near infrared spectrum, spectral reflectance approximates to spectral reflectivity. An addition of 4 leaves to the existing mat of 4 leaves, does not increase the spectral reflectance more than the change to be observed between a mat of 3 leaves and a mat of 4 leaves. 1: single leaf; 2: mat of 2 leaves; 3: mat of 3 leaves; 4: mat of 4 leaves. The vertical lines show the standard deviations (8 leaf samples); (b) for comparison, characteristic spectral reflectance/reflectivity of a deciduous broad-leaved species, *Betula sp.*; 1: single leaf; 2: mat of 2 leaves; 3: mat of 3 leaves; 4: mat of 4 leaves.

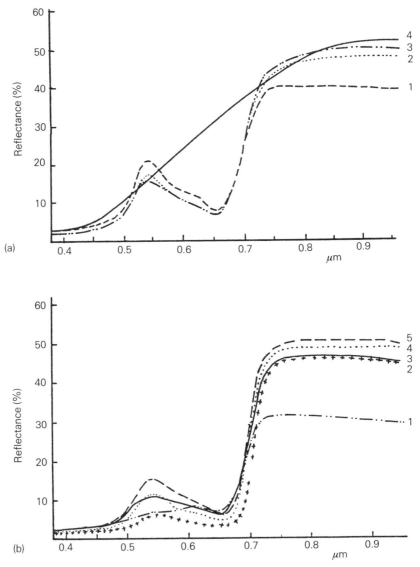

Figure 3.14 (a) Spectral reflectance curves of single leaves of oak, *Quercus robur*. The curves show the changes, in sequence, of spectral reflectance with increasing leaf age over a period of about 6 months: 1: young yellowish-green expanding leaf (spring); 2: light green expanded leaf; 3: mature green leaf at end of summer season; 4: old brown leaf about to be shed (autumn) – about 6 months old. (b) Spectral reflectance curves of single sunleaves of the broad-leaved evergreen species, *E. goniocalyx*. The curves show the changes of spectral reflectance with the increasing leafage over a period of about 18 months: 1: young orange–red tinted expanding leaf; 2: old full green healthy leaf, which will shortly be shed – about 18 months old; 3: leaf mature for about a year; 4: leaf mature in size for several months; 5: recently fully expanded light green leaf.

Leaf ontogeny

In Figure 3.14, comparison is provided on spectral reflectances according to **leaf ontogeny** (i.e. leaf ageing) of both a deciduous and an evergreen broad-leaved species. These show the considerable differences in spectral reflectance that occur with the ageing of the leaves in both the visible and near-infrared bands. In the case of the deciduous species (Figure 3.14(a)), the change is **phenologic**. That is, ageing with seasonal change. In the case of the eucalypt (*E. goniocalyx*) most leaves may remain on the tree for 18–24 months (Howard, 1969) or even longer. Knowledge of the leaf ontogeny by species and season is helpful, when planning remote sensing missions and in image analysis, in order to highlight differences between species.

As shown by curve 1 (Figure 3.14(b)), there is at first in the spring high reflectance in the orange–red part of the spectrum. As the leaf expands, the peak quickly increases in the green band (*c.* 0.537 μm); and, within a few weeks, the leaf reaches its maximum size and attains its maximum spectral reflectance in the green band of the visible and infrared spectra (curve 5). Within a few further weeks, the leaf attains a maturity of texture, colour and spectral reflectance which declines very slowly over the following months (curve 4). A similar rhythm can be observed for leaves of deciduous hardwoods; but it is condensed into one season and continues beyond the characteristics of spectral curve 2 in Figure 3.14(a). As shown by curve 4, by the time the seasonal dying leaves of a deciduous broad-leaved species are shed, there is a dramatic change in spectral reflectance in the visible spectrum.

With respect to the foliage of evergreen species, a seasonal variation in the spectral reflectance of the mature sunleaves has been recorded by the writer (Figure 3.15). This may be the explanation of some colour/tonal differences observed on satellite imagery of broad-leaved evergreen forest. In the case of disease and insect attack, the type and severity requires studying in relation to changes in the spectral reflectance. Frequently there is a change in the near-IR reflectance, particularly on the shoulder of the curve; and the gradient of the curve between the visible and near IR may change and/or shift to the right. With leaf stress occurring due to drought or bark beetles, the attack usually requires to be severe to be spectrally identifiable on the imagery.

Band ratioing

In concluding this section, mention must be made of band ratioing, due to the increasing importance attached to it under the term vegetation index, which will be considered later. An examination of the spectral reflectance curves for that part of the electromagnetic spectrum, in which most remote sensing is undertaken, will suggest that the ratioing of broad bands in the visible and

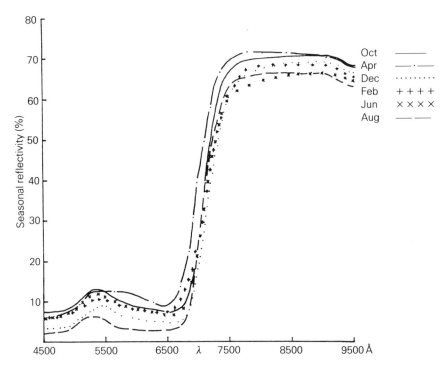

Figure 3.15 Seasonal reflectivity of the broad-leaved evergreen species *Eucalyptus obliqua*, as recorded for mats of six green leaves (i.e. LAI 6) commencing in winter (August), through spring (October) and summer (February) and up to autumn in the next calendar year (April). It will be noted that there is an increase in the spectral reflectivity of the leaves in the spring before the expansion of new foliage.

IR spectra could prove useful to imagery analysis. In analogue analysis it forms part of additive viewing, but the main incentive to its use comes from processing digital data. The ratio, for example, between the spectral reflectance of vegetation in the near-infrared and the red of the visible spectrum is useful in studies of plant cover and plant condition, by reducing the effect of variations in the intensity of solar illumination.

Attention was drawn to the usefulness of band ratioing in satellite imagery analysis by Carneggie *et al.* in 1974 in the study of Landsat (ERTS) data. The ratio of infrared/red was used to monitor the peaking of seasonal green vegetation change (i.e. maximum green productivity, not maximum biomass). Several of the more common band ratios which have been used are summarized in Table 3.1 (after Curran, 1980).

Table 3.1   Common band ratios used in remote sensing*

Name	Formula
Simple subtraction	$IR - R$
Simple division (vegetation index)	IR/R
Complex division	$\dfrac{IR}{R + \text{other wavelengths}}$
Simple multiratio (normalized vegetation index)	$\dfrac{IR - R}{IR + R}$
Complex multiratio (transformed vegetation index)	$\sqrt{\left(\dfrac{IR - R}{IR + R} + 0.5\right)}$
Perpendicular vegetation index (vegetation reflectance departure from soil background)	$\sqrt{[(R_{soil} - R_{veg})^2 + (IR_{soil} - IR_{veg})^2]}$
Green vegetation index	$-0.29(G) - 0.56(R) + 0.60(IR) + 0.49(IR)$

* Note: G represents the blue–green/green energy band, as sensed by Landsat MSS, R the red energy band and IR the two infrared bands. The same formulae can be used for other satellite sensing and airborne sensing in broad spectral bands.

3.5   THERMAL INFRARED REFLECTIVITY AND EMISSIVITY

The fact that only brief comment will be made on the topic of thermal IR reflectivity to end this chapter does not imply a lack of its scientific importance. If there was not a dearth of research on the emissivity of vegetation, the title and content of this chapter might be different. Also, the study of emissivity is more complex than that of reflectivity, and the resolution of the imagery is much coarser.

Some of the incoming electromagnetic radiation from the sun is absorbed by the earth and then emitted at longer wavelengths. For practical purposes this can be considered as occurring beyond a wavelength of $3\,\mu$m and controlled by the 'windows' in the atmosphere at 4–$6\,\mu$m and 8–$14\,\mu$m. As mentioned earlier, thermal sensing is used operationally in forestry in the 4–$6\,\mu$m band for detecting forest fires. Band ratioing, which combines thermal sensing in the 8–$14\,\mu$m and one or more of the previously described spectral reflective bands, is being studied experimentally for regional monitoring of forest cover changes. For agriculture the 8–$14\,\mu$m band is of experimental interest in crop cover studies and for monitoring the development of drought and soil moisture stress. In geology, thermal sensing in the 8–$14\,\mu$m band has been used

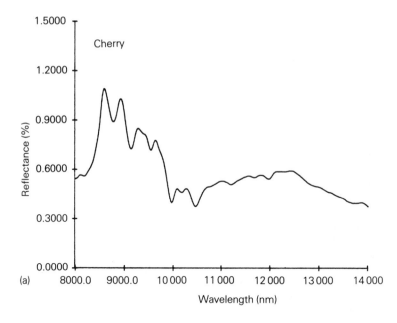

(a)

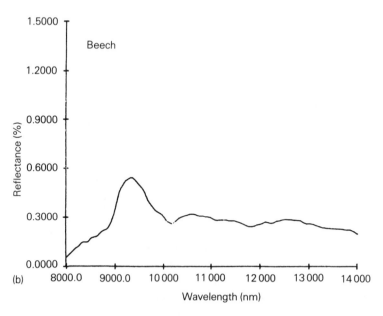

(b)

Figure 3.16 Spectral reflectance curves of the green leaves of two tree species in the far IR (8–14 μm). Note the very low reflectances and the dissimilarity in the curve shapes between species.

successfully in locating large ore bodies, and in monitoring thermal activity near the earth's surface. It is also used in the surveillance of water pollution.

Thermal IR sensing is always more complicated than reflective IR sensing, since the emissivity depends on both the surface roughness and the dielectric constants of the objects being sensed, which vary with their moisture content. Emissivity is a function of the wavelength, the temperature of the landscape surface, the look angle of the sensor and frequently the temperature gradient of the soil and sub-surface rock. Variability in the field measurements of the same type of surface object will result from the rock type, its nearness to the surface, the soil type, its moisture content, the presence of surface water (e.g. after rain) and the physiognomy, cover and condition of the vegetation.

The ratio of the radiation given off by the earth's surface to the radiation given off by a black body at the same temperature provides a measure of its emissivity; and since the emissivity of a black body is equated as 1, all objects of the landscape will have an emissivity less than 1. From the information available on the emissivity of vegetation, it can be assumed that its emissivity will usually exceed 0.90 and is commonly about 0.95 in the 8–14 μm band, but is lower in the 4–6 μm band.

In comparison, the emissivity of clear water approximates to 0.98 and is sometimes used to calibrate airborne sensors and as a reference when analysing digital data. The emissivity of soil in the 8–14 μm band ranges from about 0.85 for dry sands to about 0.97 for moist clays and loams. As reported by Fuchs and Tanner (1986), with the rapid drying out of the top 2.5 cm of a sandy soil and the decline of the moisture content from 8.4% to 0.7%, the emissivity changed from 0.94 to 0.88, which equates with a thermodynamic difference of about 9° C.

Since the emissivity of foliage is high, its spectral reflectivity in the thermal band can be expected to be low. This was confirmed by Gates and Tantraporn (1952), who presented specular reflectance data for 27 plant species with the average thermal reflectances under 4% at 10.0 μm. Later, Wong and Blevin (1967) provided hemispherical measurements for the dorsal surfaces of 47 species for which the thermal reflectances did not exceed 7%. Recently, Gates (1980) for leaves of three species, and Salisbury (1986) for six tree species, have provided specular spectral reflectance curves. Salisbury's data (Figure 3.16) draws attention to the low spectral reflectance occurring in the far IR. This also suggests that the shape of the spectral reflectance curves of green leaves varies greatly between species, which is the converse of that found for the spectral reflectance of sunleaves of many tree species in the solar spectrum.

Part Two

Collection of Remotely Sensed Data for Forest Records

INTRODUCTION

With the increasing complexity of remote sensing, there is a growing need to have an understanding of remote sensing technology used in the collection of data before proceeding to consider the analysis of the collected data and its application. Frequently, a decision on the collection and use of the remotely sensed data is made by persons not fully conversant with the wider technical issues, the planning involved and the economic consequences. The young graduate will often favour satellite remote sensing, which lends itself to computer-assisted data analysis, while the forest expert approaching retirement is likely to prefer aerial photographic methods and possibly the maximum use of field collected data. Also decisions are usually made to sub-optimize under conditions of subjective uncertainty or subjective risk and rarely under conditions of objective certainty. It is therefore essential to overcome this problem, as far as possible, by being familiar with the important sensing systems and their limitations, advantages and costs.

In the ensuing text, exact costs have not been quoted, since because of inflation and technological innovations, these may become rapidly outdated. Often top management, in reaching a decision on using remote sensing, fails to consider opportunity cost and the inclusion in the price ratio of all relevant costs of the final product (e.g. completion of a forest inventory). Often, decision-makers view the cost of procuring remotely sensed data as expensive and as an end objective. They fail to recognize that the additional cost of acquiring the data per unit of forest is small compared with the overall operation; and that the overall cost can be considerably reduced if the most suitable remote sensing system is introduced into the operations at the most appropriate time, often at the earliest stage in the operations or as a pre-project activity.

In view of the continuing importance of aerial photographs in the day-to-day work of the forester, two chapters are provided on the topic. The technical aspects of satellite remote sensing (SRS) are dealt with separately in a single chapter. This is not because of its lack of importance in the modern world, but simply because most types of space related data needed by the forester are usually better obtained from aerial photographs, since many details of satellite technology are peripheral, and because the forester has had little need for the very low spatial resolution of environmental satellite data.

4

Aerial photography

Knowledge of **aerial photography** or the taking of aerial photographs is needed for several reasons. Since aerial photographs continue to be the most widely used type of remote sensing data in forestry, an understanding of aerial photography is essential in the training of the forest image analyst. Many of the older publications remain pertinent today, and some provide the best source of information on methods and techniques. It may also be necessary to provide flight specifications for special forest photography. In so doing, a knowledge of aerial photography is necessary in order to communicate adequately with photogrammetrists; and possibly to present fully the remote sensing requirements of forestry, when photography is being planned for planimetric and topographic surveys.

Aerial photography normally includes the processing of the exposed film and the supply of one or more sets of prints and probably a print lay-down. Several major technical stages are involved in this procedure. These require knowledge of the aerial camera, choice of film–filter–lens combination, preparation of flight specifications and awareness of what is involved in film processing.

Modern airborne remote sensing of the earth relies operationally on aircraft, of which there is a wide choice. These range from conventional unpressurized survey aircraft with a ceiling of about 9000 m to high-altitude pressurized aircraft and light fixed-wing aircraft, helicopters and microlight aircraft using 35 mm or 70 mm cameras. The range of space platforms is illustrated in Figure 2.1.

Most aerial photographs are taken with the camera axis vertical to the ground. This is termed **vertical aerial photography** and is the method referred to in the text unless stated to the contrary. If the camera axis is fixed between the vertical and horizontal, the term **oblique aerial photography** is used. If the horizon is recorded, the term **high oblique** is used, and if the horizon is excluded from the photograph, the term **low oblique** is used. A hand-held 35 mm camera operated from an aircraft window usually provides high oblique photographs.

4.1   AERIAL CAMERAS

The advantage of the aerial camera is that, for a specified flying height, its capability of resolving ground objects is several times finer than that obtained by using electro-optical scanners or radar. Fundamentally, all cameras rely on the principle of the **camera obscura** in which the pinhole, through which light enters a box to form an inverted image, is replaced by the camera lens and the viewing surface by the photographic film.

Single-lens frame cameras

Most single-lens aerial cameras are constructed to satisfy a particular requirement, being initially designed for military purposes. World-wide there are over 100 different models of aerial survey cameras in use; and in addition there is continual improvement in camera construction to improve image quality. The aerial survey camera provides photographs with a format of 23 cm × 23 cm, although a format of 18 cm × 18 cm was common in the 1960s; 70 mm small-format military aerial reconnaissance cameras and 35 mm amateur cameras are often used for large-scale vertical and oblique photography in forestry.

The film in the magazine of the aerial mapping camera is advanced by an electrically operated drive mechanism providing single exposures, and controlled timewise by an intervalometer. The section of the film to be exposed rests in the camera's focal plane, where the light rays focus on passing through the lens. At time of exposure, the film is held in position by an air suction/vacuum pad; and after exposure, the film is stored on a spool contained in the magazine.

To compensate for forward movement of the aircraft, which could cause image blur, some survey cameras with focal plane shutters have a built-in image movement compensation mechanism. This operates by moving the film across the focal plane at a rate proportional to the ground speed of the aircraft and helps to improve the resolving power on the film of ground objects. To compensate for tip and tilt of the aircraft in flight, the survey camera is placed in a gyro-controlled mount, that is set to compensate for drift (also termed crab) of the aircraft. The modern mapping camera, so equipped, with high-quality lens and with a carefully chosen film–filter combination, can provide image resolution as useful to the forest photointerpreter at scales of 1 : 60 000/ 1 : 80 000 as photographs taken in the 1950s at scales of 1 : 25 000/1 : 30 000.

Other cameras

In addition to the single-lens frame camera, there are two other types of single-lens cameras; and also mutlilens camera systems, which have received the attention of the forest photointerpreter. With the exception of the following comments, these will not be considered further.

The **strip camera** is shutterless and records images by moving the film across a fixed slit in the focal plane. The speed at which the film is advanced continuously across the slit is proportional to the ground speed of the aircraft and is in the direction of flight. This type of camera was designed for low-level military reconnaissance and has had only very limited success for civilian purposes, where large-scale linear photography is needed (e.g. for road construction). The quality of the photographs is lower than that obtained with a modern survey camera, particularly at higher flying altitudes.

Panoramic cameras operate on the principle of a scanner (Chapter 6). The ground below the aircraft is covered, as the camera lens assembly rotates at right angles to the direction of flight and the images are recorded on the curved film through a narrow slit. Usually the lens assembly sweeps horizon to horizon; and after each sweep the film is advanced.

A panoramic photograph has the disadvantage of a constantly changing scale during each sweep (i.e. panoramic distortion); however, like an oblique aerial photograph, it covers a much larger area than the vertical aerial photograph taken with a survey frame camera. About the centre line of the photograph, a very high resolution of ground objects is obtained.

Multilens camera systems comprise a battery of usually two to four simultaneously operated 35 mm or 70 mm cameras (Howard and Gittens, 1974). Alternatively, a survey frame camera is used, in which the single lens is replaced by four parallel-mounted lenses. The disadvantage of these systems is the small format of each photograph, and hence the small area of the ground covered by each exposure. The advantage is that each lens simultaneously covers the ground area with a different spectral band, and with the multicamera system several different film types can be used. Excluding multiband studies using a light aircraft, airborne multispectral scanners are favoured nowadays.

Lens characteristics

As no lens of the aerial survey camera is aberration free, it is possible to bring only two wavelengths into focus at the same time, and hence the blue, green and red and near-IR wavebands of the reflected ground radiation are seen to focus at different distances along the optical axis. The distance between the position of focus of blue light at about 0.42 μm and of red light at 0.60 μm can be about 20 nm in the lens of older survey cameras. This is corrected by using a '**meniscus' filter** as a front lens; but is unnecessary with modern '**universal' lenses** when photographing in the visible spectrum. However, the camera lens will still require to have its focal length checked and, if necessary, adjusted periodically (i.e. calibrated).

Another important characteristic of the lens is that its resolving power in the focal plane of the camera is not uniform. The intensity of the illumination reaching the film declines from the centre of the lens outwards and may have to be corrected by using an **anti-vignetting filter**. This reduces some of the rays of light reaching the centre of the focal plane. The illumination on the negative falls off approximately as cosine^4 B, where B is half the angular field of the camera lens. Consequently, the fall-off in illumination is most severe with a lens having a wide angular field; and similar objects (e.g. a tree species) will have changing densities, when recorded as images across the exposed film. As each film exposure is square and the lens is circular, the fall-off in illumination is maximum in the corners of the photograph. With colour photography, particularly colour IR, this may cause overexposure or prevent interpretation at the corners of the photographs. Alternatively, the centre of the photographs will be overexposed. To minimize the problem, a longer focal length lens is often used with colour IR in forest resources studies. That is, focal lengths of about 150 mm, 205 mm or 300 mm are used with the aerial survey camera, and not a focal length lens of 88 mm, as is popular for topographic mapping.

Focal length

By knowing the focal length of the camera lens, the nominal scale of the photograph can be calculated and used to estimate ground areas and linear distances. The longer the focal length of the lens, the higher the plane will require to fly to provide photographs at the same scale. In topographic mapping, accuracy is more closely related to flying height than scale; and consequently smaller-scale photography at low altitude is preferred to larger-scale photography at higher altitudes.

The effect of using lenses of three different focal lengths over a forest is illustrated in Figure 4.1. For the same photographic scale, the base of the trees and the ground datum plane will be more readily observed on stereo-pairs of photographs using a longer focal length. With a short focal length (e.g. 88 mm), the forest floor may not be seen at all and the crown cover will be overestimated. For example, Smith, Lew and Dobie (1960), working in western Canada, concluded when measuring the crown closure and the height of trees that photographs taken with a 300 mm lens at a scale of 1 : 14 000 were

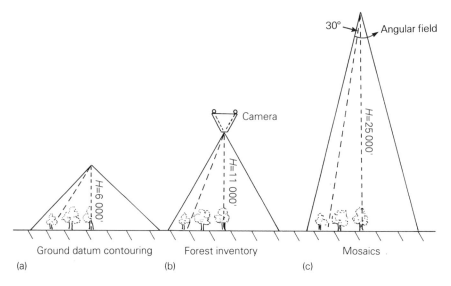

Figure 4.1   The effect of focal length on crown cover of the forest for the same nominal photographic scale (e.g. 1 : 15 840): (a) with a short focal length (e.g. 88 mm) and wide angular field lens (90°), the ground is seldom seen and crown cover will be overestimated in the stereoscopic model, but the 'firm model' is well suited to topographic mapping; (b) with a focal length lens of 150 mm, crown cover is better estimated in the stereo-model and the ground is frequently seen; (c) with a long focal length lens (300 mm) the ground is readily observed in the stereo-model, crown cover easily estimated and mosaic maps will contain less error due to radial displacement. Indicative flying heights (H) are shown in feet and not metres, which is common practice in English-speaking countries.

preferable to photographs taken with a 150 mm lens at scales of 1 : 15 600 and 1 : 28 000. Also on stereo-pairs of photographs at large scale using a short focal length lens, it may not be possible to obtain stereoscopic models of very tall trees.

F-stop

The ratio of the focal length of the aerial camera lens, always set at infinity, divided by the lens diameter is termed the F-stop or exposure number (e.g. F/5.4, F/5.6, F/8). This is used to obtain the best exposure setting of the camera in conjunction with the film speed etc. Since no lens is aberration free, it is preferable to concentrate the exposure on the centre of the lens by reducing the F-stop (e.g. F/8 and not F/5.6). Because of the ground speed of the aircraft, the exposure time cannot usually be less than 1/150 second. If an exposure meter is used directly to determine the exposure settings, the exposure time is likely to be underestimated, due to recording some of the solar radiation being backscattered in the atmosphere. It is corrected by reference to aerial photographic exposure tables, but is often based on practical experience. Black-and-white panchromatic film has a wide tolerance to over or under-exposure (e.g. two stops), but colour infrared film is intolerant (e.g. half stop).

4.2   FILM–FILTER COMBINATION

For effective forest photointerpretation, the film–filter combination requires selecting carefully. The combination has a profound effect on the tone and contrast of images on black-and-white photographs and on the hue, value and chroma and colour contrast of images on colour photographs. Within reasonable limits, the film–filter combination seems to have no effect on stereoscopic parallax and very little effect on image sharpness (Colwell, 1961).

Film–filter combinations utilizing the shorter wavelengths of the visible spectrum and fine-grained film usually provide the best details in shadow, and thus enable forest stand heights and information on the understorey to be better assessed. Some colour and all IR aerial photography are prone not to record details of shaded features. This comment also applies to shaded mountainous terrain, particularly if the shadows are at right angles to the flight path, and can result in the 'black out' of large areas on the photographic prints.

Combinations of film and filter favouring intermediate wavelengths are frequently the best for soil mapping; and long wavelengths (i.e. near IR) give the best images of water courses, shorelines and the boundaries of swampy areas, as well as providing better haze penetration. Colour photography is often essential for the identification of forest species and forest types, and where the ground is exposed soil and rock types. When forest typing involves differences in the leaf area index, forest condition and forest cover, colour IR may be preferable. Colour IR film is usually preferred in studying the incidence of disease and insect damage of forest stands.

Filters

Filters are not part of the camera, being an external accessory, but are varied according to the type of film used, and to a lesser degree according to the environmental conditions. Filters are fitted to the front of the lens to reduce the incoming scattered shortwave radiation, to minimize the effect of haze and to ensure that the recorded images on the film are formed in the spectral bands best suited to the objectives of the aerial photography.

With all types of aerial photographic film, it is normal practice to reduce the incoming solar radiation at short wavelengths by using suitable filters. A yellow-coloured (blue and UV-absorbing) filter is used with panchromatic black-and-white photography, and panchromatic (i.e. normal) colour film is usually exposed through a peach-coloured (minus-UV) filter. These filters are either made of gelatine, which is mounted between clear glass plates, or are of high-quality tinted glass.

Colour IR photographs are taken with a deep yellow or orange-coloured filter. Since all films are inherently sensitive to the blue and UV end of the solar spectrum, infrared black-and-white photography is taken with a deep red ('black') filter, which eliminates most of the light of the visible spectrum. Modified black-and-white infrared photography is obtained by substituting a minus-blue filter for the deep red filter; but its use has decreased in popularity in recent years, as also has black-and-white IR. Modified IR gives better shade penetration but less image contrast.

The function of the spectral absorption filter in aerial photography is illustrated for the most commonly used film–filter combination in Figure 4.2. The yellow-coloured filter is referred to as a **minus-blue filter**, since it absorbs most of the blue solar radiation, which, together with ultraviolet solar radiation, is heavily scattered by the atmosphere. The figure also draws attention to the fact that no filter is perfect in its function, since the cut-off of spectral energy is not sharp at 0.5 μm, but permits some transmission of the shorter wavelengths. Further information on the transmittance curves of filters is available in the many publications on the subject. These include several Kodak pamphlets.

When narrow-pass-bands or short wave-band transmission are needed, the

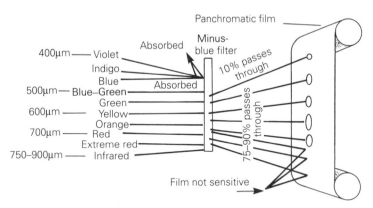

Figure 4.2 Panchromatic black-and-white film with a yellow-coloured filter (minus-blue filter) is the most commonly used combination for forest image analysis and mapping. As illustrated, solar energy on reaching the filter is absorbed at the blue end of the spectrum, but most of the remainder passes through the filter. The film, however, is not sensitive to the extreme red (far red) and the infrared solar energy, and thereby this part of the spectrum is eliminated from the photography.

spectral absorption filters are not suitable and spectral **interference filters** are used. These, however, have the disadvantage of considerable transmittance outside the pass-band. If the desired pass-band is narrow, it may not be possible to obtain sufficient solar radiation to expose correctly the aerial film without considerable loss of image resolution, because of the required long exposure time.

Films

Three important film types are widely used nowadays in forestry. These are black-and-white panchromatic film, panchromatic colour film and colour infrared (CIR). Black-and-white infrared film is increasingly replaced by CIR. Plate 1 provides a comparison of a forest area covered simultaneously by multiband photography using black-and-white panchromatic film, black-and-white infrared film, panchromatic colour film (i.e. normal colour film, e.g. Ektachrome) and colour infrared film.

As the name suggests, panchromatic film is sensitive throughout the visible spectrum; but, as illustrated by Figure 4.2, the minus-blue filter blocks out the incoming solar radiation in the blue and violet bands. Black-and-white IR film is sensitive in both the visible and near-IR spectra as far as about 0.95 μm. As pointed out by Clark (1949), there are also fundamental reasons why the sensitivity of the emulsion of aerial films cannot be increased much beyond 0.90 μm. However, using 35 mm terrestrial IR photography under carefully controlled conditions, the author has observed that images can be recorded out to about 1.1 μm.

During exposure, the silver halide crystals within the emulsion undergo a photo-chemical reaction and form an invisible image on the film (i.e. a **latent image**). The intensity of the photo-chemical change is proportional to the quantity of reflected solar radiation falling on the film. During processing, the latent images recorded in the black-and-white film are reduced to silver grains in the negative. A single 35 mm processed colour transparency will contain over 3.5 million silver halide grains or 'memory elements', while a black-and-white transparency will contain about one million to 1.5 million elements. In comparison, a solid state video-sensor has fewer than 0.3 million picture elements and a CRT video-display about 0.7 million picture elements.

The main advantage, however, in using colour film is that the human eye can interpret many more colours than shades of grey. Using the Munsell colour notation (Figure 4.3), **hue** refers to the colour (e.g. blue, green, blue–green), **value** to the colour's brightness or dullness between white and black, and **chroma** to its intensity (i.e. spectral reflectance). All hues are derived from the three primary colours of blue, green and red.

All colour films have three light-sensitive emulsion layers, with the exception of Russian **spectral zonal film**, which has only two light-sensitive

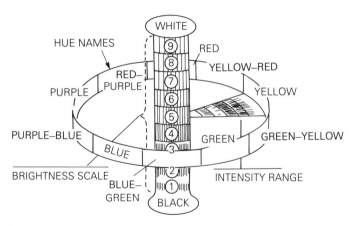

Figure 4.3 Munsell colour notation. The 'Munsell wheel' illustrates the three important dimensions of colour, namely hue, value (brightness) and chroma (intensity/reflectance), which are used in visual image analysis.

layers of emulsion, but a coloured film base. In panchromatic (normal) colour film, the film base supports the three light-sensitive layers and one spectral filter. As might be expected from previous comments, the main function of the filter incorporated in the film is to reduce the incoming blue-band solar radiation from reaching the green and red sensitive layers. The green sensitive layer, by absorbing the green-band radiation, also acts as a filter and permits the red-band radiation to reach the third layer of the film. Since the film is not sensitive to infrared radiation, no precaution is needed against radiation beyond about 0.70 μm. Instead of a single latent image being formed as with black-and-white film, three latent images are formed, i.e. one in each layer. During processing, the exposed silver halide latent images in each layer are reduced to silver grains, depending on the intensity of the exposure in each of the three spectral bands, and unexposed silver grains are washed out. In the processed negative, colour dyes replace the silver grains.

Concerning the structure of colour infrared film (CIR film), the comment of the previous paragraph on the film being in three layers applies. However, there is no minus-blue filter layer. The top layer is usually infrared sensitive and replaces the blue layer of normal colour film. As all films are inherently blue sensitive, the film remains sensitive to the blue spectral band of the incoming solar radiation. Also there have been attempts to improve the colour balance of colour infrared films by introducing a surface filter and rearranging the layers; but these types of film are not widely available for civilian use and workers in some countries may even experience difficulty in obtaining standard CIR film other than in 35 mm format.

Resolution

This refers to the ability of the film to produce spatially defined sharp images. The connotation differs, however, in relation to aerial photography, radar and electro-optical scanning. We also speak of **spectral resolution** when referring to parts of the electromagnetic spectrum, and **temporal resolution** when referring to the frequency timewise of coverage of an area by a satellite. In electro-optical scanning resolution usually refers to the instantaneous field of view (IFOV). With radar, it refers to the effective beam width and the effective width of ground objects recorded in the range direction.

In aerial photography, resolution refers to the resolving power of the photographic system comprising the camera lens and film, or separately to the film. The **resolving power** is usually expressed as the maximum number of line pairs of a target, as recorded on the film, that can be seen and measured as separate lines per millimetre in the image plane. In the USA, standard US Air Force three-bar test targets are normally used for this purpose.

Since no lens is aberration free, the resolving power of the camera–film system will always be considerably less than the resolving power stated for a specific film type, and will decrease on exposed film from the lens centre towards the limit of the angular field of the lens. The manufacturing stated resolving power of a fine-grained slow film will be greater than for a faster coarser-grained film; but in taking the aerial photographs this may be reversed, since with the faster-grained film the exposures are made through the more aberration-free centre of the lens (i.e. at a higher F-stop).

Over forest, the resolving power of the film will be further degraded because of the lower contrast between objects on the ground as compared with tests using bar charts. Assuming a film resolution under forest conditions of 0.01 cm (10 lines per mm), which is lower than would be expected from most modern films, the best resolution depending on the aerial photographic scale that could be expected on semi-matt prints would be as follows (Howard, 1970a):

Vegetation	Photographic scale
Individual large tree crowns	1 : 90 000
Conifer tops (4.5 m diameter)	1 : 45 000
Leafy branches (2 m long)	1 : 15 000
Tall seedlings/saplings/shrubs	1 : 5 000
Leaves (10 cm)	1 : 1 000

4.3 FILM PROCESSING AND PRINTING

Extensive literature exists on the subject of aerial film processing and printing, and to achieve expertise in the subject as an art and science requires specialized training and additional practical experience. Further, with the continuous increase in the depth and breadth of remote sensing, less and less attention is being given to the subject in remote sensing textbooks. This trend will be followed here by providing only the briefest introduction, although skilled film processing is recognized as essential to image analysis and for maximizing image quality. Skilled printing helps to provide the field worker with the most suitable aerial photographs and satellite hard-copy imagery. Hence, close liaison should always be maintained between field staff and the processing laboratory.

To obtain the aerial photographic prints for interpretation from the exposed negative film, two distinct operations are required, namely negative processing of the film and then the printing from the negatives. Sometimes exposed colour aerial film is processed directly to positive transparencies ready for viewing. This is referred to as 'reversal' processing, and in amateur 35 mm photography, the products are the popular 35 mm slides.

Normally, in large laboratories the developing of the exposed film to negatives is by continuous processing. The alternative is batch processing, in which the exposed film is treated as a unit in one processing solution before being transferred physically as a unit to the next solution. In black-and-white photography, the exposed film is passed through developer, washed in water, chemically fixed, washed in water to remove the excess chemicals and then dried. The developer is basically a reducing agent, which converts the silver halides of the latent images to metallic silver and permits the unexposed halides to be washed out. The developing process of the images is stopped at the desired level by immersing the film in an acid solution (i.e. stop bath).

If the processed and dried negative is examined, the tonal rendition of the images will be observed to be reversed to the reflecting properties of the objects on the ground at the time of exposing the aerial film. This is again reversed in the printing process. Light is passed through the negative on to the light-sensitive printing paper, so that the image tone correlates as far as possible with the reflective properties of the objects in the field. The printing paper then contains the latent images of the negative, which by developing, fixing, washing and drying provides the ready-for-use aerial photograph. Often the printing process includes electronic or manual dodging, in which the full intensity of the exposure light is reduced for certain areas of the print paper (e.g. shadowed forest clearings).

Usually, forest photointerpreters prefer photographs printed on a semi-matt (semi-glossy) paper as distinct from a glossy paper. The latter gives more contrast between large images but retains less of the subtle details contained within the negative. Contrasty photographs are preferred by landscape and portrait photographers. Matt prints are not suited to field use since annotations in chinagraph pencil cannot be erased. Figure 4.4 helps to illustrate the need for skilful processing, and provides a comparison between photographs obtained from the same negative and printed on 'slow' and 'fast' papers. The

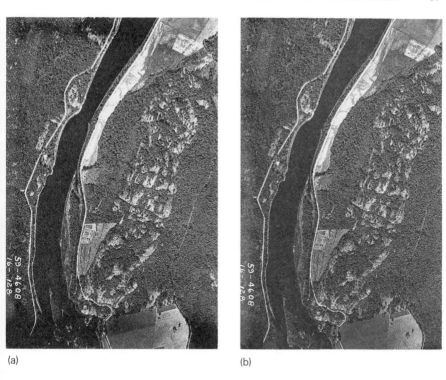

(a)   (b)

Figure 4.4 Comparison of vertical aerial photographs (part) at a scale of 1 : 15 000 from the same panchromatic negative: (a) left, contains more forest details, including tonal contrasts between the crowns of different tree species and regeneration in shadow, than the fast paper print (b) on the right. Note the photographs have been mounted with shadows towards the reader, which is the correct position for photointerpretation.

fast paper print (right) is less pleasing to the eye, and information is sometimes lost. The slow paper print (left), by retaining its tonal range, can help in tree species identification.

The processing and printing of the negative colour film introduces **subtractive** colour processing in which the true colours of the objects are rendered in the negative in their complementary colours before being rendered in their true colours in the photographic colour print. For example, after processing, the negative blue image is rendered by yellow, the green-sensitive layer contains magenta dye inversely proportional to the reflected green light of the original photographed scene, and the red-sensitive layer contains cyan dye. Thus, the primary colours of blue, green and red are represented by the dye colours yellow, magenta and cyan in the negative, and, when projected on to the three-layer colour printing paper, will produce the images in their primary colour components.

Characteristic curve

An essential task in the skilful printing and processing of the film is to relate the image density values measured on the film to the exposure levels producing them. This may be done using a step wedge (Figure 8.6) or preferably using a densitometer. Simple densitometers have a variable spot size for measuring the opacity of the light transmitted through the negative. This type of densitometer, or alternatively flat-bed or drum-scanning densitometers, are used in image analysis to identify the density values of the images on the print or negative associated with important objects in the field, such as the crowns of different tree species.

The type of density curve shown in Figure 4.5 is obtained by plotting the measured film density against the log of the exposure time (log E) and is termed a **characteristic curve**. The shape of the characteristic curve will vary with the type of film and print paper used, and the conditions under which the

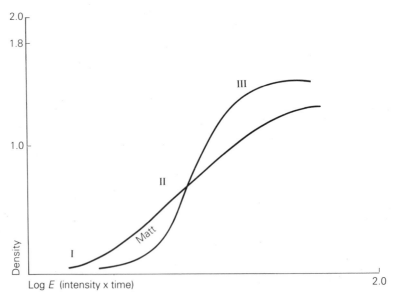

Figure 4.5   The characteristic curve (see text). I is the toe of the curve and II is the gamma (γ) or slope of the curve. Matt or semi-matt papers, having a low gamma, are likely to retain more image information than the faster print paper having a high gamma and providing contrasty images and retaining less image information along the 'gamma' between the toe and the head (III) of the curve.

camera exposures were made. The slope of the curve is termed the **gamma** (γ) and its control in printing will affect the contrast range of the images in the photograph and hence the image analysis. A high gamma will result in the photographs being undesirably contrasty. A gentle slope to the characteristic curve will usually ensure that the photographs are well suited to imagery analysis, by containing the maximum detail of objects in the shadow of tree crowns. Careful control of the gamma for each layer of colour film will help separate the vegetation types and tree species recorded on the imagery. If the image analyst does not have the necessary skills for this part of the work, then it is essential that he/she works closely with a well-qualified technician and guides the technician on what is needed from the imagery/photographs, and what is to be judged as a high-quality product. Success at this stage may save many hours of fieldwork.

5

Aerial photography: flight planning

As aerial photography is complicated and the timing of its completion depends very much on the vagaries of the weather, it is strongly recommended that the objectives and technical specifications be clearly and concisely set out in writing, and if on a commercial basis, be in the form of a signed contract. Attention needs to be given to the preparation of a flight map/flight plan including consideration of the photographic scale and photograph overlap, flight characteristics influencing the geometry of the aerial photograph, ground control, timing of the photography and its cost. This chapter introduces new terms, some of which could be equally considered later in Chapter 9.

The following comments in this chapter are directed at obtaining vertical aerial photographs using the standard format of 23 cm (i.e. 9 inch format); but before proceeding, brief comment will be made on small-format photographs (70 mm, 35 mm). The latter type of photography will be at much lower cost, can be operated from a light aircraft on local hire and can be directly controlled by the investigator. The main disadvantage is the small area covered by each exposure and the ensuing difficulties in handling the analysis of the photographs if many exposures have been made. Each 70 mm and 35 mm photograph covers a ground area of approximately one-tenth and one-sixtieth respectively of the ground area covered by a 23 cm photograph at the same scale. Small-format aerial photography is well suited to research studies and to reconnaissance surveys involving the taking of the photographs along spaced strips or at selected spots. It often forms part of a multistage sampling design in order to provide large-scale remotely sensed data to complement satellite-collected data.

The problem of the ground location of small-format photographs is overcome by using a carefully designed flight plan and by referencing suitable features on the photographs to existing maps, photographic mosaics, orthophotomaps, small-scale 23 cm format aerial photographs and earth resources satellite imagery. The accuracy of measurement using small-format photographs is usually comparable with similar measurements obtained from 23 cm photographs, and includes measurements obtained stereoscopically.

5.1   FLIGHT MAP (FLIGHT PLAN)

The flight map records on an existing map the boundaries of the survey block to be covered by the aerial photography, the location of each flight line to be followed by the aircraft and the location on the flight lines of the starting and finishing points for the taking of photographs. Part of a flight map is shown in Figure 5.1. If there is no suitable map on which to plot the flight lines as in some developing countries, then an earth resources satellite scene can be used for this purpose; and, in any case, it may be useful to have as an addition a recent satellite scene of the area to hand. This helps in checking the forest boundaries and locating other features requiring to be photographed. Terrain features are usually more conspicuous on satellite imagery.

In Figure 5.1 the flight lines are shown as north/south, which is customary in photogrammetry. Otherwise, the flight lines can be placed in the direction of the longest boundaries of the area to be photographed to minimize the unproductive time used in banking the aircraft, when photographs are not taken. In mountainous terrain with shadowed valleys, the quality of the photography may be improved by flying, if possible, in the general shadow direction, so that maximum stereoscopic coverage is obtained of the shadowed valleys.

The first and last flight lines are located within the area to be photographed to reduce the number of flight lines; and, if there are any high mountain peaks, these should, if possible, be located on a plotted flight line and not at the mid-point between lines. The former will reduce the radial displacement of mountain peaks on the aerial photograph. Before plotting flight lines on the map, a decision is needed on the photographic scale, which in turn needs decisions on the focal length of the camera lens and information on the flying height and average ground datum (i.e. ground level above sea level). By knowing in advance the sidelap and the endlap between photographs, the number of photographs to be taken can be calculated.

Aerial photographic scale

This depends on the objective of the photography. If tree heights and stand heights and tree crown diameters are to be measured, a scale larger than about

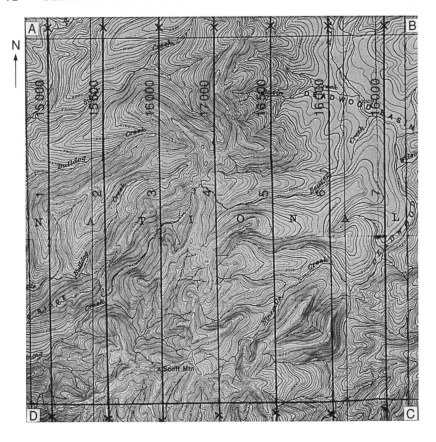

Figure 5.1 Flight plan at a scale of 1 : 125 000 of the area ABCD requiring photographic coverage at a scale of 1 : 15 840. Note that the flight lines are arranged to fall within the area ABCD and the mountain peak (Scott Mountain) is located close to the flight line (3). All flight lines are north/south, as is customary in mapping projects. Note that the flying heights have been indicated along each numbered flight line and the location of final film exposures on each flight marked with a cross (X).

1 : 20 000 will be needed. In the US Forest Service, a scale of 1 : 10 000 or 1 : 15 840 (1 inch to 1 mile) is frequently preferred. Scales larger than 1 : 10 000 are often preferred for tree species identification, unless the stand comprises a single conifer or a single broad-leaved species. For mapping the boundaries of consociations, broad forest types based on stand structure or plant sub-formations, especially in the tropics, a scale of 1 : 60 000 or even 1 : 80 000 may be sufficient; but as mentioned earlier, this will be influenced by the film–

filter–lens–camera combination. The smaller the scale, the fewer will be the number of photographs, and hence for photogrammetric mapping, the number of control points and stereo-models is reduced.

The photographic scale (S) or representative fraction is calculated using the formula

$$S = \frac{f}{H - h}$$

where H is the flying height (altitude) above sea-level and h is the ground datum above sea-level. The scale (S) is the mean or nominal scale in the vicinity of the centre of the photograph (i.e. at the principal point). All measurements used in the formula must be in the same units (i.e. metres or feet). Of the four common focal lengths of camera lenses (88, 156, 205, 300 mm), 156 (6 inches) and 205 (8¼ inches) are preferred in forestry. At large scale with coverage of very tall trees or rugged terrain, longer focal lengths may be needed to avoid excessive radial displacement of the recorded images on the photographs.

The mean ground level or datum plane is obtained by sampling along the plotted flight lines on a topographic map. The mean of samples at 2 or 3 cm intervals should be sufficiently accurate. If a topographic map is not available, then the crudest estimate of mean ground level has to be made using whatever local information is available and taking into consideration the heights of mountain peaks and major changes in the level of the terrain. A field traverse may be necessary, and may include establishing ground control points. When the photographic scale and focal length of the camera lens have been fixed, and using the estimate of the mean ground level, the flying height of the aircraft is calculated from the formula given in the previous paragraph.

Aerial photographic overlap

Aerial photographic overlap of contiguous pairs of photographs is necessary for providing stereoscopic coverage and establishing on the photographs common ground control points and photogrammetric control points. As shown in Figure 5.2, the overlap is the amount one photograph extends over the ground area covered by the adjoining photograph and is expressed as a percentage. The overlap between contiguous photographs of the same flight line is termed **endlap**; and the overlap of photographs of adjoining flight lines is termed **sidelap**. Endlap must never be less than 55% and is usually between 60% and 70%. In mountainous terrain, an endlap of 70–80% is sometimes necessary. Sidelap is usually calculated as 20–30%; but may be as low as 10%, and in mountainous terrain up to 45%. By keeping the sidelap to an acceptable minimum, the number of flight lines, and hence overall cost, is reduced considerably. Variation in the endlap, and hence the number of photographs along a flight line, will not greatly influence the overall cost of the contract for

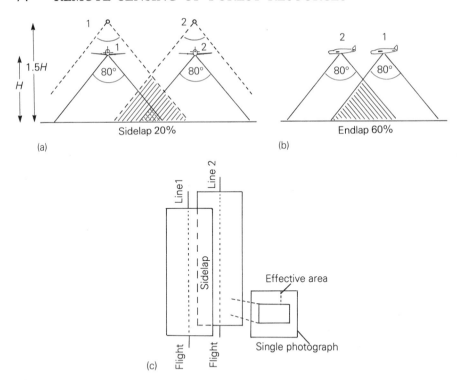

Figure 5.2   Flight planning must allow for overlap of the aerial photographs (see text): (a) shows the aircraft flying towards the reader to provide a sidelap of 20% along flight lines 1 and 2; increasing the flying height from H to 1.5, H increases the sidelap to 46% for the same spacing of flight lines; the angular field of the camera lens is shown as 80°; (b) shows the aircraft flying along the flight line with the consecutive film exposure stations (1, 2) set by intervalometer to provide a 60% endlap over level or undulating terrain; (c) represents the plan view of (a).

taking the aerial photographs; but mapping costs will be increased because of the increased number of control points and additional stereo-models.

By knowing the sidelap and endlap of the photograph, the central area of the photograph defined by the photographic overlap can be calculated. This is termed the photographic **effective area** (Figure 5.2(c)), and is used to calculate the total number of photographs required to cover the area of the flight plan. Within the effective area delineated on an aerial photograph, radial displacement will be minimal and the change in film density caused by tree shadows etc. will be reduced. An aerial photographic mosaic is basically an assemblage of effective areas.

5.2 IN-FLIGHT FACTORS

Several in-flight factors will influence the geometry of the aerial photographs, and hence must be controlled during the taking of the photographs. These include the flying height as affecting the photographic scale, aircraft tilt, aircraft drift and aircraft swing. An on-board altimeter and other navigational aids are used to control the flying height.

Aircraft tilt

This is the combined effect of the lateral tilt of the aircraft wings in flight and the occasional dipping fore and aft of the aircraft, which is termed longitudinal tilt or **dip**. In photogrammetry, lateral tilt is referred to as x-tilt and longitudinal tilt as y-tilt. The maximum combined tilt requires to be clearly specified in the flight contract and should not exceed about 2° for the overall photographic coverage and should not exceed 5° for any one photograph. Elliel (1939), for example, pointed out that with a 205 mm focal length lens, an error of about 1% in scale is introduced at the edge of each 23 cm (9 inch) photograph for each 1° of tilt.

The effect of lateral tilt on the aerial photograph is illustrated in Figure 5.3. G_1 and G_2 are two identical trees of the same height, h, on the ground and d_1 and d_2 are their image displacements on the aerial photograph, when the camera is tilted $t°$. The diagram shows that the images have been distorted by the tilt. The image of G_1 has been displaced outwards from the centre of the photograph, and the image of G_2 has been displaced inwards towards the centre of the photograph and is compressed. If points on the ground replace the base of the two identical trees, it will be observed that these points on the photograph are also displaced by tilt. Further examination of the diagram will suggest that the nominal scale of the photograph will vary on the two sides of the axis of rotation caused by the tilt; and that the safer way of calculating scale, after the photographs are available for fieldwork, will be to use mean ground distances (D) on each side of the flight line and compare these with the mean of the measured distances on the photograph (d), i.e.

$$S = \frac{f}{H-h} = \frac{d}{D}$$

Drift and swing

Sometimes in flight, due to side winds, the aircraft will drift away from its flight track, as established on the flight plan. This is corrected by the pilot adjusting the direction of the 'nose' of the aircraft. At the same time this

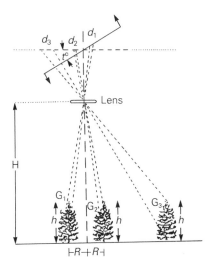

Figure 5.3   The effect of lateral tilt of the camera, mounted on the aircraft, at the time of taking the photographs. On the exposed film, the image displacements of three trees of identical height (h) are compressed or extended according to the direction of the aircraft tilt. Thus, on the side of the photograph of upward tilt, the image of the tree is expanded (d_1) and compressed on the side of downward tilt (d_2). d_3 is greater than d_2, since it is further radially from the geometric centre of the photograph, although the tree heights are the same.

horizontal displacement must be corrected by adjusting the orientation of the aerial camera. Failure to do so will result in the sides of the vertical aerial photographs not being parallel to the baseline plotted later on the aerial photographs.

Drift, sometimes termed **crab**, should be specified in the aerial photographic contract, as not to affect more than 10% of the photographs' width in relation to the planned ground coverage. **Swing**, sometimes termed **yaw**, refers to the rotation of the aircraft about its vertical axis and is observed occasionally on aerial photographs as a rotation of the plotted baselines about the principal points.

5.3   GROUND CONTROL

This refers to the acquisition of accurate data of the horizontal and/or vertical positions of identifiable points on the ground. As ground data is expensive to collect, unless available from existing maps, the responsibility for establishing ground control points requires to be clearly stated in the flight specifications. Under difficult circumstances, the cost of this field activity may exceed that of the aerial photography.

Ground control points

Ground control points (GCP) are needed in establishing geometric accuracy, when the aerial photographs have been taken and are being used in the preparation of maps, photographic mosaics and ortho-photomosaics. Usually in forestry, planimetric control (i.e. horizontal GCPs) will suffice; but if

(a)

(c)

(b)

Figure 5.4   Ground control points (GCPs) are established in the field before taking the aerial photographs, and should be sufficiently contrasty against the forest background to be clearly visible on the photographs; GCPs also help the aircraft crew to locate the area to be photographed: (a) pannel boards in the form of a cross used as a simple GCP and (b) locally cut logs used as a GCP; (c) in larger aerial surveys, satellite transmitted signals of the Global Positioning System (GPS) are used accurately to locate ground control points.

contour maps are to be prepared then the much more expensive vertical control (i.e. topographic GCPs) will be needed. Since each control point is established on the ground ahead of flying, the materials used should not be attractive to thieves and must be of sufficient size and contrast against the forest background to record on the aerial photographs. A cross of white-painted logs or of contrasting coloured plastic sheeting or panel boards (e.g. 1 m × 2.5 m) (Figure 5.4(a)) may be sufficient, or a more substantial GCP may be used (Figure 5.4(b)).

The minimum number of GCPs to be established in the field ahead of the aerial photography will depend on several factors including purpose of the survey, area being covered and terrain conditions. GCPs should be linked by survey with existing geodetic beacons. The minimum number for planimetric and thematic mapping (Chapter 11) will be four or five. For example, Kelsh (1940) carried out a study in Maryland (USA) of a forested area of 140 square miles and covered photographically at a scale of 1 : 20 000. With four corner GCPs and one at a mid-point, the average planimetric error was approximately 6 m and without the mid-point the average error was 14 m.

Traditionally in forestry in many parts of the world, the distance between the ground control points has been measured using a chain and compass, with a hypsometer to measure slopes and an altimeter for changes in altitude. Nowadays, for speed and accuracy, a geodometer can be used to determine the distance between two GCPs, although high forest will limit its use. The geodometer permits a ray of light to be emitted and reflected back and, on the basis of the time interval, the distance between the two points is calculated. For further information on geodometers and tellurometers (using radio waves) a modern survey textbook should be consulted.

In addition, consideration can be given to adopting maritime navigational satellites for establishing the distance horizontally between GCPs when small-format aerial photographs are being taken from a light aircraft or helicopter. Recently in Australia, for example, the Global Positioning System (GPS) has been used to direct the aircraft along the flight lines, trigger the cameras at fixed distances and record the photographic co-ordinates. Some 90% of the photographs were determined to be within 100 m of their map positions (Biggs, Pearce and Westcott, 1989). The cost of operating the system is low. During the recent 'Gulf war' about 10 000 low cost GPS pocket navigators were issued to the US desert forces. At high cost and high precision, as required for small and medium-scale topographic surveys, the Transit Satellite System is increasingly used in developing countries for establishing ground control points (e.g. Brazier, 1985).

It may be expected that in the next few years the establishment of GCPs on forested land will become common practice using satellite positioning, and that accompanying low cost satellite fixes and the geodetic rectification of satellite imagery, the number of GCPs per satellite scene to meet the accuracy standards of planimetric mapping will be considerably reduced. For example,

Welch *et al.* (1985), using first degree polynomials to rectify Landsat TM data of Iowa were able to obtain sub-pixel accuracy (e.g. residual mean square error $\pm$ 0.26) with four ground control points for quarter and full scenes. This is well within the accuracy standard for 1 : 50 000 planimetric maps.

5.4 SEASON AND TIME OF DAY FOR PHOTOGRAPHY

These are both important in the planning of successful aerial photography in order to ensure that the imagery obtained is the most effective and carried out at the minimum cost. The two factors, season and time of day, are closely linked, since having determined the best season meteorologically, this may be further shortened by daytime constraints, such as shadow length and reflex reflection. Even then, there may be a further constraint on the timing of the photography by the state of the vegetation best suited to image analysis, which can lead to a conflict of interests. For example, photogrammetrists and often geologists prefer the photography to be taken when the deciduous trees are leafless. Foresters and photogrammetrists may prefer a fairly high sun angle, while geologists and archaeologists prefer a low sun angle.

Season

The need to consider the seasonal weather conditions and the phenology of the vegetation has been commented on in Chapter 3. Sources of meteorological data should be checked, and, for large contracts, a study of the periodicity of cloud cover as shown on environmental satellite imagery (preferably over several years) is recommended. Studies of environmental satellite imagery, by the author and colleagues at the FAO Remote Sensing Centre in Rome, showed that periods best suited to cloud-free aerial photography can often be pin-pointed. The technique was used in 1974 and 1978 in planning high-altitude aerial photography in Sierra Leone and Liberia, where otherwise SLAR would have been recommended, and was further developed later for mainland Malaysia. In Liberia, the number of suitable flying days between mid-November and March was estimated as 18 days $\pm$ 7 days at the 90% confidence level, and was based on the study of the daily NOAA imagery covering a three-year period.

Dust haze in addition to cloud cover can be a limiting factor in the taking of aerial photographs, and can occur even if the country is distant from arid and semi-arid lands (e.g. in Sierra Leone from April onwards). Smoke from the seasonal burning of grasslands and forested lands is a further constraint. In the tropical woodlands of East and Central Africa, grassland fires occur annually and the haze is often so dense that aerial photography is to be avoided in August and September. In south-eastern Australia in bad fire years, smoke haze can prevent photography throughout most of January and February.

Whenever the terrain is mountainous, the use of colour photography may be restricted by the season, because of long shadows cast by the mountains, and can result in shaded valleys being 'blacked out'. In Victoria, Australia, this problem often precludes winter colour photography in the East–West oriented Great Dividing Range. At high latitudes (e.g. in northern Europe and Canada), the **shadow lengths** cast by trees may restrict the timing of all photography to 2–3 summer months. Welander (1952), working in Sweden, concluded that the constraint of excessive shadow favours a shadow length not exceeding 1.5 times the height of the trees; and 1.0 times provided photography ideal for forest interpretation.

In most circumstances in the planning of aerial photography for forestry, and other studies of the natural environment, consideration requires to be given also to the seasonal state of the forest and grasslands. This extends to the seasonal occurrence of pests and disease and to monitoring the effect of 'acid rain' in Europe and North America. The influence of leaf phenology on the spectral reflectance of foliage was commented on in Chapter 3. If maximum difference between species is required, then for temperate zone deciduous forests late spring or early autumn photography may be preferred; and in the tropics, where there is a long dry season, photography at the end of the rainy season or early dry season may be the most favoured (if cloud cover is not a problem).

In Wales, the late flushing habit of oak favours its identification on photographs taken in mid-spring; in North America the very distinctive colouring of some species in autumn (e.g. sugar maple, red oak) favours photography at that time of the year; and in Tanzania, muninga (*Pterocarpus angolensis*) can be most readily identified on large-scale photographs taken early in the dry season when the large yellowish dry fruits are conspicuous and the foliage is sparse. In central Europe, the mapping of crown dieback/defoliation caused by acid rain is best observed on imagery taken between mid-July and mid-September; and obviously if the main need is to separate stands of cool temperate conifers and deciduous broad-leaved species (angiosperms), then winter photography will be preferred. There are, however, many exceptions; and it is important therefore to spend time, when planning the photography, in studying carefully the seasonal behaviour of the vegetation; and, if available, its appearance on old photographs.

Time of day

In considering the timing of aerial photography within the selected season, there will be further constraints. As already indicated, shadow length is important; and thereby for forest studies, early morning or late evening photography is excluded. For example, Mikhailov (1961), working in the USSR, recommended that colour photography should not be used for forested lands unless the sun angle is more than 20° above the horizon, since the colour balance of the film is adversely affected; and Maruyasu and Nishio (1962) in Japan recommended a solar elevation of at least 45° when using colour film.

Particularly at lower latitudes, attention requires to be given to the sun's angle to the horizon (or zenith) to avoid recording the reflex reflection zone on the film; or at least to plan the taking of the photographs so that the (diffuse) reflection zone is confined to the corners of the photographs (Figure 5.5). The occurrence of the reflex reflection zone on the aerial photographs can be confusing to the interpreter. When viewing a pair of contiguous photographs stereoscopically, the reflex reflection zone is observed with one eye and the other eye sees a normal image. Within the zone, tree heights cannot be assessed accurately using a parallex bar and the shadows of the trees will be observed to be directed towards its centre. This is due to the radial displacement of the tree crowns exceeding the radial displacement of their shadows. The only way to minimize this effect, which is particularly severe on photographs of high forest

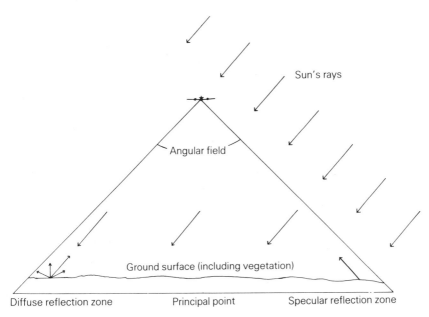

Figure 5.5 Diffuse reflex reflection zone - sometimes termed no-shadow point. Diagram illustrates the geometry of flying conditions under which the zone is recorded on the aerial film and hence aerial photograph. As shown above, the angular field of the aerial camera lens is sufficiently wide for the reflex reflected sun's rays to be recorded on the film, as it is exposed in the aircraft. Note that over water surfaces the specular reflection zone will be recorded on the film, being on the opposite side of the ground principal point (ground nadir). The aircraft is flying towards the reader.

at lower latitudes, is to use the longest focal length lens possible and to avoid flying at certain times of the day and year.

5.5 COST OF AERIAL PHOTOGRAPHY

Brief comment will be made now on the cost of aerial photography and its relation to other operations. These observations will be made in general terms, since to quote exact costs would relate to specific examples which could easily be misleading and which might also be rapidly outdated. The most important cost components are the positioning of the aircraft, unproductive grounding of the aircraft during bad weather, the actual flying time and aircraft maintenance during the taking of the photographs.

For many developing countries, the **positioning fee** for bringing the aircraft from its usual base to where the photographs are to be taken is the major cost component of mobilization. This problem also exists for some large developed countries. In Australia, for example, survey aircraft are usually based in the south; and for work in the north, this could require flying 3000 km or more. The only way to reduce this fixed cost is to combine the specific flight with other work, or to specify to the contractor several months or even a year in advance that he is to combine the work with other contracts. In general the shorter the notice when the flying is to be undertaken, the higher will be the tender price. If special equipment or a special aircraft (e.g. high-altitude jet) are needed, this will add to the positioning fee, since they may not be locally available, and also fewer companies can tender. The ultimate conclusion is, if possible, to keep the equipment and aircraft simple, but this may result in the use of older equipment and in lower quality photographs. If high-altitude aerial photography is favoured, the types of civilian aircraft to undertake the flying are few (e.g. Mystère, Learjet) and likely to be based commercially in Europe or North America.

If the area to be covered by aerial photographs is small, then the cost per photograph or per unit of land area can be very expensive, since the positioning fee has to be added to the cost of taking only a few photographs. In the USA, the cost of photographing a small area in Texas would not usually justify bringing a survey aircraft from California or the North-East. The only solution is to try to find others nearby who are planning aerial photography and to co-operate in sharing the cost of positioning the aircraft, even if this delays the work by a flying season. The alternative, if the area is very small, may be 70 mm miniature camera photography, as has been used in Argentina; but this will greatly increase the number of stereo-models and photographic control points.

In the actual taking of the aerial photographs, the major cost is the flying time of which up to 40% may be unproductive. To this must be added the time taken for refuelling, unsuitable times of the day for flying and unsuitable days when the aircraft is on the ground (i.e. standby costs). Flying time will be decreased by ensuring that the photographic scale is the minimum for the objective of the photography and ensuring that the sidelap (i.e. not endlap) is also minimal (section 5.1). By doubling the scale (e.g. 1 : 30 000 replacing 1 : 15 000), the cost can be expected to be halved.

So far, no mention has been made of the cost of the film, since this will probably contribute less than 1–4% to the total cost; but it is essential to ensure that all the exposed film is processed promptly and before the aircraft leaves the target area, so that any reflying, due to poor quality photographs, can be undertaken quickly and at minimum cost to the contractor. It will be appreciated that the choice of the film type will not add much to the total cost; but what will add considerably to the cost when colour is used at the same scale, particularly colour IR, will be the cost factor the contractor adds for uncertainties, including fewer flying hours, fewer flying days and, possibly, difficulties in colour film processing immediately following flying. Colour processing and printing is several times more expensive than black-and-white processing and printing. In combination, all these factors can increase the tender price for taking colour photographs by 20–40% or even more under very difficult or unknown conditions.

When the total cost of the aerial photography is considered as part of the overall cost including the end product (e.g. planimetric, forest stock maps), it will probably not exceed 10–33$\frac{1}{3}$% of this. For topographic maps, the ground control may cost up to twice the cost of the aerial photography; and the photogrammetric preparation of the maps may be three times or more the cost of the aerial photography. Excluding ground control, the preparation of ortho-photomaps can be expected to be more expensive than the aerial photography. Included in the cost of the aerial photography is usually the supply of the negatives, one or two sets of prints and a **print laydown**. The latter, sometimes termed an index mosaic, is a photographed assembly of all contiguous aerial photographs, so that the index of each photograph is clearly shown.

If planimetric mapping, topographic mapping and ortho-photomosaics are dispensed with and the end product is interpreted photographs with field checking, the cost of the aerial photography would be expected not to exceed per hectare the cost of the photointerpretation. Stellingwerf (1963) in Holland placed the cost of photointerpretation at a little less than the cost of the aerial photography. In conclusion, it is emphasized that the cost of the aerial photographic coverage will form but a small part of the total cost of a land resources or forest inventory.

5.6 TECHNICAL SPECIFICATIONS OF FLIGHT CONTRACT

In concluding this chapter, the technical specifications requiring to be considered when preparing a flight contract are summarized below. These reflect the contents of a typical US Forest Service contract and the comments given on flight planning in this chapter. They do not take into account the skills of the aircrew/contractor needed to cover the survey block nor the req₁ ired air navigational aids including gyro compass, doppler fix and possibly ground-based radio beacons to assist in ground positioning of the aircraft during flight. Other sources of flight specifications include national topographic survey departments and the *ITC Journal* (1985) on guideline specifications.

Technical specifications

1. **Confidential nature of the aerial photography.** This should be stated, including not divulging information except with written permission.
2. **Ownership of the photographic negatives and storage.** These are the property of the purchaser initiating the work and no sales or reproductions can be made without written permission. Storage arrangements should be specified.
3. **Responsibility for damage, losses, etc.** The contractor is fully responsible.
4. **Area to be covered (survey block).** Show clearly on an attached flight plan.
5. **Scale.** Requires to be maintained for stated ground elevations and as calculated with a permissible variation of 5%. Must be stated in contract.
6. **Overlap.** Along the line of flight this shall average 60% and shall be between 55 and 65%. Between flight lines, it shall average 30% and be between 15 and 45%. More precise details are sometimes given.
7. **Broken flight lines.** Two exposures of the new photography shall overlap the last two exposures of the old photography.
8. **Tilt.** This should average less than 2° and should not exceed 5°.
9. **Crab.** This should not affect more than 10% of the width of the photographs.
10. **The make, model and serial number of camera and lens and last date of lens calibration.** These should be specified. Recalibration of focal length to be specific for infrared photography.
11. **Film and filter combinations.** These must be specified, including the film type and its manufacture, and the filter type by preferably quoting the Wratten number.
12. **Photographic quality.** This must conform to the submitted samples. All photographs to be free from blemishes or cloud shadows. Cloud cover of 1% may be acceptable.
13. **Season of photography.** The date of completion of photography and delivery of materials should be given, as also should the earliest commencement date.

14. **Flight log.** This should be supplied to the purchaser and should give time, dates, etc. for each strip of photography.
15. **Delivery.** Standard delivery consists of two sets of contact prints and one set of index mosaics. Both sets of prints should be free from blemishes and preferably be semi-matt.
16. **Radial line plots.** It should be clearly stated who will furnish ground control.
17. **Business arrangements.** These include prices, delivery schedule, payments, cost of additional materials and cancellation provisions.

6

Electro-optical sensing and side-looking airborne radar

As indicated in Chapter 2, the two important airborne methods of remote sensing in addition to aerial photography are electro-optical sensing, well established through the use of satellites, and radar. In addition, video cameras are increasingly popular for collecting data at low level from light aircraft to supplement aerial photographs and satellite imagery. Television systems, although used on satellites (RBV), have the disadvantage in airborne sensing of lower spatial resolution than is provided by aerial photography and electro-optical scanning.

Side-looking airborne radar functions at wavelengths many times longer than aerial photography and electro-optical scanning. While electro-optical scanners operate at wavelengths of up to about 10^{-3} cm ($10\,\mu$m) SLAR operates best at wavelengths longer than about 3 cm. As a consequence of this difference in wavelengths, the microwave reflection of objects provide images which bear little relation to images formed by passive sensing in the photographic and reflective infrared spectra. Some objects, which are diffuse reflectors to the human eye and appear lightish toned and coarse textured on the imagery recorded at the shorter wavelengths, will behave specularly when actively sensed by radar, and appear smooth and darkish on the imagery. In addition to active sensing at longer wavelengths using radar, it is practical to passively sense in this microwave part of the electromagnetic spectrum, but the available image-forming energy is very low, and consequently the image resolution is very coarse.

As illustrated by Figure 6.1 and further discussed in this chapter, the principle of remote sensing by aerial photography, electro-optical scanning

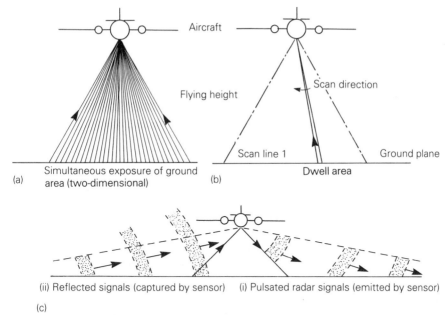

Figure 6.1 Diagram illustrating the difference in principle of sensing passively: (a) using aerial photography; (b) by electro-optical scanning; (c) sensing actively by side-looking airborne radar (SLAR). The arrows indicate the direction of energy-flow. Technical constraints prevent recording immediately below the aircraft using SLAR.

and side-looking radar differs. An aerial photograph (Figure 6.1(a)) provides simultaneously a perspective view of a relatively large area. An electro-optical scanner (Figure 6.1(b)) records within its instantaneous field of view a very small area, and builds up a line of recorded signals of these small areas (pixels). Each line is sigmoid in shape until geometrically corrected. SLAR (Figure 6.1(c)) emits a slit-like burst of energy, some of which is reflected back by the ground objects and recorded sequentially according to the traversed distance. Both electro-optical sensing and SLAR rely on the forward movement of the aircraft to build up the lines of signals. Aerial photography and electro-optical sensing provide examples of passive remote sensing, since they rely on an environmental source of energy (e.g. solar radiation); radar represents an active system, since the energy source required for the sensing is generated by the sensor.

6.1 ELECTRO-OPTICAL SENSING

Following the *Manual of Remote Sensing*, the above umbrella title is used in this section to cover frame sensors, pushbroom sensors and mechanical scanners.

Generally speaking, electro-optical image sensors can be divided into two main categories, namely electronic and mechanical. However, in practice there is a mix of technology. Frame sensors, developed from the television camera, have not achieved the same operational use as pushbroom sensors and mechanical scanners. Frame sensors include the return-beam vidicon camera (RBV) used on the earliest Landsat satellites, and in miniature are best represented by the amateur video camera used in airborne videography. Frame sensors are suited to imaging in broad bands of the visible spectrum; they do not require the support of a processing laboratory; and the video-film can be viewed immediately.

Both the pushbroom sensor and the mechanical scanner have important advantages over the single-lens aerial camera. They have the capability of sensing simultaneously in two or more spectral bands between about $0.3\,\mu m$ and $3\,\mu m$, while the aerial survey camera senses in a selected single broad band between about $0.38\,\mu m$ and $0.95\,\mu m$. Furthermore, narrower spectral bands can be used with an electro-optical sensor than with small-format aerial cameras; and, since the data is acquired in discreet picture elements (pixels), the data (like radar data) can be processed directly in digital form using an increasing array of computer-assisted programmes and without loss of radiometric accuracy. The ground area covered by each picture element will depend on the instantaneous field of view (IFOV) of the electro-optical sensor and the altitude of the space platform. However, the resolution of the aerial camera is always considerably finer for the same flying height, and this for most purposes outweighs the advantages of airborne electro-optical scanning.

Table 6.1 Multispectral scanning (MSS): example of sensor bands (University of Michigan/Bendix Corporation)

Band number	Wavelength (μm)	Location in electromagnetic spectrum
1	0.38–0.44	Visible spectrum
2	0.44–0.49	Visible spectrum (blue)
3	0.49–0.535	Visible spectrum
4	0.54–0.58	Visible spectrum (green)
5	0.58–0.62	Visible spectrum
6	0.62–0.66	Visible spectrum (red)
7	0.66–0.70	Visible spectrum (far red)
8	0.70–0.74	Far red/near IR
9	0.76–0.86	Near IR
10	0.97–0.106	Near IR
11	9.75–12.25	Far IR (thermal)

Mechanical scanners

These have been widely tested and used on both aircraft and satellites, and their success is probably best demonstrated by the Landsat programme started in 1972 (Chapter 7). The airborne instruments may be single channel, as often used in thermal sensing, or may be designed to sense in several channels simultaneously, when the term **multispectral scanner** (MSS) is used. Four-channel multispectral scanners were used on the Landsats, while airborne MSS may have eleven channels or more (Table 6.1). As multispectral scanners are designed with a small instantaneous field of view (e.g. 2.5 milradians) to sense in several wavebands, the amount of energy available for capture per spectral band is small, the noise produced by other factors can be limiting, and it is not presently feasible to improve greatly the resolution to make the resolving power comparable with aerial photography at the same flying height. Airborne MSS systems are generally operated at altitudes between about 300 m and 10 000 m.

Choice of spectral bands

The major advantage of mechanical scanners is their ability to provide synoptic data simultaneously in several bands including the mid-IR and far (thermal) infrared. The careful choice of spectral bands can provide information not otherwise obtainable by sensing in the photographic spectrum. The choice of spectral bands will be limited by atmospheric absorption, which is maximum at about 0.95, 1.1, 1.4, 1.8, 2.6, 2.8 and $6.2\,\mu$m for water vapour, 4.3 and $15\,\mu$m outwards for carbon dioxide and 2.9 and $9.6\,\mu$m for ozone. Transmission through the atmosphere will be greatest between these absorption bands (Figure 6.2). The wavebands within which the transmission occurs, and which are consequently used in remotely sensing the earth's surface, are frequently termed 'windows of the atmosphere'.

Studies in Michigan in the mid-1960s indicated that the thermal IR might be used to identify tree species, but there has been little progress in this respect. Later, Olson (1972) and Weber (1971) found that pine under stress and bark beetle attack have higher crown temperatures; and obtained differences between healthy and unhealthy trees, as high as $7°$C; but the average differences were only about $1°$C. Possibly environmental factors may have a greater influence. These include nocturnal air circulation (e.g. Fritchen, Balick and Smith, 1982). Recently Sader (1986), working in Oregon, has observed that aspect and slope gradients have a greater influence on thermal emission of young stands than older stands, due to the greater green biomass and canopy closure.

At present, operational thermal sensing in forestry is confined to fire detection, and was developed for this purpose by the US Forest Service. The thermal band between about $4.5\,\mu$m and $6\,\mu$m is used to locate, through

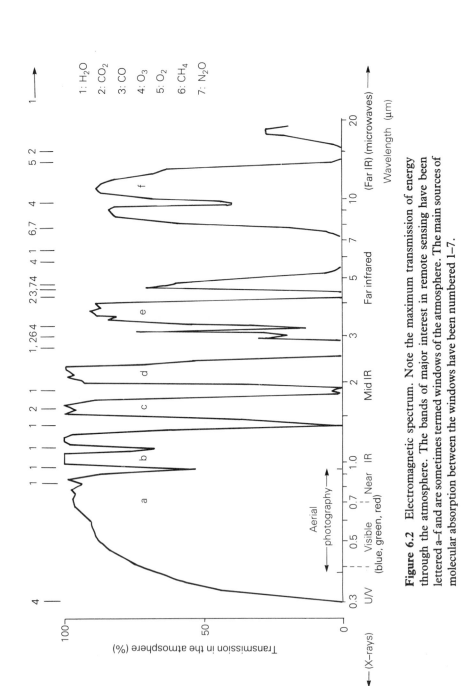

Figure 6.2 Electromagnetic spectrum. Note the maximum transmission of energy through the atmosphere. The bands of major interest in remote sensing have been lettered a–f and are sometimes termed windows of the atmosphere. The main sources of molecular absorption between the windows have been numbered 1–7.

smoke haze, the boundaries of forest fires. The noise factor produced by the earth's emission restricts using the $10-12\,\mu m$ band for this purpose, except to provide simultaneously a 'terrain map' to assist in aircraft navigation etc.

Scanner operation

The heart of the scanner is the detector, which transduces the incident reflected or emitted signal into an electrical signal. This is then recorded on magnetic tape and later computer processed into imagery. In older-designed equipment, the signal is amplified and displayed on a cathode ray tube; and, as with older radar, the permanent record is obtained by photographing the face of the cathode ray tube. The film advances across the CRT to record each single successive line scan; and resulting from the forward movement of the aircraft, each line scan will be S-shaped until electronically corrected.

A degree of error will arise in the imagery because of the aircraft tilt in flight and aircraft drift or crab, and results in the compression–extension of the images near the edge of the film. The processed imagery is not a series of separate perspective views from the nadir as the centre point, as with vertical aerial photographs, but a continual film strip built up from the scan lines. Unlike radar, the scanner operation records the ground below the aircraft, and thus provides a record of the flight track.

In flight, the scanning rate must be set so that, by the time scan line is completed, the aircraft has advanced by a distance equivalent to the ground area covered by the scan. Excessive tip of the aircraft will introduce gaps between the contiguous scan lines recorded on the film strip. This is termed **underlap**; and the converse is **overlap** (line compression), which is also used in electronic processing to increase the signal-to-noise ratio. When present, the reflex reflection zone will be recorded parallel to the flight track as a continuously over-exposed line of pixels.

As the aircraft moves forward (i.e. towards the reader in Figure 6.1(b)), the rotating mirror in the scanner moves smoothly across the field of view at right angles to the flight direction; and the recorded signals will produce a strip map of the ground scene (e.g. line 1). By adjusting the scanner speed to account for the flying height and ground speed of the aircraft, the film strip is built up line by matching line.

Within each scan, the rotating prism mirror with several faces dwells long enough on a ground area covered by the instantaneous field of view of the scanner for a meaningful signal to be recorded. The scan line is thus built up of a contiguous series of spectral radiance measurements, of which one is represented within the **instantaneous field of view** (IFOV) in the figure (dwell area). The IFOV refers to the narrow angular field of view of the sensor's detector, which during its scan of perhaps 120° at right angles to the flight track, will dwell momentarily over each small ground area in order to provide a distinctive recorded signal. This signal in its processed form is termed a **picture element** or **pixel**.

Since the angular resolution of the airborne scanner is fine (e.g. 1–2 milradian^2) for normal flying heights, the image of each large ground object (e.g. a tree) will be represented by a number of individual radiance measurements. If the instantaneous field of view (resolution angle) is one milradian, ground resolution will be approximately $0.1\,m^2$ for a flying height of 300 m.

If one small object in the instantaneous field of view provides a very powerful signal, then this may be sufficient to so change the averaged recorded signal that its pixel will have a grey scale value sufficiently distinct for it to be exactly located in the digital processing of the recorded data. Thus a small air navigational light on the ground may be used in airborne thermal scanning at night to provide a ground control point; and a 60 cm mirror reflecting the sun's rays is sufficient to saturate a single Landsat MSS pixel covering a ground area of about 0.44 ha. This can then be used as a ground control point.

Pushbroom sensors

These have become important operationally since the launch of the Satellite Probatoire pour l'Observation de la Terre (SPOT). Unlike mechanical line scanners, this type of sensor has no moving parts, which results in a longer life for the sensor, and relies not on a single detector but on an array of detector elements, which are termed **charged couple detectors** (CCDs). In the design of SPOT-1, which is considered in the next chapter, there is an array in the sensor of 3000 CCDs for fine sensing in the visible spectrum as a single band.

The disadvantage of the pushbroom sensor as compared with the mechanical scanner is that suitable detector elements are not generally available for sensing much beyond about 1 μm and calibration of the individual CCDs is more difficult. Identical with the mechanical scanner is the use of the forward movement of the aircraft to provide the scan in the flight direction (i.e. the x-direction). The scan at right angles to the flight direction (i.e. the y-direction) is achieved simultaneously by the fixed wide-angle array of CCDs. Usually the signal from each detector is amplified separately before being sampled sequentially to provide a serial representation of the ground scene within the angular field of the CCDs. Resulting from the system design, the dwell time of the detector elements on the earth's surface is longer and this is an advantage over the mechanical scanner, since the noise factor is reduced.

Video cameras

With the increasing cost of collecting field data in many countries and of verifying fieldwork, and the need often to supplement the coarse resolution of satellite data, interest is expanding in using video cameras in some of the tasks pioneered by small-format aerial photography. This is particularly encouraged by the increasing range of amateur video cameras, which are decreasing in cost and are improving in image quality.

An example of the use of videography in the tropics is provided by the national forest resources inventory of the Philippines (Schade, 1987). A three-stage sampling design was used involving on the ground 6-point relascope clusters to provide timber volume estimates. Landsat MSS false colour composites were used to separate forest cover from 'other land use' and to stratify the forest into closed broad-leaved forest and brushland. Flight strips were positioned in these strata; and imagery, accompanied by voice recordings during flight, were taken by a vertically-mounted video camera in the aircraft. The strip sampling yielded 2900 pairs of video and Landsat observations, which after visual and computer analysis provided the ground area of three forest types (i.e. mossy forest, dipterocarp old growth and dipterocarp residual forest) in addition to brushland and 'other land use'. A previous simulation study showed that sub-sampling Landsat MSS imagery using video camera-covered strips of 500 to 1000 m width, gave similar results to sub-sampling with 1 : 60 000 scale black-and-white aerial photographs; but was considerably faster and at a much lower cost of analysis.

Videography has the advantage over small-camera photography in that the imagery can be viewed immediately after acquisition, the time delay of film processing is avoided and there is no need to consider setting up a field/local film processing and printing laboratory and the employment of a skilled technician. It is a fact that these factors sometimes restrain the use of small-format aerial photography in forestry. The main disadvantages to using video cameras are low spatial resolution, which can be expected to provide image quality similar to commercial TV, the lack of flexibility of film–filter combinations, the lack of stereoscopic viewing and the lack of experience in analysing video data for forest purposes. Video cameras are available for providing black-and-white imagery in the spectral range of panchromatic photography, in colour using either a striped colour filter or a higher quality optical system to separate the incoming reflected solar radiation into its three primary colours, and can be modified by replacing the photomultiplier tube(s) to record in the near-infrared out to about 1.3 μm. Typical resolution is 240 lines per frame using 2 cm magnetic tape as compared with more than one million grains on a 35 mm aerial photograph and assuming a resolution of about 40 lines per millimetre.

A video display of 240 scan lines per field of view with currently available video cameras is usually sufficient to avoid flickering caused by image motion in flight, when the analogue scenes are being individually examined on a TV-type monitor. As video cameras have extremely short focal lengths (e.g. 8 mm, 18 mm), the flying altitude is low (e.g. 500–2000 m) in order to obtain imagery at usable scales and covering an adequate ground area. This can be a limitation to the technique in difficult flying conditions (mountainous terrain) but permits flying under high cloud cover.

As yet there is a dearth of publications on the use of video cameras in aerial work related to forest resources. Basic principles, for example, have been covered by Meisner (1986), Measuronics Corporation in Montana (USA), and the general applications in remote sensing by Vlcek (1983) and Meyer *et al.* (1983). Videography has already been found to be well suited as a support to aerial reconnaissance related to damage assessments and animal counts, and as providing a record of clear felling operations and the progress of annual plantings.

6.2 PASSIVE MICROWAVE IMAGE SENSING

Although passive microwave sensing functions in approximately the same wavebands of the electromagnetic spectrum as K- and X-band radar, its operating principle is not that of this active remote sensing system but the same as used in thermal (IR) sensing. Its energy source is emitted electromagnetic radiation by the earth's surface, and is the same source, which peaks in the thermal band at about 10.6 μm and which tails off rapidly towards the microwave band. The very low energy source has been the major constraint to the civilian development of passive microwave sensing.

Instrument design imbues the technology of both thermal sensing and radio astronomy. The microwave imaging radiometer consists basically of a large paraboloidal mirror (one to several metres in diameter), at the focus of which is placed a small detector to collect the microwave energy and to feed it to an amplifier. As the spatial resolution is governed by the size of the mirror, which is limited on the aircraft, the dwell time on each ground element has to be long. Also, the spectral band needs to be broad, and each captured signal is weak and therefore requires to be greatly amplified before images can be recorded of ground objects. Thereby, the noise factor is a problem since strong signals are received from four sources, namely emitted energy by the ground objects/ landscape, emitted energy by the atmosphere, transmitted energy from the terrain subsurface and reflected microwave radiation by the atmosphere.

In combination, these factors favour passive microwave sensing as providing very coarse resolution imagery useful only at the regional and global levels and collected by satellite. There are no known forestry applications of passive microwave sensing, but in the medium term this type of sensing may prove useful for the regional assessment of soil moisture and contribute to the mapping of rock types. In the fields of synoptic meteorology, climatology and oceanography, the usefulness of passive microwave sensing is already recognized. More information will be found in the *Manual of Remote Sensing* (American Society of Photogrammetry, 1983).

6.3 RADAR

There are three well-established radar systems, namely plan position indicator radar (PPI), real aperture radar (RAR) and synthetic aperture radar (SAR), and, of these, the last two are used in forestry. In everyday life, radar is to be observed on a TV screen, when a passing aircraft causes flickering by introducing a time delay between signals.

The great advantage of radar over other remote sensing methods is its near-all-weather capability of providing imagery, including the penetration of complete cloud cover, light to moderate rain and snow. Airborne radar can scan sideways distances exceeding 30 km. Until recently, its main disadvantage has been the low resolution of the imagery as compared with aerial photography and, to a lesser degree, electro-optical scanning; and consequently, in forestry, it has been restricted operationally to reconnaissance surveys.

In general, the price of radar coverage per square kilometre tends to be competitive with aerial photography provided fine-image resolution is not the primary consideration. For aerial photography the price will be influenced by several major factors including problems associated with cloud cover and number and distribution of flying days a year; and, for radar, the geographic location of the SLAR equipment and the purpose of and type of information required from the aerial survey.

Real aperture side-looking airborne radar (RAR/SLAR; brute force radar)

Real aperture radar depends on a fixed side-facing slit antenna mounted below the aircraft and serving as both transmitter and receiver. The slit, being fixed, cannot scan mechanically as with electro-optical systems, but emits a narrow spreading path of pulsed radio waves of known wavelength. The active emission of the radio waves spreads out in a narrow path at right angles to the aircraft direction of flight (Figure 6.1(c)). This is known as the **range** direction, and depending on the radar system may be from one side of the aircraft or both sides (as shown). Some of the waves reflected by objects on the ground are captured by the antenna as the aircraft moves forward (i.e. towards the reader in Figure 6.1(c)–ii). Scanning is continued by the aircraft's forward movement. The time for the return of the signals from the ground objects depends on the slant range distance between the ground and the antenna mounted on the aircraft.

In order to be imaged separately, two ground features that are close together must reflect differently in time, so that their signals are received separately by the antenna. Primarily for this reason, radar, unlike electro-optical scanning and the vertically mounted aerial camera, does not record the scene immediately below the aircraft. Within shadows on SLAR, unlike those on aerial photographs, no details are recorded. Depending on the design and versatility of the SLAR system being used, the **depression angle** (vertical angle of illumination, look angle) can be adjusted to best suit the objectives of the survey, flying height and terrain conditions. Since the scanning is continuous at right angles to the flight direction, the images on film will be in the form of a continuous strip.

Geometric principles

The geometric principles involved in SLAR imagery are shown in Figure 6.3. The angular beam width β represents the swath of pulsed radio waves emitted from the transmitter in the aircraft, some of which, on reaching the ground, are

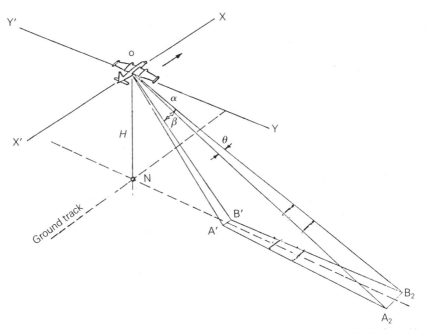

Figure 6.3 Diagram illustrating the major geometric principles of side-looking airborne radar. The flight direction is indicated by an arrow. H is the flying height; α is the depression angle of the pulsed signals transmitted from the aircraft; β is the width in the vertical plane of the pulsed signals; NA_2 is the ground range – the distance NA' is not covered by the signals; oB_2 (or oA_2) is the slant range; oB' is the near range; $A'B'B_2A_2$ represents the image swath. The recorder of the reflected signals on the aircraft 'measures' in terms of slant range and not ground range. This causes spatial distortion of the raw signals in the horizontal plane, which can be computer corrected during image processing. The image resolution declines outwards, away from the flight track, in the direction of $A'B'$ to A_2B_2.

reflected back and captured as the image-forming signals. The depression angle of the antenna is represented by α and is set to avoid receiving signals at or near the vertical, which also results in the ground nadir (N) and flight track being outside the imagery. A comparison of the **near range** and **far range** indicates that all ground objects are not beamed at the same angle and radio wave intensity. Consequently identical objects, including similar forest types, may return different signal strengths. The direction o to B is the **range direction** and this distance as measured to a target is the **slant range**. The distance along the ground from N to B_2 is the maximum **ground range**.

The ground resolution of the SLAR system will be controlled by two independent parameters. As the angular beam β fans out on reaching the ground at increasing distances from the aircraft, the distance between objects has to increase for them to resolve separately on the imagery. This is termed **azimuth resolution** (i.e. in the flight direction) and differs from the **range resolution**, which occurs in the range direction and depends on the period of time that the antenna emits its burst of energy and the slant distance between the two objects. Usually the nominal 'across the flight track resolution' of real aperture radar is finer than in the azimuth direction (e.g. Westinghouse 25 m, 8 m; Texas Instruments 46 m, 15 m). The two are similar for synthetic aperture radar (e.g. Goodyear 10 m; Goodyear Gems 15 m; Macdonald Dettweiler IRIS 6 m).

Object/image characteristics

The intensity of the image-forming radio pulses, reflected back to the antenna, depends on the object's electrical properties and geometric properties. In the microwave part of the electromagnetic spectrum, the electrical properties of natural surfaces are usually expressed in terms of their (complex) dielectric constants, which are indicative of their reflectivity and conductivity. Most natural dry surfaces have dielectric constants in the range of 3–8, but water on the other hand has a dielectric constant of about 80. Hence the amount of water/moisture in soil or vegetation greatly changes their dielectric constants and their reflected signals would be expected to vary accordingly. This fact is useful in discriminating between dry and moist natural surfaces, including dry soil and green vegetation, and water and dry land. On the other hand, this can result unfortunately in the same natural surface having more than one distinctive signal.

Concerning geometric properties, both the micro- and macro-features of the landscape are important in the image-forming process. When the radar resolution is finer than about 10 m, many individual trees can be discriminated on the imagery, and more forest types recognized. As mentioned previously and depending on the choice of the radar wavelengths, some micro-surfaces will behave as specular reflectors, which in the solar spectrum would behave as diffuse reflectors. Heavily grazed dry grassland, for example, may behave specularly to the radar, while on aerial photographs it would be recorded as a diffuse surface. As was indicated in Chapter 3, for reflection in the visible spectrum, reflection can be characterized as being reflex, specular, diffuse and spread; but the component captured by the radar sensor is normally the reflex component which is referred to as **backscatter**.

An exception occurs when the specular or spread component is reflected back to the radar sensor by a second surface approximately at right angles to the first surface. This is referred to as **corner reflection**, and most commonly occurs in urban areas, where buildings and the paved road surfaces act as dihedral or trihedral reflectors. The compounded signals can result in small whitish point 'targets', which serve usefully as ground control points. Small man-made dihedral or trihedral targets can also be used for this purpose. Corner reflection is produced by a cliff face, and can be very conspicuous on imagery along the line formed by cliffs merging with the sea. It has also been observed in seasonal swamp forest in Australia, when the corner reflection is caused by the water surface and tree boles (section 15.6).

The physiognomy of the woody vegetation strongly influences the tone and texture of the recorded radar images, but usually cannot be predicted ahead of the imagery acquisition. Frequently the image boundaries of plant formations, and sometimes the boundaries of plant subformations or forest types, can be accurately identified and delineated; but the imaging characteristics of radar can be counter-productive. Depending on the radar wavelength, the recording can be a mixed signal produced by the surface roughness of the crowns of the trees, the understorey and the texture of the terrain, in which changes in the main canopy layer do not have the greatest influence. There may be abrupt and unpredictable changes in the dielectric constants of the terrain surface, which emphasize geomorphic features or local soil moisture and not the vegetation or soil boundaries. The reflective properties will be very much dependent upon the water content of each of the surfaces, so that reflection will be highest from the surface with the greatest moisture content. Furthermore, even when the forest cover and terrain are uniform over extensive areas, the slant range will influence the texture and tone of the recorded images.

In considering macro-features of the landscape, differences in the ground relief strongly influence the geometry of the imagery according to the slant range and the angle of the incident signal on the ground. Whereas the relief displacement on vertical aerial photographs is radial outwards from its principal point, the relief displacement of SLAR images is perpendicular to the flight line/ground track (Figure 6.4(a)). In rugged terrain, the relief displacement is reversed, due to the fact that the image-forming signals from the top of the object are received first by the antenna (Figure 6.4(b)). This results in the recorded images leaning towards the flight line and not away from it, and is termed **layover** or radar layover. Layover is most severe at near range and least conspicuous at far range, more noticeable the more the object tends to have a vertical facing slope to the radar signals and is accompanied by lengthening black shadows in the direction of the far range. In Figure 6.4(b), the layover is maximum in the vicinity of point 1. No ground details are recorded within the strongly shadowed areas.

Choice of wavelength

Side-looking airborne radar is usually designed to operate at a specific wavelength/wave frequency, or occasionally at two or three wavelengths. The use of civil multiband radar is experimental. As radio waves are used, the frequencies at which radar operates are allocated by the International Telecommunications Union (ITU). By convention, military letter codes are used to designate each of the several wavelengths, which consequently are not related to their civilian use.

Table 6.2 lists the approximate wavelengths and their frequencies, which have been used in civilian SLAR studies, and shows the corresponding code letters commonly used to designate the wavebands. Wavelength (λ) is converted to frequency (F) using the formula: $F = c \times \lambda^{-1}$, where the constant $c = 2.998 \times 10^{14}$. Wavelengths shorter than about 3 cm are not favoured, since adverse atmospheric conditions can attenuate the signals. However, short wavelengths (e.g. 1 cm) are used in ground-based PPI systems at airports to map areas, where rain is falling. An additional airborne or satellite problem when operating at shorter wavelengths is that more energy is needed to power the systems.

Although radar signals exceeding about 3 cm are little affected by cloud, haze, fog and smoke, only long wavelengths can adequately penetrate heavy precipitation; and, at shorter wavelengths, the radar reflection by the water droplets may be substantial and provide an excessively high noise factor. The longer wavelengths will provide much less information on surface roughness of vegetation than the shorter wavelengths, but more information on the terrain. In forestry, the shorter wavelengths are preferred, while soil scientists and geologists will usually obtain more information from imagery acquired at the longer wavelengths. In early SLAR studies, K_a-band was used and gave some of the more interesting published results on vegetation mapping using radar (e.g. Morain and Simonett, 1967). Most modern equipment operates in the X-band, followed by L-band and experimentally by C-band. The Environmental Research Institute of Michigan (ERIM) has equipment adaptable to C-, L- or X-band by changing the antenna. At the University of Kansas (USA), several experimental studies have indicated that C-band is to be preferred in agricultural studies, including agricultural crop identification.

(a)

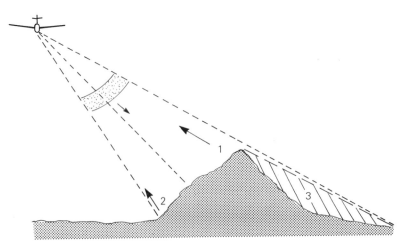

(b)

Table 6.2 Wavelengths used in radar showing their corresponding code numbers

Wavelength (approximate)/ waveband (cm)	USA	UK
0.4	–	O
0.6	Q/W	V
1	K_a, K	Q, K
2	K_u	J
3	X	X
5	C	C
10	S	S
20	L	L

Polarization

Irrespective of the waveband, the direction of the vibration of the emitted signals of the SLAR system is controlled, so that the signals are emitted in a single plane (i.e. **plane polarized**). Usually the transmitting signals are horizontally plane polarized (H) and the return signal is filtered to be horizontally received (H). Likewise it can be vertically transmitted or received and this gives four notations (i.e. HH, HV, VH, VV). Experimental results by varying these combinations suggest that the discrimination on the imagery between vegetation types can be improved if two different combinations are used over the same target area.

Figure 6.4   (a) The relief displacement on SLAR imagery is at right angles to the flight line. In the above photograph strip of continuous recording on 70 mm film enlarged to a scale of 1 : 400 000, the high mountains have been displaced towards the ground track, which is outside the top edge of the photograph. Note that towards the bottom of the photo-strip, the hill shading is away from the ground track of the aircraft. Texture on the steep mountain slopes is mainly provided by mature tropical rainforest. (b) Schematic drawing of a mountain as recorded towards the bottom edge of (a). The aircraft is flying towards the reader and the radar signals are at right angles to the flight direction. The reflected signals from the top of the mountain (1) will be recorded before the signals from the base of the mountain; and hence the mountain will be recorded as leaning towards the flight track. The mountain slope (3) which is more or less parallel to the pulsed signals will reflect little or no energy towards the aircraft and hence records as shadows on the imagery.

Synthetic aperture side-looking airborne radar (SAR; coherent radar)

The use of longer wavelengths and quality improvement of the imagery requires increasing the antenna length; and obviously there is a physical limit on the length of the antenna that can be accommodated on the side of the aircraft using real aperture radar (RAR). To overcome this constraint, side-looking airborne synthetic aperture radar or SAR has been developed and is now widely used. For example, a synthetic aperture antenna of 2 m can function as equivalent to a real aperture radar antenna of about 600 m. SAR has the added advantage of providing uniform azimuth resolution, irrespective of range distance; while with RAR, the resolution in the azimuth direction is increasingly coarser the greater the range distance. Otherwise, the comments given in the previous section apply to SAR.

The basic concept of SAR or coherent radar is to record the change in amplitude and frequency of the reflected signals, as the antenna (fixed on the aircraft) moves forward in the azimuth or flight direction. A change in the observed frequency of electromagnetic waves and sound waves caused by the relative motion between source and observer is termed a **Doppler frequency shift** or Doppler effect, and can be illustrated by the change in tone of sound waves caused by the whistle of a train, as it passes by the observer. The recording and primary processing of SAR signals, however, is much more complicated than for RAR and for information, reference should be made to *Imaging Radar for Resource Surveys* (Trevett, 1986) and the *Manual of Remote Sensing* (American Society of Photogrammetry, 1983).

In RAR systems, the images of the ground objects are continuously displayed on a cathode ray tube with their brightness varying according to the strength of the reflected signals of the ground objects. The permanent record is usually obtained on negative film, which is passed across the cathode ray tube proportionally to the ground speed of the aircraft. Normal photographic techniques are used to process the film, to obtain prints at contact scale (e.g. 1 : 250 000, 1 : 400 000) and enlargements of up to about 1 : 50 000.

In SAR systems, the signals can also be recorded optically on film, and the signal film (raw data film) later used on a specially designed optical correlator to produce normal imagery film. Usually, however, the signals are recorded on to high-density digital tape (HDDT) and later transcribed on to computer-compatible tape (CCT) for use with a digital analysis system or converted using a photo-writer into analogue images on standard film strips.

The main advantage of electronic recording and processing SAR data is that the dynamic range of the signals is retained and permanently recorded and handled digitally; while in photographic recording systems, the signals are limited to a grey-scale range in black-and-white of about 50 to 100 shades; and the slant view inherent in the imagery will generate some of the problems associated with oblique aerial photography. However, in electronic processing, there is the problem of handling large amounts of data with the number of pixels (picture elements) varying as the square of the resolution. With a resolution of 50 m the number of pixels is only 400 per square kilometre, but with 30 m resolution about 1100 and with 10 m resolution over 8000.

7

Satellite remote sensing

The collection of data about the earth from outer space became feasible with the launching of the USSR satellite, Sputnik-1, in 1957; and the usefulness of recorded satellite data was clearly demonstrated by the interpretation of colour photographs with a spatial resolution of about 100 m, which were acquired during the USA missions of the early 1960s (e.g. Mercury, Gemini).

As pointed out in Chapter 2, satellite remote sensing (SRS) in forestry can be considered for practical purposes to have started with the launching in 1972 of the US earth resources technological satellite (ERTS-1), which was later renamed Landsat-1. The major value of the Landsat programme to forestry has been the continuity of satellite data supply for nearly 20 years; being able to plan with the knowledge that the same area of the earth's surface is covered repetitively, albeit cloud cover can be a serious constraint; that the resolution of the satellite data has been adequate for the development of practical applications in forestry; and that the data is increasingly used in combination with airborne data as part of multistage sampling.

Nowadays, there is an ever-widening choice of data source, data products and imagery analysis techniques. In the following sections, satellite technology relevant to forestry will be considered under two broad groupings: (a) earth resources satellites, which are of major importance, are sometimes called earth observation satellites (EOS), and include the popularly termed second generation satellites of Landsat TM and SPOT and (b) the environmental satellites, which provide much lower ground resolution imagery than the earth resources satellites, and the data from which are primarily used in meteorology.

Further subdivision of each can be made technically according to the type of sensor and performance characteristics of the satellite platform. On the basis of sensor the US Landsats use electro-optical scanners and earlier also return-beam videcon cameras (RBV). The French SPOT, as mentioned previously, uses fixed couple charged detector (CDDs) (Brachet, 1986); the USSR Salyut/Soyuz frame cameras (Oesberg, 1987) and the European (ESA) satellite and the planned Canadian one use synthetic aperture radar, as was also used on the short-lived US Seasat in 1978. Little information is available on the imaging satellite programme of the People's Republic of China, which launched its first earth resources satellite in 1975. India also has its own imaging satellite programme and with USSR launching assistance placed its first satellite, Bhaskara, in orbit in 1979. Comparative spatial resolutions are Landsat MSS (80 m), the Japanese MOS (50 m), Landsat TM (30 m), SPOT (10, 20 m), USSR/GDR MFK 6-channel camera (10–20 m) and the USSR metric camera (5 m). Landsat TM is the only earth resources satellite system that operates not only in the visible and near-IR spectrum but also in the mid-IR and thermal IR. The spectral bands of the Japanese Maritime Observation Satellite (MOS) are similar to Landsat MSS.

The environmental satellites on the basis of orbit can be grouped according to their being polar orbiting, or stationary relevant to the rotation of the earth (i.e. **geostationary**). The former include the US NOAA series of environmental satellites and the USSR Meteor imaging satellites with RBV (1 km resolution) and thermal IR scanner (8–12 km resolution). The earth resources satellites are also polar orbiting and are unmanned with the exception of USSR Salyut/Soyuz series and several US short-period experiments. The latter include Skylab in 1973 and the ongoing Space Shuttle programme, which was not operational in 1986–88. The experimental flights of the reusable Space Shuttle have included synthetic aperture radar (Sir-A, Sir-B), metric camera photography and the first sensing using charge couple detectors.

With the exception of military satellites, which may be made to dip to below 100 km above the earth's surface and thereby have a shorter life expectancy, the earth resources satellites with the exception of Salyut/Soyuz at about 260 km, orbit at altitudes between about 600 km and 950 km. Landsats 1–3 orbited at about 900 km, Landsat-4 onwards at about 707 km and SPOT at a nominal altitude of 830 km. The environmental satellites either orbit at higher altitudes (e.g. NOAA satellites at about 1450 km) or are geostationary at even greater distances from the earth (e.g. GOES, Meteosat at 36 000 km). The USSR Meteor series orbit lower at about 600 km altitude.

Resulting from the acquisition of enormous quantities of data by these satellites, there is a growing need for new methods of archiving data, new retrieval techniques and much faster data processing and imagery analysis systems. As indicated later, a single Landsat multispectral scene in four bands consists of about 30 million pixels of data; and with the trend towards the acquisition of imagery with finer resolution by SPOT and earth resources radar satellites, a rapid expansion of acquired data can be expected in the years ahead. This will necessitate expanded training programmes in satellite applications and the establishment of more regional data processing facilities.

Unfortunately, not all the recorded data has been archived and much of the archived data is on photographic film and not on digital tapes. Archives of wide interest exist in several countries (e.g. Japan, UK, USA, USSR, Canada, Brazil, India), in regional centres (e.g. in Europe at ESRIN/ESA) and in the United Nations family at the FAO Remote Sensing Centre in Rome, Italy. The most extensive available collection of environmental satellite data is held by the National Environmental Satellite Data and Information Service of NOAA in Maryland (USA). This latter archive dates back to April 1969 and contains data of wide scientific interest obtained by the sensors on board more than 40 environmental satellites.

7.1 EARTH RESOURCES SATELLITE SYSTEMS

Over 120 member states of the United Nations have used earth resources satellite data to assist developmental activities. The principal reasons are economic and the availability of up-to-date data, although often current aerial photography would be preferred if it was available. The major global suppliers of current data are the Earth Satellite Corporation (Earthsat) with its headquarters in Maryland (USA) and SPOT Image with its major facilities at Toulouse in France.

In addition, there has been a worldwide expansion of ground receiving stations, which acquire and process data to meet regional and national needs. There has also been the development of national data contact points and for the USSR Salyut/Soyuz regional contact points (e.g. Finmap, Helsinki for Western Europe). Landsat receiving stations now cover most of the world, but exclude part of subsahelian Africa, Central America, north-west South America, Antarctica and part of Siberia. However, not all these stations have the capability of receiving both Landsat TM and MSS data. North America has the highest density of receiving stations with stations at Fairbanks, Goldstone and Greenbelt in the USA and at Prince Albert and Shoe Cove in Canada; and Europe has stations in Italy and Sweden.

A network of SPOT regional centres is also being developed, which includes data receiving stations in Canada for North America, Brazil for Latin America and Bangladesh and India in Asia. Also to be determined are the station localities of the polar-orbiting European land applications satellite system (LASS), the Japanese earth resources satellite (ERS), the Canadian radar satellite and the Indian satellite for earth observations (SEO). Existing facilities are also being modified to receive MOS (e.g. in India and Canada). Brazil and Indonesia have expressed interest in launching equatorial orbiting satellites.

There is little doubt that the data of the USSR earth resources satellites would have been much more widely used if bureaucratic procedures had earlier permitted its more ready availability and speedy delivery, since the resolution of the imagery, as fine as 5 m, makes it suitable for many natural resource applications. Some countries, including the USSR, maintain that imagery with a resolution finer than 50 m can only be supplied to the country it covers and the imagery has to be ordered by nationals within the country concerned.

Orbits of Landsat and SPOT

The satellites are **polar orbiting**. That is, all the orbits pass over or pass close to the North and South Poles (i.e. Landsat orbits at about 9°N). Both are **sun-synchronous**, which implies that the timing of their passes across the equator keeps precise pace with the earth's westerly rotation and is at the same sun-time each day. In consequence, shadows under clear skies on the imagery will match over a short period of time, but will be incongruous when compared on a seasonal basis, due to the change in time of sunrise.

Whatever the most suitable shadow length in aerial photointerpretation, the analyst has to accept satellite imagery at a sun-synchronous local time throughout the year of approximately 1100 hours and 1030 hours for Landsat and SPOT respectively. The only alternative is to choose imagery on a seasonal basis according to the time of sunrise, but then the condition of the forest is likely to be different. Resulting from convergence of the orbits at the poles, there is minimal lateral overlap (sidelap) of the imagery at the equator and increasing sidelap polewards (e.g. about 85% for Landsat at 80°N).

Landsat passes over the same area of the earth's surface every 16 days (previously 18 days) and SPOT every 26 days. As SPOT can also record in a sidewards direction, oblique views of the same ground areas at different look angles can be obtained every 2–5 days depending on the pass and latitude (Figure 7.1). However, it cannot be assumed that imagery can be obtained in accordance with the orbiting cycles, since in practice clouds and possibly heavy haze may prove to be major constraints. In monsoonal countries, maximum cloud cover usually coincides with the growing season of agricultural crops.

Even after nearly 20 years, cloud-free imagery has not been obtained for some regions, particularly the climatic zones of tropical rainforest. On oblique satellite imagery, due to the lower look angle, even moderate cloud cover may record as 100% on the imagery. That aerial photographs can be acquired seasonally in a region does not imply that earth resources satellite imagery can

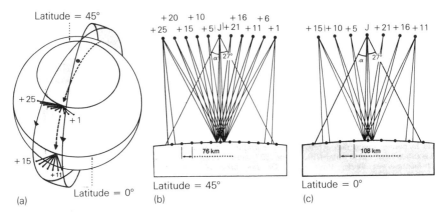

Figure 7.1 Characteristics of SPOT. As indicated in (a) with the satellite being in polar orbit, the ground tracks converge at the poles and are maximum spaced at the equator (cf. (b), (c)). The same ground track is covered every 26 days; but as the look angle (27°) of the twin electro-optical sensors on board SPOT can be varied, there are six opportunities in each 26-day cycle to record obliquely the same ground area (i.e. days 15, 10, 6, 21, 16 and 11 in (c)). Stereo-pairs of photographs are obtained with a scene size of 60 km × 60 km and a ground base-to-orbiting ratio (i.e. base : height ratio) of 0.1 to 1.2.

also be acquired, since aerial photography can be obtained at different times of the day and over a sequence of several days and also under high cloud cover.

Similarly to aerial photography, satellite sensing incurs errors due to instability of the (outer) space platform during orbit. In the case of aerial photography, quality control of the imagery is exercised through the flight specifications, through the pilot/navigator/photographer, checking the aerial photographic proofs, and during film processing and photographic printing. With satellite sensing, corrections are made by the ground monitoring station and during the primary processing of the acquired data. In-flight corrections include adjusting for the drag effect of the atmosphere, so that the satellite crosses the equator exactly at the same times; ensuring that the satellite maintains its scheduled altitude (i.e. in the z-direction), as otherwise the imagery scale and pixel size would be changed; and preventing drift of the satellite off its flight track. Swing (yaw) and tilt (roll) are also corrected, since these adversely affect the x and y geometric accuracy of the imagery. Occasionally, the result of the latter will be observed in analysis, when a redundancy of pixel data (overlap) or an absence of pixel data (underlap) causes a **line striping** across the imagery. Repetitive occurrence of the striping would suggest a faulty detector. For further information, reference is recommended to the *Manual of Remote Sensing*, chapter 12.

Spectral characteristics

The sensors of both SPOT and Landsat record in the visible and near-infrared parts of the electromagnetic spectrum; but the electro-optical mechanical scanners, as used on recent Landsats, also provide the opportunity of recording in the mid-infrared and the far-infrared (i.e. thermal). In contrast, the current charged couple detectors (CCDs) of the pushbroom scanners used on SPOT do not permit sensing in the near-infrared beyond 1 μm. Of historical interest is the television-type 3-band return-beam videcon camera (RBV) operated on Landsats 1 and 2; and by which, due to system failure, less than 2000 scenes were recorded. The spectral bands in which the current sensors of Landsat and SPOT were designed to operate are shown in Table 7.1.

As the recordings take place in parallel in several spectral bands, several simultaneous brightness readings are recorded. These signals are converted into grey-scale values and radio-transmitted to the ground receiving station. The data pixels are directly transmitted to a regional ground station, or stored on tape on the satellite for relay to the ground station when in range; or as has been used experimentally with Landsat, data is transmitted directly via a tracking data relay satellite system (TDRSS) to a single station in the USA (i.e. White Sands, New Mexico). X-band is used for the transmission of Landsat TM and SPOT data; and x-band and s-band for Landsat MSS data, since older stations receive only in s-band.

Table 7.1 Spectral characteristics of Landsat and SPOT (in micrometres)

Landsat TM	Landsat-5 MSS	SPOT* MS	Remarks
0.45–0.52			Band improves water penetration
0.52–0.60	0.50–0.60	0.50–0.59	Vegetation reflectance peaks c. 0.54
0.63–0.69	0.60–0.70	0.61–0.68	Maximum chlorophyll absorption c. 0.66
	0.70–0.80		Near IR begins
0.76–0.90	0.80–1.10	0.79–0.89	Contrasts living/dead vegetation; sensitive to surface soil moisture
1.55–1.75			Mid-IR atmospheric window
			Sensitive to crop and soil moisture conditions and stand density differences
2.08–2.23			Mid-IR window – useful in geology
10.4–12.50			Far IR window – sensing thermal phenomena

* Note: SPOT panchromatic records in a single spectral band at 0.51–0.73 μm with double the spatial resolution of SPOT MS.

The transmitted grey-scale range varies with the satellite and sensor. In the case of Landsat MSS, a grey scale of 64 pixel values is used, which, during raw data processing on the ground, is transformed into a grey-scale range of 124 pixel values. With respect to the raw data of Landsat TM, this is recorded, transmitted and processed in a grey-scale range of 256 pixel values in each of its channels. On-board calibration of the satellite sensor is made between scans against direct solar radiation, grey-scale illumination wedges or reference lights. No on-board correction during sensing can be made for that part of the incoming signal comprising reflected solar energy from the earth's atmosphere.

Spatial characteristics

Each recorded radiance measurement of the satellite provides a single picture element or **pixel**, the size of which is determined by the instantaneous field of view of the sensor. The spatially ordered assembly of pixels then provides the individual scan lines, and their assembly, when processed, provides the user of data with hard-copy imagery.

As illustrated in Figure 7.2, Landsat with an instantaneous field of view of 11.56° and a nominal altitude of 705 km covers an area of 79 m by 79 m at each dwell area (dwell point); but in the rapid on-board analogue-to-digital conversion, the pixel size is reduced to 56 m by 79 m. That is, the grey-scale value of each pixel of 56 m by 79 m (0.44 ha) represents the reflectivity of an area of the earth's surface of 79 m by 79 m (0.62 ha). In literature the pixel size is sometimes given as 80 m by 80 m and the satellite spatial resolution as being 80 m.

For Landsat MSS, the scanning, which images six lines simultaneously, covers a cross-track swath of approximately 185 km of the earth's surface, and the scanning in the forward direction (x-direction) is provided by the continuous forward movement of the satellite. Thus, there is no endlap, which is provided later in the laboratory as part of image processing. Each processed Landsat MSS scene, covering a ground area of approximately 185 km by 185 km, is made up nominally of 2340 scan lines with 3240 pixels in each scan line. That is, about 7 580 000 pixels in a single spectral band or 22.5 million pixels in three bands of a standard 23 cm colour print at a scale of 1 : 1 million and covering a ground area of approximately 34 000 km^2. In comparison, an aerial photograph at a scale of 1 : 120 000 covers an area of about 760 km^2.

With the commonly termed second-generation earth resources satellites, Landsat TM, SPOT etc., their geometric accuracy has been improved by reducing the pixel size. Thus each pixel of Landsat TM covers a ground area of approximately 30 m by 30 m; and this necessitates transmitting from satellite to ground station and the later processing of more than five times the number of

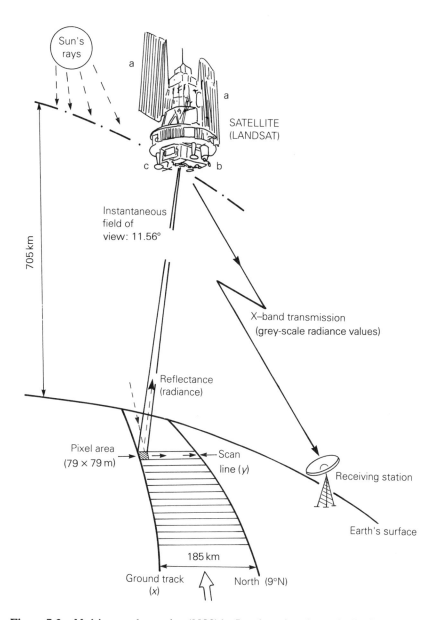

Sun's
rays

a

a

SATELLITE
(LANDSAT)

c   b

705 km

Instantaneous
field of
view: 11.56°

X–band transmission
(grey-scale radiance values)

Reflectance
(radiance)

Pixel area
(79 × 79 m)

Scan
line (y)

Receiving station

Earth's surface

185 km

Ground track
(x)

North (9°N)

Figure 7.2 Multispectral scanning (MSS) by Landsat. As schematically shown, some of the solar radiation incident on the earth's surface is reflected upwards towards the sensor on the satellite, which records (captures) this energy in units as the sensor scans crosswise to the satellite's ground track. After on-board processing into discrete grey-scale numbers, the digitized data is transmitted by radio to the antenna of the receiving station. Note on the satellite: (a) solar panels generating power for the on-board operations and navigation; (b) the multispectral scanner; and (c) the relay antenna sometimes used to transmit data received from data collection platforms on the ground.

pixels to cover the ground area of one Landsat MSS pixel. The Japanese MOS-1 satellite is intermediate between Landsat MSS and TM with a pixel size of 50 m. The increased data flow and its primary processing can result in production problems, including slow delivery of data to users, as occurred with a data flow of 117 megabits per second for the short-lived Seasat (cf. Landsat MSS, 15 megabits per second).

Spot carries two identical CCD sensors, which provides the satellite with versatility not available to Landsat. Besides providing 'twin' imagery with a swath of 117 km (cf. Landsat, 185 km), imagery can be acquired simultaneously with a swath of 60 km using both the panchromatic sensor with a spatial resolution of 10 m and the multispectral sensor with a spatial resolution of 20 m. By operating the in-built sensor steerable mirror, off-nadir recording can be made with look angles from 0° to 27° (Figure 7.1(b),(c)); and by recording at different viewing angles and a different orbit, stereoscopic pairs of imagery are obtained. With vertical (nadir) viewing, the pushbroom scanner records in the single scene (60 km × 60 km) 6000 × 6000 pixels in the panchromatic mode and 3000 × 3000 pixels in each band of the multispectral mode. In recording oblique imagery, the length of the scene in the flight or ground-track direction is always controlled in processing as 60 km, but the swath or cross-track direction is a function of the look angle and will range between 60 and 80 km.

7.2 PRIMARY DATA PROCESSING (EARTH RESOURCES SATELLITES)

This, also termed **preprocessing**, covers receipt at a ground station of raw data from the satellite, storing the data, corrective processing of the raw data and recording the corrected data on master high-density digital tapes (HDDTs). The HDDTs are used to supply primary data products (hard copy, CCTs) to national contact points and sometimes to other users. Usually the processing is entirely digital, and fully computerized.

Both radiometric and geometric corrections are made to the raw digital data. With the increasing demand world-wide for satellite data products of high quality, there is a trend to provide users with a wider choice of digital data and analogue hard-copy products. These range from computer-compatible tapes with the minimum of radiometric corrections to photographic colour negatives and prints, in which the scenes have been rectified by reference to carefully identified ground-control points on existing maps. Radiometric corrections include for SPOT the equalization of the response of the CCD detectors; and for the data of Landsat, the removal of sixth-line striping resulting from the earth rotation and displacement of the scanner sweep with the forward movement of the satellite, and minimizing the effect of haze/airlight on the grey-scale values of the pixels.

Software programmes are available for use by the imagery analyst to reduce the haze effect. A rough estimate of this can be made by searching digitally for pixels having the lowest grey-scale values, assuming or checking that these are for surfaces on the earth with little or no spectral reflectance (cf. water, Figure 3.12) and then subtracting the average of these values from all the other pixels to arrive at new grey-scale values. This type of correction, as also with geometric correction, is needed when monitoring changes in land use and forest cover, which demands the comparison of satellite data acquired on two or more different dates or by two different platforms.

As pointed out by Bernstein and Ferneyhough (1975), there are several important geometric distortions that must be corrected during the primary processing of the data, and which occur due to changes in the altitude, attitude (i.e. roll, pitch, yaw) and velocity of the satellite and eastward rotation of and curvature of the earth. In hard-copy format, it is important for many purposes to have the images displayed in a usable map projection (e.g. Universal Transverse Mercator). This may be derived using either platform orbital parameters, to establish the exact location of each pixel, or polynomial equations, to model the distortion of ground control points located in the imagery. The latter uses regression equations in which the coefficients have been determined by interrelating geographic and image co-ordinates. Imagery is available, for example, from SPOT Image in which registration accuracy is within 0.5 pixels, i.e. 5–10 m (Brachet, 1986).

Resulting from the increasing cost of data acquisition and the increasing sophistication of satellite data users, changes are occurring in the processing and supply of primary data products. In the first few years of Landsat data acquisition, complete hard-copy scenes in colour ($c.$ 34 000 km^2) were often preferred at scales of 1 : 1 million or 1 : 500 000 or sometimes 1 : 250 000. Analysis by the user included density slicing or additive viewing with spectral band transparencies at a scale of 1 : 3 340 000.

Suppliers now favour computer-compatible tapes (CCTs) of part of the scene covered by the satellite sensor. There is also a trend for the purchase price of the data to be related to the number of pixels or unit ground area (e.g. per square mile or square kilometre). Landsat TM data is sold of areas of one-quarter of a Landsat MSS scene. A SPOT hard-copy scene (MS) represents an area of 3600 km^2, as compared with 34 000 km^2 in a full Landsat MSS scene, which makes the purchase price based on ground area and not scene price much more expensive. Table 7.2 lists the sources of historic and current satellite data of interest for forest resource studies.

7.3 ENVIRONMENTAL SATELLITE SYSTEMS

Although the data of this family of satellites has greatest use in meteorology, the term 'environmental satellite system' is preferred to meteorological

Table 7.2 Summary of principal sources of satellite data for monitoring forest lands by earth resources satellites*

Data-orbiting earth resources satellite	Year(s)	Spectral bands (μm)	Resolution (m)	Time interval between repetitive coverage, etc.
Landsat-1 (ERTS-1)	1972–78	4 MSS bands 0.5/0.6 0.6/0.7 0.7/0.8, 0.8/1.1 0.5/0.75 RBV	80 40	Non-operational. Direct relay at 18-day intervals
Landsat-2	1975–Nov. 1979 Reactivated June 1980	4 MSS bands (same as Landsat-1) RBV 0.5/0.75	80 40	Non-operational
Landsat-3	1978–80	4 MSS bands (same as Landsat-1) 1 MSS thermal band 10.5/12.4 (failed) June 1978 RBV 0.5/0.75	80 240 40	18 days direct relay and on-board tape recording Non-operational

			Resolution (m)	
Landsat-4	1982–84	MSS as Landsat-3 Thematic mapper (6 channel visible/near-IR, 1 channel thermal)	80 30 visible IR 120 thermal	Lower orbiting altitude – 16 days direct relay and on-board recording Non-operational. May be reactivated
Landsat-5	1984+	MSS as Landsat-3 Thematic mapper as Landsat-4 (6 channel visible/IR) (1 channel thermal)	80 30	16 days direct relay + on board recording. Relay as Landsat-4
Soyuz/Salyut	1975+	Visible/near IR (6 channel scanner) Multiband camera MKF-6 Photogrammetric camera	120 20 10–20 5	Films – returned to ground periodically
SPOT-1	1986+	0.5/0.6 0.6/0.7 0.8/0.9	20	26 days, any point on the globe is accessible obliquely every 2–4 days – also stereo
SPOT-2	1989	0.5/0.73 (panchromatic)	10	
ERS-1	1991	C-band radar	30	SAR imagery and cloud/sea-state data
MOS-1		MSS 4 channels (cf. Landsat MSS)	50	

* Note: Short period sources of outer space data include Skylab and the Space Shuttle.

satellite system, since there is a spectrum of unrealized uses. In this respect, foresters have been slow in adapting environmental data to day-to-day needs. There are a number of potential uses in forestry, including applying imagery in the vegetation index format to the regional monitoring of forest cover, supporting fire protection and developing models for the introduction of exotic tree species. Environmental satellite data can be used to complement sparse or unreliable rainfall and temperature data in climatological studies, to provide information useful to food security on a national and regional basis, to yield information on sea-surface conditions useful to fisheries, to provide coarse resolution data for global environmental studies, etc.

The very fact that data is acquired by an environmental satellite repetitively of the same ground area at intervals ranging from every half hour to twice daily makes the data unique and often complementary to the data provided by earth resources satellites, despite the spatial resolution being at least about ten times coarser. It was estimated by Barrett and Curtis (1976) that a single earth-orbiting environmental satellite yields more than ten million points of data daily. As a consequence of the frequency of the observations, cloud cover (unless continuous) is unlikely to be a major constraint in using the data, although, in some regions (e.g. monsoonal), the continuous cloud cover may occur at the most critical time of the year for its agricultural applications; but this would not be a constraint in forestry.

The resolution of the data transmitted by the environmental satellites and available to civilian users ranges from 0.6 km to 5 km; but many local low-cost ground stations, as used in weather forecasting and some major airports, receive the data at much lower resolution (e.g. 50, 100 km). This is sometimes a point of confusion with would-be environmental users requiring the finest resolution. Photographic products, but not tapes, with a spatial resolution of about 0.6 km are available for some parts of the world from 1973 onwards through the US Defence Meteorological Satellite Programme (DMSP). Also, as part of the US Nimbus programme, a 6-channel coastal-zone colour scanner with a spatial resolution of 0.825 km and operating in six spectral bands was mounted on Nimbus-7 to study sea-water temperature, sediment, chlorophyll, coastal wetlands, etc.

Geostationary satellites

The comprehensive coverage of the earth is achieved by a network of five geostationary satellites at a distance of about 36 000 km and synchronous with the earth's rotation at the equator (Figure 7.3). In theory, three satellites would suffice to provide complete global coverage; but there are technical constraints, including increasing the distance between earth and satellite and the low image-forming quality of oblique signals.

Coverage is provided by the US GOES, the Indian Insat over the Indian

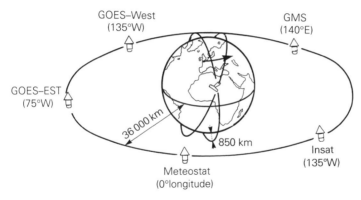

Figure 7.3 Continuous global coverage is provided by five geostationary environmental/meteorological satellites at a distance of 36 000 km from the earth. Meteosat, operated by the European Space Agency, is stationed at 0° longitude on the equator above the Gulf of Guinea, West Africa. Also, shown schematically is an earth resources satellite, polar orbiting the earth at an altitude of 850 km. Polar orbiting environmental satellites operate at intermediate altitudes (e.g. NOAA series at about 1450 km.)

Ocean, the European (ESA) Meteosat over the Gulf of Guinea (West Africa) and the Japanese GMS over the Pacific Ocean. GOES and Meteosat, for example, generate imagery in their respective fields of view every half hour in the far-IR (10.5–12.6 μm), and in the visible spectrum (0.55–0.70 μm) during daylight hours. The nominal spatial resolution at nadir is 1 km and 2.5 km in the visible spectrum, and 8 km and 5 km in the far-IR spectrum for GOES and Meteosat respectively. Both satellites have the capability of relaying numerical data from ground-based data collection platforms.

The low resolution of the imagery from geostationary satellites precludes its direct use in local/district forestry, but the data may be useful in providing estimates of fire risk, improving climatological records and monitoring the location of regional storm damage in near real time. Further information on geostationary satellites will be found in the *Manual of Remote Sensing* (chapter 14) and in several articles, including Hirai *et al.* (1975) on the Japanese Himawari(GMS), Morgan (1978) on Meteosat and Diesen and Reinke (1978) on USSR imagery.

Earth-orbiting satellites

Of increasing interest to foresters is the use of data acquired by the polar-orbiting US environmental satellites and the USSR Meteor. These orbit similarly to Landsat and SPOT, but at higher altitude and twice daily. The US

Table 7.3 Spectral characteristics of NOAA AVHRR-2

Band number	Bandwidth (μm)	Remarks
1	0.58–0.68	Visible spectrum (reflected)
2	0.725–1.10	Near IR (reflected)
3	3.55–3.93	
4	10.3–11.3	Far IR (thermal)
5	11.5–12.5	Far IR (thermal)

earth-orbiting TIROS-1 was launched in 1960, was followed by the ESSA satellites in 1966 to provide an operational system, and then in 1978 by the third generation of near polar-orbiting environmental satellites, designated TIROS-N (NOAA-6 onwards).

The on-board advanced very high-resolution radiometer (AVHRR) senses with a resolution of about 1.1 km and simultaneously in several spectral bands (Table 7.3). Approximately 200 Landsat MSS pixels represent the same ground area as covered by one NOAA AVHRR-2 pixel. One pixel covers approximately 0.005 km^2 for Landsat and 1.21 km^2 for NOAA AVHRR.

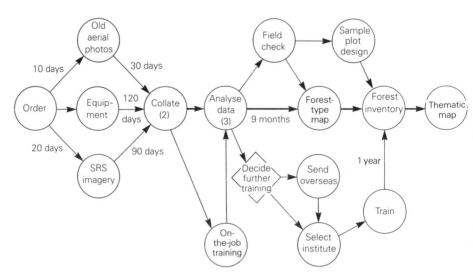

Figure 7.4 Critical path analysis (simplified). Events are represented by circles and activities by arrows. Completed events are flagged. (2) is a merge point; (3) is a burst point. As work progresses, the event forest inventory would be expanded.

As the AVHRR data are recorded simultaneously in the visible and near-IR bands of the electromagnetic spectrum, the data can be processed and printed in a vegetation index format (section 3.4), which is well suited to monitoring extensive areas for changes in greenness, living biomass and woody vegetation cover. There is, however, a major constraint on the availability of direct readout of AVHRR data, since parts of the earth's surface are not covered by ground receiving stations. The on-board tape recorder for delayed relay is limited presently to 10 minutes' recording per orbit of data in the low resolution mode of 4 km nominal ground resolution (swath width: 4000 km).

Both NASA and FAO have been using experimentally in Africa NOAA AVHRR data at full resolution (1.1×1.1 km). This type of data has been termed local area coverage (LAC), and at lower resolution is termed global area coverage (GAC). A GAC pixel represents an area covered by 15 LAC pixels, but its radiance value is provided by only the average of three resampled values. The use of GAC reduces considerably the computer processing time, as does the use of LAC data compared with processing Landsat data; but the procedure used in deriving the radiance values of the GAC pixels does not lend itself to field checking the data and the use of correlation and regression analysis techniques in image analysis.

7.4 PLANNING FOR SATELLITE DATA ANALYSIS

Unlike the taking of aerial photographs, which may be under the direct control of the forester and land-use planner or influenced by their requirements, satellite remote sensing application involves initially choosing from an increasing range of marketed products. This task can be time consuming, may involve unforeseen delays and can become complicated. To avoid disappointments and frustration, it is well worth while therefore to consider each event and each requirement separately and to integrate them into a critical path diagram.

In Figure 7.4, events are represented by circles and activities are represented by arrows. Each event indicates the completion of one or more activities and occurs before other dependent activities begin. As shown in more detail to the left of the diagram, the duration and dates of the activities can be added and the completion of the events flagged. The critical path analysis covers not only the ordering and receipt of materials, but also the ordering of equipment, the maintenance and repair of equipment and the recruitment and training of staff. When an event cannot be completed until two or more activities are brought together, this is represented by a **merge point**; and if two or more activities depend on finalizing an event, this is represented by a **burst point**. When an event is also a decision, it is represented by a diamond-shaped **decision box**. As the flow chart of events and activities is developed, it will often be found that it consists of two or more **sub-networks**. The **critical path** is the longest time sequence through the network, since this is the path that controls the timing of the completion, and would be represented in the diagram by a double activity line.

Critical path analysis applied to the planning of satellite analysis, whether digital (computer based) or analogue (photointerpretation), will ensure that activities are not overlooked, that the timing is more correctly assessed and that the costing is more precise. The method is equally applicable to the planning and execution of aerial photography and image analysis involving fieldwork and establishing a remote sensing system for forest resources (Chapter 13).

Decisions will be needed on whether or not to rely on existing equipment and expertise, or to obtain funds for specific new equipment and at the same time to give selected staff additional training. These are brought together at a merge point, and we cannot proceed until this occurs. In the example provided by Figure 7.4, the decision was made to proceed first with ordering image analysis equipment and to supplement this with SRS imagery and existing old aerial photographs. This requires technical skill and experience in the ordering of specialized equipment (Chapters 9 and 10) and choice of data source, data product and scale of the imagery.

First enquiries or ordering may indicate that the imagery is not available for the year and season required nor readily available in hard copy etc. These delays must be allowed for, since a planned period of one to two months may extend to several months; and, as indicated in the example, the most optimistic and pessimistic estimates of the slack between initiating the planning and being ready for imagery analysis associated with fieldwork could be four months.

Choice of satellite data

Depending on the data source, the choice of products ranges in hard copy from black-and-white prints and positive and negative transparencies in a single spectral band and at one or more scales to colour prints and transparencies combining three spectral bands. Similarly, there exists a choice of computer-compatible tapes (CCTs), formatted as the end product of primary processing from high-density digital tapes (HDDTs). These will have been variously corrected for errors. For example, SPOT Image provides photographic products at several scales, quick-look scenes as paper prints or in video-cassettes and small areas on floppy disc and CCTs with 6250 or 1600 bits per inch (bpi). One full SPOT scene is recorded on a single 6250 bpi tape or on two or three 1600 bpi tapes, each with a maximum capacity of 32 million bytes. CCT processing of SPOT data is available to four levels of radiometric and geometric fineness.

Irrespective of the processing techniques and the increasing cost of the satellite data product with radiometric and geometric corrections, its price per unit of ground area will always be many times cheaper than aerial photography. The cost of aerial photography, for an area equivalent to one Landsat MSS scene (about 34 000 km^2), could be expected to be at least 100 times more expensive than the sale price of the imagery. The resolution of ground objects on aerial survey photographs would be considerably finer than the highest resolution obtainable by earth resources satellites used for civilian purposes, and very much finer on the larger-scale aerial photographs preferred in most forest studies.

With the cost ratio so much in favour of satellite remote sensing, it is preferable therefore to favour whenever possible satellite remote sensing in place of aerial photography in forestry, provided the lower resolution is acceptable; and if not to consider a multistage design involving both types of imagery. Frequently, the high cost of aerial photography precludes its use in monitoring forest changes on an annual or periodic basis, but this is not so applicable to SRS data.

When monitoring gross changes over very large areas at short intervals of time, the comment on the cost of aerial photography may be equally valid for earth resources satellite sensing in comparison with environmental satellite sensing; and despite the further reduction of the information content of the latter, its possible use should be critically examined. Repetitive coverage, for example, of vast areas by Landsat to support desert locust surveillance, is too costly, but the use of NOAA AVHRR data in vegetation index format is acceptable even though the information content is reduced.

Satellite telecommunications aids

Of future importance in forestry will be the use of global positioning systems (GPS). Already, ships at sea are obtaining accurate fixes from the US Navy Navigation satellites (Navstar) and the Decca GPS using compact signal processing and Doppler signal receiving equipment. The US Navstar global positioning system, with a global network of 18 satellites at a distance of 20 000 km from the earth, opens up satellite positioning at low cost to forest survey. The degraded signals for civilian use can be expected to give an accuracy instantaneously of 100 m horizontally and 160 m vertically as compared with an accuracy of 15 m in three dimensions for military application.

For geodetic survey, satellite positioning is now widely accepted and provides positioning accuracies of ±1 m horizontal and vertical. However, each geodetic positioning takes 3–4 days and costs in the field up to US$ 10 000 per station. Recently, Indonesia was covered in two years with a first order network of 238 stations.

A further aid is ground-based instruments in combination with a telecommunications/relay system on board some satellites. Transmitters aboard the satellite relay directly in real time the signals transmitted from the ground-based data collection platform (DCP) to the ground receiving station. The satellite component of this type of data collection system (DCS) has been available on Landsat, NOAA GOES and Meteosat, but has been little used operationally. Examples of operational use include Canada for water flow monitoring and Chile for coastal current and mountain stream monitoring.

The sensors for an unmanned ground-based data collection platform are not expensive to purchase and to maintain, but must be satellite compatible in the data transmission mode and not subject to vandalism. The types of data that can be ground sensed include stream flow, water depth, snow depth, wind speed, precipitation, soil moisture, soil temperature, temperature at ground level and hours of solar radiation. The use of DCPs in forestry in isolated areas warrants further consideration, provided there is a major ground receiving station within DCS range, or the establishment of a DCP network justifies establishing a small data receiving station.

Part Three

Image Analysis in Forestry

INTRODUCTION

Modern image analysis using remotely sensed data involves, and should integrate, the methods and techniques used in aerial photointerpretation (API), photogrammetry and computer-assisted digital analysis; it should be supported by ancillary scientific skills, and the background provided by a relevant field-oriented discipline. Photogrammetry is the oldest discipline, dating from the mid-nineteenth century, and has made rapid advances in recent years using computer-assisted techniques. Aerial photointerpretation following rapid development in the Second World War acquired world-wide academic recognition in the 1950s, continues to form an important element in the professional and practical training of those working in scientific field disciplines, particularly in forestry and geology, and has expanded in recent years to include the visual interpretation of satellite imagery, radar imagery, etc. Otherwise the state-of-the-art has advanced little in the last 30 years and its skilful practice may unfortunately have regressed. In contrast, digital image analysis continues to expand rapidly through satellite remote sensing (SRS) and the development of low-cost mini and personal computers.

In choosing the sequence of the next three chapters, it is recognized that this could be differently arranged depending on the interest, training and experience of the reader. Those who have had experience recently in aerial photointerpretation will know that this does not infer an abrupt separation between visual interpretation and computer-assisted digital analysis, but a trend towards their integration, which must also include applying elements of photogrammetry. From the writer's experience, if the training does not include an introduction to photogrammetry, the photointerpreter/analogue image analyst will not make full use of measuring techniques, and will often avoid using photogrammetric parameters and the preparation of simple thematic maps. Similarly, the digital image analyst without training in the elements of photogrammetry may place too much reliance on the analysis of spectral signatures and not recognize the importance of quantitative visual image analysis, and that visual analysis enables diagnostic elements of the images to be extracted, which is not possible using computer-assisted techniques.

8

Visual image analysis

Visual image analysis or photointerpretation can be defined as the visual act of examining the images of ground objects contained within the photograph for the purpose of identifying the objects and judging their significance. The added word visual (or manual) is needed to distinguish between visual interpretation and digital analysis. The shorter term image analysis embraces both.

Visual analysis relies on the powerful faculty of binocular vision provided by the human eyes. This involves the use of the eyeball, including the fovea, retina and possibly parafovea, the optic nerve and the visual centres of the brain. Rays of light pass through the refracting media of the eye, in a similar way to light passing through the camera lens, and form an image at the back of the eyeball on the retina. The stimulus is then transmitted to the visual centres of the brain and is co-ordinated with a similar perception from the other eye. The fovea is most sensitive to green light at $0.555\ \mu$m, which is close to the peak of spectral reflectivity of green vegetation in the visible spectrum (Chapter 3).

A single eye is able to determine accurately the direction of an object, but is only able to gauge distance qualitatively. This is illustrated by holding two pencils, with the points about 30 cm apart at nearly arm's length, and with one eye closed trying to bring the point of one in contact with the top or tip of the other. It will be found that the top of one pencil will pass usually behind or in front of the tip of the other pencil. If the exercise is repeated with both eyes open, it is easy to bring the pencils into contact. The reason is that a pair of human eyes possesses the faculty of binocular vision by which it is possible to obtain a conception of relief in space (i.e. a three-dimensional model).

To illustrate the relief effect that is achieved by binocular vision, and so important in photointerpretation, hold a single pencil end-on in front of one eye (with the other eye closed). Then view it with both eyes open. It will be observed, when it is viewed with the two eyes, that an impression of its relief/ shape in space is obtained. Making use of this faculty of binocular vision, Wheatstone, in the mid-nineteenth century, showed that when two photographic images of the same object (taken at two different stations) are placed in front of the eyes, an impression of the relief of the object in space is provided by the eyes and brain in the form of the single three-dimensional model. Wheatstone then proceeded to accommodate this faculty by designing the mirror stereoscope (see Figure 2.2).

8.1 TESTING STEREOSCOPIC VISION

The importance of the stereoscope and the application of stereo-vision in forest photographic interpretation cannot be stressed too strongly. It is therefore important to test for stereo-vision before beginning to study aerial photographs, and this should be followed by testing for colour vision. The need for stereo-tests also applies to satellite imagery, if the overlap of contiguous pairs of the imagery provides a stereoscopic model. Unfortunately, some people, possibly 5%, do not have the faculty of stereoscopic vision and no amount of training will give it to them. There is no known physical aid to

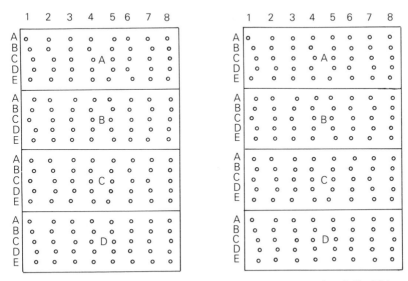

Figure 8.1 Moessner test for stereoscopic vision (see text section 8.1). This test involves using a pocket stereoscope.

developing stereoscopic sight in a person who does not possess it naturally; but exercises can help those with weak stereoscopic vision. The imagery analyst, when beginning a day's work, will probably find that stereoscopic vision improves for the first 20–25 minutes of work. A minimum practice period of 5–10 minutes is advisable before candidates are tested for stereoscopic acuity.

One of several simple tests of stereo-vision is that of Moessner (1954). This consists of two banks of 160 circles with each circle approximately 0.75 mm in diameter, as shown in Figure 8.1. Twenty-five of the circles selected randomly in the left bank have been shifted slightly to the right by distances varying from 1.125 mm at the top to 0.0125 mm at the bottom. The test, using a pocket stereoscope, is to ascertain which circles appear to be floating. Being able to select the floating circles down to a parallax difference of 0.05 mm provides a rating of 80%, and can be considered as satisfactory. An advantage of the test is that it is relatively easy to concentrate stereoscopic viewing on the small area in the vicinity of a pair of circles; and thus it serves as a training exercise as well as a test.

8.2 STEREOSCOPES

This topic is introduced as the next step of the chapter to emphasize the importance of the stereoscope as a major tool of the forest aerial photointerpreter in examining the diagnostic elements of the photographic image. Otherwise it would be better placed in section 8.5.

Stereoscopes can be grouped conveniently into three basic types, that is lens, mirror and prism. They may be also non-scanning or scanning, and preferably should be chosen for use with a parallax bar or parallax wedge, since inevitably the forest imagery analyst must be able to measure apparent heights of individual trees and to estimate forest stand heights, using stereoscopic pairs of aerial photographs. The analyst may also be required to measure spot ground heights and to calculate the slope of the terrain. The geologist in addition needs to determine strike and dip of rock structures using the difference between two spot heights.

By far the most popular is the simple lens stereoscope, which is also called a hand stereoscope or pocket stereoscope. It comprises two semi-convex simple lenses mounted in a frame, and is light and compact and ideal for fieldwork (Figure 8.2(a)). Magnification is usually two times. It is recommended that the eye base is adjustable (55 mm/75 mm). An average eye base for Europeans is 60 mm. Although cheap to purchase, pocket stereoscopes have the disadvantage that only a small section of the photograph can be viewed at a time. Except for this narrow strip, viewing requires flipping up of the side of one photograph. Several manufacturers (e.g. Abrams, Zeiss) developed the pocket stereoscope into a bridging stereoscope so that flipping is not necessary.

(a)

(c)

Figure 8.2 (a) Simple lens (or pocket) stereoscope, which is extensively used by foresters in fieldwork. Note the small carrying case for the folded stereoscope. A stereoscopic pair of photographs, taken by the German Metric Camera on-board the ESA–NASA Spacelab at a sun elevation of 29°, are being examined to assess the forest cover, snow cover, slope, aspect and access in the Himalayas of Nepal. (b) Mirror stereoscope (1) and ancillary equipment set up for interpreting a stereoscopic pair of aerial photographs (4). Note the binoculars (7) to improve the magnification of the images and the overhead light to provide side illumination towards the interpreter. 2: part of drafting arm; 3: flight lines common to the contiguous pair of photographs, and used to align the photographs and parallax bar; 5: parallax bar; 6: fibreboard working top, painted (matt) white; 8: engineer's scale or ruler, graduated in millimetres and used to align photographs. (c) Zoom stereoscope. Several manufacturers produce this type of prism stereoscope with magnifications up to 20–30 times.

Further modifications include the addition of a parallax bar fixed to the stereoscope.

The second type of stereoscope uses mirrors to supplement the simple lens, so that a relatively large area of each pair of photographs is viewed (Figure 8.2(b)). This avoids frequent moving of the photographs and flipping up of the side of the photographs. Viewing is usually through magnifying 'binoculars', although viewing of larger areas can be carried out without the 'binoculars'. The latter is particularly useful if a synoptic view of the landscape is needed. As the silver of the mirrors is on the outer surface of the glass to avoid refraction, it can easily be damaged by careless fingering. Magnification is two, three, four, six or eight times, being fixed by the 'binoculars'. Four times and six times magnification are popular in forestry. A separate parallax bar is a key accessory to the mirror stereoscope (Figure 8.2(b)). As an addition, a freely moving photographic mount in the x- and y-directions, but with the mirror stereoscope fixed (or vice versa), makes the instrument suitable for scanning.

In the third type, viewing is achieved via mirrors and prisms. External levers control the positions of the prisms and internal mirrors for scanning. Several manufacturers now produce prism stereoscopes, which zoom from a reduction ratio of 2–4 up to 20–30 times magnification (Figure 8.2(c)). The image analyst will find that the maximum magnification is limited, however, by the 'graininess' of the photograph, being possibly ten times for black-and-white prints and about 20 times for colour diapositives.

For viewing colour transparencies through the stereoscopes described above, it is necessary to provide illumination below the transparencies in the form of a light table. The brightness of the diffused fluorescent lighting of the illumination table, on which the transparencies are placed, should be similar to that used initially in processing. Often viewing is improved if a black paper mask is placed over the photographs, except for the small area under close scrutiny, since this reduces glare and eye-strain.

When examining a pair of photographs under a stereoscope, the photographs must be correctly oriented by ensuring:

1. that the flight lines are correctly aligned by checking with a ruler's edge that the two principal points and the two transferred principal points are in the same straight line (section 9.1);
2. that the distance between the principal points is suitable for stereoscopic viewing;
3. that the shadows on the photographs are towards the interpreter and the photographs are in their flight sequence, otherwise a pseudoscopic illusion may occur; in **pseudoscopic view** hill tops may appear to be valleys and vice versa; rivers will appear to flow on the top of hills!

Several hours of training with stereoscopic pairs of aerial photographs is needed before attempting fieldwork; and several days of training, preferably broken into fairly short periods, is needed in acquiring proficiency in measuring apparent heights with a parallax bar. A helpful introductory exercise to aerial photointerpretation using single photographs is for the student to assemble photographs which have been cut into pieces similar to those in a jigsaw puzzle (Photographic Jigsaw Test, Howard, 1970a).

8.3 DIAGNOSTIC ELEMENTS IN VISUAL IMAGE ANALYSIS (PHOTOINTERPRETATION)

Visual interpretation is a problem-solving activity involving the detection and identification of ground objects on the photographs, by recognizing them by their principal spatial and spectral elements and sometimes in forestry by their temporal condition. As pointed out by Estes and Simonett (1975), publications differ on the number of basic diagnostic elements, but there is general agreement on six primary ones. These are tone or colour, size, shape, texture, shadow and pattern; to these may be added height, site and association (Howard, 1970a), and also temporal difference(s) when multidate imagery is being examined.

Tone and colour form the basis of digital image analysis by sensing in different spectral bands and by recording in a grey-scale range of 64–256 in each band. In contrast, the human eye is able to discriminate in black-and-white photographs only a few differences of shades of grey between black and white; but it is highly sensitive to colour differences and is readily able to use the other diagnostic elements. In the *Manual of Remote Sensing* (American Society of Photogrammetry, 1983), tone/colour is placed first as the basic element of photointerpretation in the ordering of the six primary elements, and is followed by the remainder as representing the spatial arrangement of tone.

The importance placed on height depends very much on the purpose of the study, including the discipline (e.g. forestry, geology, geomorphology, pedology) and the characteristics of the landscape. In forestry, height assumes major importance, since it not only helps to characterize qualitatively the forest stand and the terrain as observed in a stereoscopic image, but is also an essential element when using stereoscopic pairs of aerial photographs to estimate stand volume. Height is also used to estimate forest site quality by knowing the relationship between height and age, which is usually obtained from a previously prepared graph based on field measurements. Sometimes a person, unfamiliar with stereoscopes, will view the need to study the stereo-model (and hence height differences) as unnecessary, and will illustrate this by reference to work being done on single aerial photographs, ortho-photographs or photo-mosaics. Such a statement is usually erroneous in forestry and geomorphology, and the quality of the work would be greatly improved by using the third dimension.

Tone and colour

The definitive contrasts of tones and of colours of the images within the photograph are important to their identification; and, without contrast, the diagnostic elements of size, shape, texture and pattern would not be relevant. Significant to visual interpretation is the tonal edge gradient that defines the boundaries of larger objects recorded on the photograph. The edge gradient can be varied by the film processing techniques. Edge enhancement is popular in digital imagery analysis.

Tone

Tone refers to the various shades of grey to be observed on a black-and-white photograph and can be given logarithmic density values between black and white by reference to a grey scale (Figure 8.6). Losee (1951), using a circular aperture of 0.5 cm to study Canadian forests, concluded that an experienced interpreter requires a difference in density of 0.21 or greater. For example, he noted for a single set of photographs the following density values: 1.25 (black spruce), 0.84 (white spruce), 0.80 (aspen popular), 0.52 (red and white pines), 0.4 (tamarack) and 0.32 (birch). An example of tone being used in the separation of conifers is shown in Figure 8.3. Mixed stands, if present, would have a speckled appearance and thus also provide an example of texture.

Wet surfaces normally appear darker than dry surfaces on infrared and panchromatic photographs (cf. spectral curves, Figure 3.12). Krinov (1947) concluded in the USSR that the reflectivity of wet surfaces and Sphagnum was about 2.7 less than that of dry material. A dark smooth bitumen road may appear light on panchromatic photographs, due to the fairly high reflectivity of the surface. Dirt roads and tracks appear very much darker when wet than when dry. Subsoil when exposed usually appears light on panchromatic photographs compared with the surrounding living vegetation, but will probably record darker on infrared photographs. Dry grass, snow and sand record whitish. With some film–filter combinations insect damage and recently burnt ground vegetation appear darker than undamaged areas. In areas with sparse vegetation, the darker tones provided by fire patterns on the ground are often conspicuous for several years.

The location of an image relative to the principal point of the photograph influences its tone; and hence the tonal contrast between image classes is very important. Trees of the same species may appear progressively darker in tone the further they are from the principal point, or on one side of the photograph the forest will be recorded in darker tones than the other side. This is due to the shadowing effect of the trees as related to the flight direction and the sun's zenith and azimuth angles (Figure 8.4).

Tonal contrast is important in the visual interpretation of images recorded by radar and electro-optical scanning. However, in electro-optical scanning,

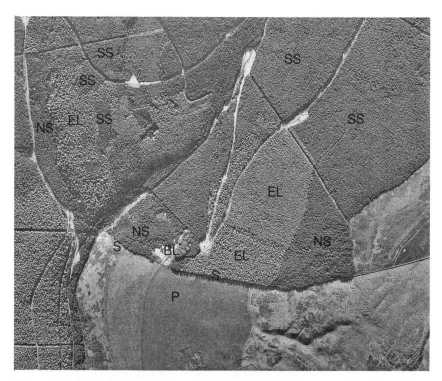

Figure 8.3 In the cool temperate man-made forests, tone/texture can be used often to separate conifers from hardwoods (angiosperms) and some conifer species. Above, on part of a panchromatic aerial photograph, the boundaries of the subcompartments between larch (light toned) and Norway spruce and Sitka spruce (dark toned) are recognizable. The size and shape of the crowns of the small stand of broadleaved (BL) hardwoods help in their identification, but not the species. The spruce are finer textured than the larch and the young larch saplings are finer textured than pole-sized larch. The shadows of the larch and spruce help in providing information on the crown shapes. EL: European larch; NS: Norway spruce; SS: Sitka spruce; S: shadows of tree crowns; P: pasture/upland grazing. Note the principal point, flight line and access roads (photographic scale: 1 : 12 000, reduced to 1 : 16 000).

since most instruments are multi-channel, the grey-scale values of each spectral band are usually colour coded. In contrast, since radar imagery is normally black-and-white, tonal contrast is important. Usually the imagery will have been recorded with a wide range of controlled grey-scale values and consequently photographic processing should attempt to preserve this range. The individual images should be initially examined at the largest workable scale in order to detect important tonal differences within and between images.

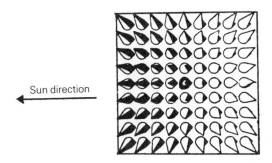

Sun direction

Figure 8.4 Diagram illustrating the influence of tree shadow on image tone on a large-scale aerial photograph of a stand of uniform-sized trees. The tree at the principal point has been scribed with a thick line. The shadow effect changes outwards from the principal point and is influenced by the sun's position (i.e. to left in diagram). As a result, one side of the photograph appears darker than the other side. Thus, tone is best used comparatively (i.e. as tonal contrast) and not as an absolute measurement in image analysis.

It will be recalled from Chapter 6 that the recording of radar imagery is quite different from aerial photography; and in consequence the shadowing effect is different and will contain no tonal detail.

The black-and-white interpretation of satellite imagery using tonal contrast is similar to aerial photographic interpretation; but with respect to single trees and small groups of trees, there is no opportunity to examine their within-image tonal differences. It will be recalled from Chapter 7 that, as each discreet pixel covers a considerable area, its grey-scale value represents the average of different types of reflectance, including shaded and non-shaded trees.

Colour

Most of the above comments are equally or even more relevant to interpretation using colour contrast between images. Colour can be represented by three variables (i.e. hue, value and chroma) and classified in different ways (cf. Munsell colour wheel, Figure 4.3). A slight change in colour is much more readily recognized by the human eye on colour prints and colour transparencies than are tonal differences on achromatic photographs. For example, Becking (1959) was of the opinion that the primary colours (blue, green, red) provide up to 200 000 observable colour differences.

Resulting from the advantage of using colour in visual image analysis, many of the techniques used today involve the colour coding of grey tones; and provided the image-forming data is in two or more spectral bands, high-quality natural or false colour products are obtained. It is also possible to render single-band data in colour by associating a restricted range of the grey-scale values with each primary colour, as is employed in colourizing old commercial black-and-white films for public television. However, the interpretative quality is low. Colour coding is used extensively in providing colour photographs from multispectral scanner data; and it is the viewing technique most favoured in the interactive digital analysis of satellite data.

Size

Sometimes the three diagnostic elements, size, shape and position (site), are combined under the term **contextural information**. Size has two aspects and usually requires the use of a stereoscope for three-dimensional viewing. The size of the object often helps to identify it, which includes the recognition of old over-mature trees, maturing trees, saplings, seedlings and shrubs. Secondly, size represents the area occupied by a group of individuals and this can be translated into timber volume or biomass per unit ground area or into stand types within the forest according to their size, shape and position.

Shape

Size and shape are often closely associated. Frequently on large-scale aerial photographs, the identification of a tree species depends on both characteristics, since the crown shape changes with age. Shape relates to the configuration or the general outline of objects as recorded on imagery. The shape of valleys often provides an important clue as to their weathering process and age, and can also be indicative of the rock type. The characteristic shape or physiography of the landform helps in the identification of land facets and land systems (Chapter 14). The characteristic shape, for example, of oxbow lakes on aerial photographs is intimately associated with river and flood-plain development and with local tree species, and is useful for environmental planning.

As regards tree shape being helpful in forest photointerpretation, there are many examples. Two of the commonest are the cone-like crowns of young broad-leaved species and most conifers, and the dome-like crowns of mature broad-leaved species and some conifers. Their crown shape changes with age. Scots pine (*Pinus sylvestris*) in the New Forest, Hampshire, for example, has a conical crown in early life and a flattened crown towards maturity. Besides age, disease and fire damage will also influence the crown shape; and, in stereo-vision, the image of the crown may appear to be different in shape at the centre and towards the edge of the photographs.

To be a successful photointerpreter, it is necessary to be familiar with the crown shape by species in the field according to age, health and site and with its crown shape according to location on the aerial photograph, time of the year of the photography, the photographic scale and type of the aerial photography.

Obviously the lower spatial resolution of satellite imagery excludes the diagnostic element of shape in studying many smaller objects, including tree crowns; and on the imagery of airborne radar at finer scales, it is only the crowns of the largest trees that can be studied unless in the content of texture.

Texture

Textural differences are recognizable in aerial photographs at all scales, and with the improved spatial resolution of the new satellites texture is increasingly important to supplement digital spectral techniques. Texture is also important in the analysis of radar imagery.

Texture was described by Colwell (1952) as the frequency of tone change within the images on the photograph. It is produced by an aggregate of unit features too small to be clearly discerned individually. It is the product of tone, size, shape, pattern, shadow and reflective qualities of the object and varies with the photographic scale. As the scale is reduced, so the texture becomes finer; and what may be considered as pattern and tree shape on larger-scale photographs, provides texture on very small-scale photographs and some satellite imagery. On very large-scale photographs, e.g. 1 : 2000, individual groups of leaves contribute to texture. At 1 : 5000 leaves and branches and small shrubs contribute to the texture; at 1 : 20 000 tree crowns provide the texture; at 1 : 80 000, it is the larger trees and groups of small trees and at 1 : 120 000 groups of trees that contribute to texture.

A fifty–fifty division of black-and-white into extremely small units provides a fine texture. The larger and more irregular the units, the coarser the texture will be. At scales of 1 : 5000 to 1 : 10 000, willows, for example, with their small narrow numerous leaves, provide a smooth texture when photographed. Immature oaks, having bunched leaves, may provide a coarse texture and old stag-headed oaks, having only a few main branches, give a very coarse texture on small-scale aerial photographs. Eucalypts, as the leaves are commonly bunched towards the ends of branches on older trees, also often provide a coarse texture. Scattered trees recording in dark tones on small-scale photographs tend to give coarse texture. For example, mangoes (*Mangifera indica*) provide black dome-shaped spots, which tend to break up or distort the texture provided by other tree crowns. Ground features, such as gilgais in Australia and termitaria in Africa, will influence the general texture of the landscape.

Shadow

This sometimes contributes to species identification, influences the image texture and, as small shadows, the tone within the tree images. Shadow is probably the major constraining factor for automatic image analysis. As associated with shape, for open grown woodlands, deciduous forests when leafless and some conifer forests (particularly with snow on the ground), shadows can assist in identifying tree species (Figure 8.3). Narrow, lightly crowned, open grown trees and tall shrubs are sometimes better examined from the shape of their shadows. In East Africa, the identification of the almost crownless *Acacia pseudofistula* (gall acacia) and of the palm-like *Borassus palmifera* was possible only by their characteristic shadows (Howard, 1965).

Several writers have commented on suitable shadow lengths. According to Welander (1952), the shadow length at the time of taking the photographs should be between 1.0 and 1.5 times the height of the tree. Seely (1942) in Canada observed that the potential error in measurement of tree height from their shadows becomes significant when the angle of the sun's rays to the horizontal is less than 35°. At 35° the length of the shadow is 1.4 times the height of the tree on level ground. Nyyssonen (1955) in Finland obtained unfavourable results with an angle of 27°. At 27° the length of the shadow is about 1.96 times the tree's height on level ground. In the north of Finland, where the stands are more open, the measurement of tree heights by shadow length was comparable with heights determined by parallax measurement; but this was not so in the south of Finland where the canopy closure is greater.

Tree shadows are measured on photographs using either a bar scale or a shadow wedge. The shadow wedge is made by scribing a V-wedge on clear transparent film and having the distances between the two arms of the V measured at fixed intervals. The wedge is placed over the shadow on one photograph of a stereoscopic pair, viewed under the stereoscope and read to the nearest 0.05–0.10 mm. A magnifying lens with attached bar scale (Figure 8.6) can be used on single photographs to read off measurements to about 0.05 mm (0.002 in). If a large number of heights are to be measured from shadows, time will be saved by preparing shadow conversion graphs as used by the US Forest Service.

Pattern

This is a macro-characteristic used to describe the spatial arrangement of images, including the repetition of natural features. These patterns can often be associated with geology, topography, soil, climate and plant communities, and their understanding helps in evaluating the quality of the land and in assessing the site quality of the forest. Particularly in studying land use, it is important to distinguish between natural patterns and those produced by human activity. In archaeological studies and mineral exploration, an illogical change in the natural pattern of the landscape may signify an area of interest. The mapping of land systems on satellite imagery and aerial photographic mosaics is facilitated by the prevailing topographic and drainage patterns. Often geomorphologists and soil scientists prefer to work from mosaics than from stereopairs of photographs, since the major topographic and drainage patterns are more readily observed. Pattern recognition is a key factor in the study of radar imagery.

Site

This term has two distinct meanings. First, it is widely used in the study of aerial photographs to describe the ground position of the image under observation in relation to its surrounding features. Its more important meaning connotes the sum of the factors of the environment influencing tree growth. Aspect, topography, geology, soil and the characteristic natural vegetation are all important factors, when examining the **forest site** on imagery. The first three may be classed as macro-characteristics; and the last two, with soil moisture and tree size, may be considered as micro-characteristics. The relative importance of each varies with local conditions. By studying sites from the quality of the forest stands on aerial photographs, some key factors of the environment can often be appraised; and conversely by recognizing site quality, the interpreter may be able to eliminate the local occurrence of certain tree species.

Referring back to site as implying the location or position of the image on the photograph, it is important always to consider the image in relation to its surroundings. For example, an open space by itself is usually meaningless, but the presence of certain urban development in its vicinity, and possibly the association of cars on the space, immediately suggests a car park. Similarly, in the country but near a town, a large open space which is T-shaped or X-shaped would suggest an airfield; and in the Amazonian rainforest, recent clearings would suggest new areas of settlement.

Site is an essential diagnostic element in image analysis. For example, on satellite imagery, the location of the tropical coastal zone enables the mangrove formation to be identified and mapped. In Western Tanzania, a familiarity with the woodland ecotypes on aerial photographs will show that the site of the *Brachystegia–Julbernardia* woodland is above the herb-line (Table 8.1 and Figure 8.5). Similarly in the Snowy Mountains of eastern Australia, the interpreter is able to locate the boundary between *E. pauciflora* (var. *alpina*) and *E. delegatensis* by knowing that the former is sited at elevations above the latter and by the former being in shape and size the common short-stemmed tree species immediately below the tree line and associated with elevations between 1500 and 2000 m.

Association

This is another term with two distinct meanings. In its ecological context it refers to a plant community of definite floristic composition, presenting a

Table 8.1   Selective photo-key – tropical woodland (Tanzania) (Howard, 1959)

HILLTOP

I.   Woodland – 50% or more of the ground covered by the crowns of medium and large trees. Stand height is usually 35–55 ft.

    A.   Hill-top woodland (H/W) – on hilltops. In Tabora District: *Brachystegia tamarindoides* woodland.

    B.   *Brachystegia – Julbernardia (Isoberlinia)* woodland (BJ/W) – In Tabora District: *Brachystegia spiciformis – Julbernardia globiflora* woodland. Productive of sawlog-size *Pterocarpus angolensis*.

HERB-LINE

(H/L) – delineated on photographs as a thick line.

    C.   Valley woodland (V/W) – always below the herb-line. Often as islands in the wooded grassland. In central Tabora District: *Brachystegia boehmii – Julbernardia globiflora* woodland. Sometimes *B. longifolia* is locally common.

II. Wooded grassland – under 50% of the ground covered by the crowns of large and medium-sized trees.

    D.   Wooded grassland without frequent termitaria – trees scattered (W/G) – tree height is usually 30–45 ft.

    E.   Wooded grassland with frequent termitaria – trees grouped (W/GT) – tree height is usually 30–45 ft.

III. Grassland.

    F.   Grassland with bushes and dwarf trees. Usually gall *Acacia* and *Combretum* spp. (AC/G) – under 10% of the ground is covered by large and medium-sized trees. Small trees sometimes attain 2 ft girth and a height of 30 ft.

    G.   Palmstand grassland (P/G) – usually single/groups of *Borassus flabellifer*. It is not an important sub-type.

IV. H.   Riverine (Riv.) – not always present in the catena and especially absent in hill areas. It is usually best described as riverine thicket; but riverine forest may occur along perennial streams.

VALLEY BOTTOM

V.   Other main types/sub-types.

    I.   Regeneration/thicket (R/T) – stand height is under about 30 ft and usually large trees are absent.

    J.   Induced vegetation – normally absent from forest reserves.

    K.   Permanent swamp – considered absent from forest reserves.

---

uniform physiognomy and growing in uniform habitat conditions. Alternatively, the term association is applied when some objects are so closely associated with others that each helps to confirm the presence of the other. The term correlation is sometimes used instead of association. Frequently, the shape, tone, pattern, texture, area, height and/or site are associated with a class

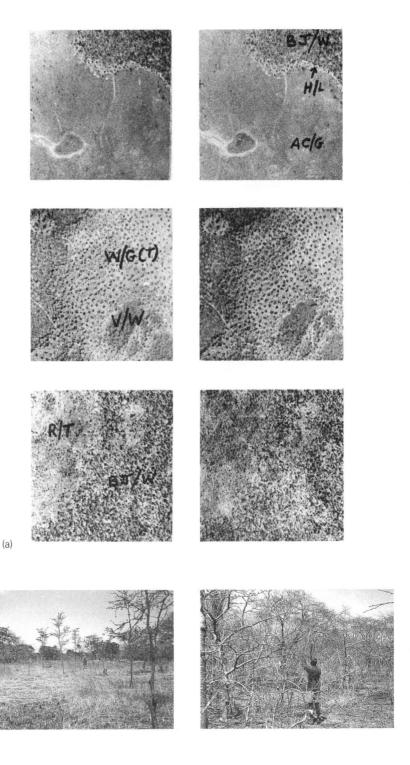

(a)

(b)

of object not recorded or not clearly recorded on the imagery. Studying one or more of these image characteristics, which have been observed to be associated with the object not clearly seen on the imagery, enables the latter to be evaluated. Thus, the savanna woodland types of Africa and tropical Australia and Latin America are closely associated with the soil types. With the aid of a catena diagram, the soil types can be delineated by the boundaries of the forest types as seen in stereo-vision.

Height

This is one of the most important diagnostic elements in forest aerial photointerpretation. It is more correctly termed **apparent height**, since a number of factors will result in the measured or calculated height of the

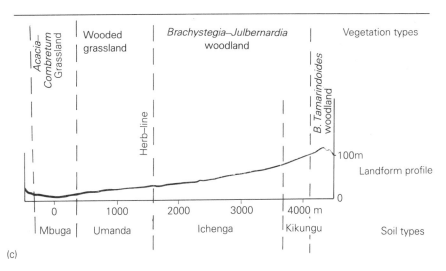

(c)

Figure 8.5 (a) Stereograms of three vegetation types occurring in the savanna woodland formation of east and central Africa (nominal aerial photographic scale: 1 : 30 000). The stereograms were annotated in the field using a felt pen. BJ/W: *Brachystegia–Julbernardia* woodland; H/L: herb-line; AC/G: *Acacia–Combretum* grassland; W/G(T): wooded grassland with termitaria; R/T: regeneration/thicket; V/W: valley or floodplain woodland. (b) Supporting field (ground) photographs to the stereograms, which assist the photointerpreter with the analysis of the aerial photographs. Left: *Acacia–Combretum* grassland with wooded grassland in the background. Right: thicket. (c) Part of a land catena showing the transect vegetation types from the Acacia grassland on the black cracking cotton soil (mbuga) in the valley bottom to *Brachystegia tamarindoides* woodland on the red lateritic soil (kikungu) on the hilltop. The figure also illustrates the close association between natural vegetation types and soil types of the tropical woodland.

photographic image(s) being different from the actual heights measured in the field. Even the heights of trees measured by hypsometer in the field may not be exact, and may be found to differ from those of trees that are felled and measured. It cannot always be assumed that the measured heights of standing trees in the field are more accurate than those obtained by parallax measurements. Therefore, it is misleading to term the unverified field measurements as 'ground truth'.

When measuring the apparent height of trees on photographs and determining the elevation of ground points, heavy shadows can cause an overestimate of height due to the floating mark of the parallax bar 'sinking' into the ground (as seen in the stereoscopic model). Underestimates of the height of trees occur due to difficulties in identifying the tops of trees, particularly cone-shaped crowns which may not fully resolve on the photographs, due to the absence of shadow on the photographs and due to tall ground vegetation providing a false datum plane in the stereoscopic model. In the practice of photointerpretation as an art, the image analyst with local experience may be successful in simply visually gauging the height of the forest stands in broad classes from the stereoscopic model. In this respect, the crude estimates of stand height are being made in a similar way to making crude occular estimates when walking through the forest. Emphasis is placed on the tree size and/or shape of the tree crowns in the stereoscopic model.

Both the focal lengths of the stereoscope and of the camera, and magnification produced by the lens of the stereoscope, influence the relief exaggeration of the tree images when stereoscopically viewed. Topographic differences in height are more conspicuous when viewed on a stereoscopic pair of aerial photographs than from a vantage point on the ground or from an aircraft. In stereo-vision, a slope may be exaggerated two times or more; and the difference in height between dominant trees and the main canopy layer and between the co-dominant trees and the understorey is similarly exaggerated. This exaggeration has been termed **appearance ratio** or **exaggeration ratio**, and one of the several formulae used in its calculation is as follows (Stone, 1951):

$$\text{Exaggeration ratio} = \frac{B}{H} \times \frac{f}{(f_s)(E)}$$

where B is the airbase, H is the flying height above ground level, f is the camera focal length, f_s is the stereoscope's focal length and E is the inter-pupillary distance set on the stereoscope.

On satellite imagery the height of forest stands cannot be measured stereoscopically, although SPOT imagery is available which provides a stereoscopic view of the terrain and an impression of its relief. Some outer space photographs (e.g. MOMS, Salyut/Soyuz) and some airborne radar imagery have also been taken with stereoscopic overlap, which enables the terrain to be viewed and parallax measurements to be made in three dimensional space.

An important fact overlooked in the interpretation of satellite imagery is that the spectral reflectivity of the forest stand, as recorded by the sensor, is influenced not only by the density of the forest cover/leaf area index, but also by the stand structure, especially its height. Two different grey-scale levels portrayed in the imagery may result not from changes in the forest cover but from differences in stand height. As observed by the writer, under favourable conditions and with local knowledge, it is possible to use the spectral grey-scale values of the satellite images to categorize the forest in broad stand height classes. It requires emphasizing therefore that, without field checking, it is no more correct to interpret changes in tone or colour of the image to be due to forest cover than it is to attribute them to stand height.

Boundary delineation

In the process of examining the diagnostic image characteristics, the delineation of boundaries on the imagery to separate the different classes often improves the efficiency of the interpretation. Thus, in the tropics the delineation of the boundary of the mangroves helps in the classification of the lowland tropical forest and the mapping of the boundaries of riverain (gallery forest) helps in the classification of adjoining savanna woodland. It is convenient to assume that forest types have recognizable sharp boundaries in the stereoscopic model. The boundaries so delineated will be found not to coincide exactly with the boundaries as observed on the ground or map but only to approximate to them. It is the process of drawing a line which is ideal. Vink (1964) referred to the technique as idealization. The boundary, as observed on the photographs, may be a combination of the tree crowns, shrubs and their shadows and exceed the width of a drawn pencil line (i.e. about 0.2 mm). The tree crowns on the single aerial photograph will also be displaced radially outwards from the principal point (section 9.2).

8.4   PRELIMINARY AIDS TO VISUAL IMAGE ANALYSIS

Several aids will be briefly commented on as background to and improving the quality of image analysis. These are as follows.

Maps

Reference should be made initially to existing topographic, planimetric and thematic maps. The latter include those based on geology, terrain or geomorphology, soils, vegetation, land use and forest stock.

Terrain maps

Terrain maps, if available, are useful in showing geomorphic units, which by subsequent fieldwork may be subdivided into smaller units showing soil types and/or vegetation/forest types.

Vegetation maps

Vegetation maps, showing the normal range of plant species, of forest types, of crops, etc. are helpful in equipping interpreters with local knowledge, enabling them to eliminate certain land types or tree species and also to substantiate their interpretation as being correct. Thus, in Malaysia, it is possible from land-use maps to locate areas of mangroves, rubber and oil palm, rice growing, etc.

Forest stockmaps

Forest stockmaps show the distribution of different forest or stand types, which have a bearing on forest management. In the United Kingdom, for example, the Forestry Commission prepares stockmaps showing compartments by species and year planted at a scale of 1 : 10 000. These serve as a valuable photointerpretation aid. Thus, when preparing a working plan of a typical area in west Wales, the aerial photographs were checked against the stockmap (Howard, 1961). After a few minutes' study of the photographs, it was usually possible to identify on the photographs the tree species, by using texture, tone, size and the stockmap data, to delineate areas where the species had failed to grow and to correct the compartment/sub-compartment boundaries for errors in the existing stockmaps.

Ecological field keys

An ecological key can be designed to aid the photointerpreter in the rapid and accurate identification of the images recorded on the photographs. Keys fulfil three important functions and are preferably accompanied by stereograms. They serve as a reference file for the trained photographic analyst, so that he/she can carry out the interpretation with a minimum of field visits. They provide a means of standardizing the classification by different interpreters at different places and at different times. Finally, keys serve as a useful training device for students by helping them to recognize and to classify objects correctly.

To obtain the maximum value from keys, they should provide (a) a brief description in words of important recognition features; (b) a stereogram illustrating the features (Figure 8.5(a)); and (c) when possible one or more terrestrial photographs of the type of features covered by the stereogram (Figure 8.5(b)). It is recommended that ground-based photographs should be taken and tied in with the key. Ground photographs are not widely used, which probably results from the fact that ecological keys were mainly developed during the Second World War, when ground photographs were not obtainable.

Both selective and dichotomous keys are used. A **selective key** gives a brief written description with or without accompanying photographs of the principal vegetation types likely to be encountered, and from these the photointerpreter chooses the most appropriate. Table 8.1 is an example of this type of key. A **dichotomous key** (Table 8.2) fulfils a similar function to a taxonomic key as used by the botanist. It starts with a separation into two

Table 8.2 Dichotomous photo-key – East African vegetation (Langdale-Brown, 1967)

1a. Upland sites (hills and hillsides and other well-drained areas), 2.
1b. Lowland sites (valleys and flood plains), 7.
2a. Trees, shrubs or bushes present, 3.
2b. Trees, shrubs and bushes absent or very widely spaced. Grassland.
3a. Ground completely obscured by trees, shrubs or bushes 20–200 ft high, 4.
3b. Ground or grass visible between trees, shrubs and bushes up to 50 ft high , 6.
4a. Open or barely continuous cover of trees which form a single layer not more than 50 ft high. Woodland.
4b. Continuous cover of trees and shrubs 20–200 ft high, 5.
5a. Continuous, uneven cover of trees 40–200 ft high with crowns touching, intermingling and overlapping. Layers of different heights often visible, individual tall trees ('emergents') common. Forest.
5b. Continuous, even, interlocking cover of trees and shrubs seldom more than 40 ft high. Thicket.
6a. Tree–shrub–grass mixture, trees and shrubs from 20 to 50 ft high cover more than half the area. Woodland.
6b. Tree–shrubs–grass mixture, trees and shrubs up to 50 ft high cover less than half the area including low bushy growths of no apparent height. Savanna and Scrub.
7a. Trees absent or widely spaced, 8.
7b. Trees present, usually abundant, 9.
8a. Herbaceous vegetation of no apparent height on seasonally waterlogged sites. Seasonal Swamp.
8b. Herbaceous vegetation up to 15 ft high on permanently waterlogged sites. Herbaceous Swamp (e.g. Papyrus).
9a. Scattered or continuous tree or palm cover 10–150 ft high. Tree Swamp.
9b. Even-banded tree cover 10–100 ft high in tidal areas, water often visible. Mangrove Swamp.

groups, for example wild vegetation and cultivation or upland and lowland and then proceeds to divide these into broad groups before dividing them further into specific groups.

Catena diagrams

Milne (1936) introduced the concept of the soil catena or 'soil chain' into his fieldwork in Tanzania, in which the topographic sequence of soil types is recorded. Howard (1970c), when developing aerial photographic land-unit classification in south-eastern Australia, recognized the need to introduce an intermediate landscape unit between the land facet and land systems (Chapter 14), and termed this the land catena. A **land catena** diagram represents in profile the topographic sequence of land facets which are geographically related (e.g. valley bottom to hilltop to valley bottom) (Chapter 14). Figure 8.5(c) shows part of a land catena diagram to support associated stereograms. Land catena diagrams are useful to the image analyst, since the known sequence of units helps in the identification and mapping of those units of the sequence which are not so readily observed on the photographs or mosaic; and also helps to associate the vegetation with the soil types and landforms.

The comparable diagram in plant ecology is the **vegetation profile**, which portrays the characteristic internal structure of the plant formation/plant sub-formation along a line or narrow strip transect. Unfortunately, vegetation profile diagrams are usually unavailable, are time consuming to prepare and usually require intensive fieldwork. The profile diagrams are useful in photointerpretation when studying seral vegetation, vegetation succession and forest stand structure. Whenever possible, catena and vegetation profile diagrams should be supported by stereograms and photographs taken in the field.

Stereograms

Aerial photographic stereograms (Figure 8.5(a)) form an important aid to photointerpretation, particularly as part of forestry training, forest inventory and forest mapping. Each stereogram is a subjectively selected sample representing a major ground feature including forest types and volume classes. It is provided by the relevant part of a stereoscopic pair of aerial photographs, and is usually mounted for viewing with a pocket stereoscope. To accommodate the limited viewing of a simple lens stereoscope, the two contiguous stereoscopic samples are mounted about 60 mm between common points and each sample is about 5 cm wide. By viewing the left sample with the left eye and the right sample with the right eye, a three-dimensional view of the landscape is obtained. Each stereogram should be accompanied by a brief but adequate description of the landscape/forest features, and is used to identify similar sites, forest types and volume classes on stereoscopic pairs of photographs. The stereograms, if possible, should be prepared from the same set of photographs as will be used later in the field, in order to avoid the effect of differences in photographic scale, season, etc.

Land information systems

These may be manual or digital, and, nowadays, are best represented by (digital) geographic information systems (GIS). In its simplest form, a land information system will consist of a set of maps representing geology, geomorphology, soils, forest stock, etc., a card index or personal computer containing the relevant information and suitable equipment for overlaying maps and photographs. In digital form a land information system can serve as a powerful tool to improving the efficiency of visual and digital imagery analysis. If a comprehensive GIS is not contemplated, but the area to be covered by imagery analysis is large, consideration should be given to setting up a simple land information system. The digital approach to establishing such a system is outlined in Chapter 12.

8.5   EQUIPMENT

Equipment for visual interpretation is needed for four main purposes: (a) viewing the photographs; (b) making measurements on single and stereoscopic pairs of photographs; (c) transferring data from the imagery to base maps, planimetric maps and thematic maps and from aerial photographs to satellite imagery and (d) storing interpreted information in digital format in the memory of a computer system. The first was discussed in section 8.2 under stereoscopes and (c) and (d) will be discussed in Chapters 11 and 12.

Measuring aids

In the taking of measurements from photographs (b), simple equipment suited to fieldwork is available for measuring height in the z-direction), for linear measurements in the x- and y-directions and for measuring areas. Height measurements normally require stereo-pairs of photographs and are made using a parallax bar (Figure 8.2(b)). Simpler, robust and much less expensive is the parallax wedge (Figure 8.6). This is made of transparent plastic material, and well suited to determining the height of trees on stereo-pairs of photographs in the field. Linear measurements taken on single photographs require an engineer's scale graduation in 1/50 inch or 0.1/0.2 mm (Figure 8.6). A school ruler graduated in millimetres can be read to about 0.3 mm. Finer readings of short lengths are made using a monocular magnifier

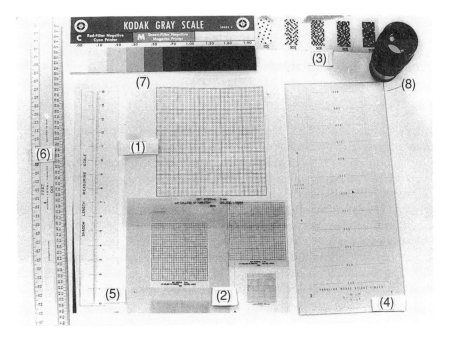

Figure 8.6   Simple measuring aids; 1: dot grid with 2 mm between dots; 2: three grids with dot intervals of 1 mm, 0.5 mm and 0.25 mm; 3: crown closure (crown density) scale; 4: parallax wedge; 5: shadow wedge or crown diameter wedge; 6: engineer's scale graduated in 1/50 of an inch so that readings can be made to 1/100 inch; 7: Kodak tonal scale; 8: monocular magnifying scale.

incorporating a bar scale, graduated in 1/10 mm (Figure 8.6). The width of tree crowns and tree shadow lengths can be determined on stereo-pairs of photographs with a crown/shadow wedge (Figure 8.6).

Areas can be measured with a transparent dot grid (Figure 8.6). The dots are equally spaced and suited to the nominal scale of the imagery or maps. When an area is being measured, the grid is placed on the surface of the photograph and the number of dots falling on the object to be measured (e.g. clear cut forest) is counted. By knowing or calculating the total ground area (A), the area (a) occupied by the object can be determined as follows:

$$a = A \times \frac{\text{number of dots on object}}{\text{total number of dots in area}}$$

It is preferable to use grids with coloured dots (e.g. orange/red) for measurements on black-and-white aerial and satellite photographs and black dots for measurements on colour imagery. Depending on the scale of the photographs, purpose and area of the survey and accuracy required, literature provides the density range of the dots varying between two and 200 per square centimetre. Spurr (1960) suggested for forest inventory, when using photographs at a scale of 1 : 15 840, about 7 dots/cm^2.

Although the dot grid can be made locally using transparent film and is very low in cost to buy, counting becomes tedious if the area is large. Alternatives are provided by manually operated **polar planimeters** and electronic digitizers. A polar planimeter calculates the area as the marker on the planimeter arm is traced around the periphery of the area. With an **electronic digitizing planimeter** (Figure 8.7), the area is determined similarly by operating manually the tracing arm; but its x-, y-co-ordinates are fed continuously into a microprocessor, which is programmed to express the results as area measurements.

If the intention is to estimate the percentage (or density) of the ground covered by the tree crowns of a forest stand, and not the ground area, a micro-dot grid can be used with larger-scale photographs. However, it is easier to use a crown closure/crown density scale, which has 10% or 20% texture classes as representing crown cover (Figure 8.6). The forest stand is examined on a contiguous pair of photographs under the stereoscope, preferably mirror-type, and cross-checked against the crown closure scale. The texture of the scale needs to be suited to the scale of the photographs.

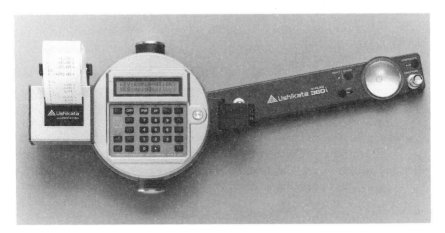

Figure 8.7 Modern electronic digital planimeter. The tracing lens is to the right. To the left is a 16-digit thermal printer, which provides a permanent numbered record. The mounted electronic calculator contains a 16-digit two-line liquid crystal display (LCD) and provides measurements of linear distances, curves and areas. Values can be accumulated and the instrument can be interfaced with a personal/mini computer.

Interpretation aids

These include those illustrated in Figure 8.6, as being well suited to field use. Also, the photointerpreter may wish to transfer data to base maps, or to revise existing thematic maps as outlined in the next chapter, to study grey-scale and colour contrast differences and possibly combinations of the several spectral bands. For the latter, three types of analogue equipment have been used: (a) density slicers, (b) additive viewers and (c) diazo colour printers. Analogue density slicers are little used nowadays, being replaced by digital imaging equipment. This is due primarily to problems in standardizing colour coding on the viewing screen and the high cost of analogue density slicers. The aerial photograph or satellite imagery has, however, to be digitized.

As mentioned in section 4.3, non-imaging densitometers are important in film processing, and are also used to improve diazo printing and in research related to film density and the grey-scale values of images. Standard commercial grey scales are also used for the same purpose (Figure 8.6).

Additive viewers were developed principally in the 1960s to be used with multispectral black-and-white diapositives; and, although expensive to purchase, were frequently used in the 1970s and early 1980s for the analogue analysis of Landsat imagery (e.g. Howard and de Kock, 1974). Basically, an additive viewer requires three coincident projectors for each of the three primary colours (blue, green and red), but was in general constructed with four projectors to accommodate the four spectral bands of Landsat MSS. A rotating variable colour filter operates in front of each lens and four separate resistances control the brightness of each projected black-and-white scene.

The alternative, the **diazo colour printer**, is well suited to satellite spectral imagery analysis and is simple to operate, and the purchase price is little more than that of a mirror stereoscope with parallax bar. Colour transparencies in a range of hues are obtained at contact scale from the black-and-white transparencies, and viewed as a single scene on a light table or projected on to a map base for the transfer of thematic detail. Two disadvantages of diazo photographs are the difficulty in using the imagery in the field and the fact that a strong light source is needed for viewing.

9

Elements of photogrammetry

Unless the image analyst/photointerpreter has had an introduction to photogrammetry, it is unlikely that maximum use will be made of stereoscopy and the taking of precise measurements from the imagery. In this chapter several selected elements of the broad subject of photogrammetry are considered. These have been chosen to provide an understanding of the geometric characteristics of the vertical aerial photograph, parallax measurement of height using stereoscopic pairs of vertical aerial photographs, and the geometry of imagery from outer space.

In preparing topographic and planimetric maps, the photogrammetrist uses well-established methods and techniques. These include the rigorous subject of analytical photogrammetry, which is beyond the scope of this book, and which provides the basis to the design of modern first and second order analytical plotters. Also, the computer processing of x-, y- and z-co-ordinate data has in recent years been closely integrated into photogrammetry. Nearly all topographic and planimetric maps are now prepared using aerial photographs, and maps at small scale are increasingly revised using satellite data.

9.1 SCRIBING VERTICAL AERIAL PHOTOGRAPHS

Vertical aerial photographs are used singly, in pairs or as an assemblage in photointerpretation and to prepare planimetric and topographic maps. Aerial photographs are frequently used in the forest as sketch maps. A temporary assemblage of photographs (e.g. print laydown) or a permanent assemblage (photo-mosaic or an ortho-photo-mosaic) are important in providing a synoptic view of the landscape for the purpose of forest management and planning and to complement and correct the data contained on old maps.

Ideally the caption along the bottom edge of the aerial photograph should contain the following information. This can be used to locate visually the photographs on an existing map, to calculate the photographic scale (section 5.1), and to identify the season in which the photographs were taken:

1. a locality reference (e.g. Thetford);
2. photograph and flight line numbers;
3. date of flying;
4. time of the exposure;
5. altitude of aircraft above sea level at the time of the exposure;
6. focal length of the camera lens;
7. type of film, if not panchromatic black-and-white.

N.B. If available, the flight plan can be used to obtain the geographic location of the photographs and the flight lines.

Scribing procedure

Before proceeding to take stereoscopic measurements from contiguous pairs of photographs it is necessary to scribe on each photograph photogrammetric control points (Figure 9.1). On the marked-up photograph, the geometric centre is termed the **principal point**. This is located at the intersection of the two diagonal lines from the corners of the photograph, or at the intersection of lines between the mid-points on the sides of the photograph. If the stereoscopic measurements are limited to tree heights, it is only necessary to establish the principal and transferred principal points and flight line. The principal point is marked on the surface of the photograph with a pin, with a faint white or red ink circle about 0.5 cm in diameter surrounding the pin-point. The centre point is then transferred stereoscopically to the adjoining photograph of the same flight line by pricking through its transferred position and scribing it with a circle of identical diameter. This point is known as the **transferred principal point** or **conjugate principal point**.

By extending a faint ink line from the edge of the circle of the principal point to the circle of the conjugate principal point, the **flight line/flight track** of the aircraft is established on the photograph. This is also called the **base line** or **photographic air base**, when applied to the principal points of two consecutive photographs in the same flight line. The air base, as recorded on the photograph, is used for lining it up in relation to similarly marked photographs, in determining stereoscopically tree and spot heights and for preparing maps. If the same air base as recorded on two adjoining photographs is measured, these will usually be found to differ slightly in length. This is due

Figure 9.1 A vertical photograph of 23 cm format (reduced), on which the location of the principal point, the two conjugate principal points, the flight lines between these points, the six wing points and one ground control point (triangle) have been scribed in white ink (see text).

to resultant tilt and changes in the flying height of the aircraft above the ground datum plane. The line passing through the principal point and coinciding with the flight line, is known as the **x-axis** and the line perpendicular to it passing through the principal point is the **y-axis**.

After marking the principal point and transferred principal points on each photograph, the next step, when preparing a map, is to choose six points towards the edge of each photograph. These points must be common to all contiguous photographs of the same flight line and adjoining flight lines. Normally photographs are taken so that the corresponding principal points of the photographs in adjoining flight lines are opposite to each other. If a single strip of photographs is being used, then each point will be common to only three photographs. These additional points are termed **wing points** or **pass points** or **photogrammetric control points**. As the last name suggests, the points are used in a manner similar to ground control points to provide orientation between photographs in mapping. The wing points, in combination with the transferred principal points, provide a total of eight peripheral points on a photograph, as shown in Figure 9.1.

9.2   GEOMETRIC CHARACTERISTICS OF THE VERTICAL AERIAL PHOTOGRAPH

The aerial camera records a perspective view, and thus the images of ground objects are displaced radially outwards on the photograph. The radial displacement of an image is also due to the tilt of the aircraft at the time the photograph was taken and to changes in the ground elevation (i.e. topographic displacement). It is important to remember that the displacement due to topography is radially outward from the nadir point. The ground **nadir** is that point on the ground vertically beneath the perspective centre of the camera lens. On a truly vertical aerial photograph, the principal point and the nadir point coincide, which is convenient, since the principal point, as mentioned in the previous section, is readily located, as the geometric centre of the photograph. This assumption is usually accepted in aerial photointerpretation irrespective of the fact that tilt may be present in the vertically taken aerial photographs.

The combined effect of a little topographic displacement and a little tilt in a nearly vertical photograph will emphasize the displacement outwards of an image from the principal point, unless they are in opposite directions. Extra care must therefore be taken to ascertain the tilt in photographs of rugged terrain, if image measurements including distances are being made on the photographs.

Topographic displacement

The effect of changes in the ground elevation on radial displacement in a truly vertical photograph is illustrated in Figure 9.2. The hilltop (G_1) is displaced radially outwards from the principal point (P) on the photograph due to the topographic relief of the hill above the datum plane being nearer the camera. A similar point (G_2) in the bottom of a valley, and exactly the same radial distance (R) from the ground nadir (N), is displaced radially inwards towards the principal point (P), due to the topographic relief of the valley bottom being further from the camera. Additional consideration of the diagram will help in

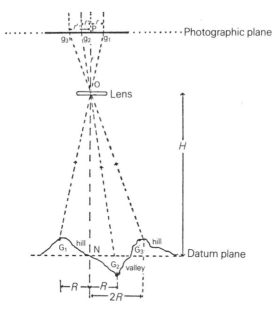

Figure 9.2 Topographic displacement. The diagram illustrates the effect of changes in the ground elevation of points G_1, G_2 and G_3 on the radial displacement of the same points recorded as images (g_1, g_2, g_3) on the aerial photograph (see text).

formulating the following generalizations: that the topographic displacement is radial from the centre point (P) and that all objects above the datum plane are displaced outwards and all objects below the datum plane are displaced inwards.

The displacement of a further hilltop (G_3) on the same datum plane as the first hilltop (G_1) and of the same height, but at a distance $2R$ from N, will be displaced on the photograph more than G_1 and proportional to its distance from the centre of the photograph. Images at the same distance and on the same datum plane, irrespective of which side of the principal point they are located, will be displaced the same distance from the centre of the photograph. If the flying height above the ground datum at which the photograph is taken is increased, then radial displacement is decreased; and if the flying height is decreased, then the radial displacement is increased. In other words, topographic displacement is inversely proportional to the height of the photography above the datum plane for a specified focal length. As large-scale photographs are normally preferred in studies of forest resources, topographic displacement assumes greater importance than in small-scale planimetric mapping projects.

Height displacement of tree images

The measurement on single large-scale photographs of the heights of tall objects provides a useful introductory classroom exercise in image measurements on aerial photographs. The displacement of the top of a tree, building or television mast from the bottom, as shown on a photograph, is similar to the displacement of the top of a hill above the datum plane. Height displacement is again radial from the nadir, and as tilt is slight, it is assumed to be radial from the principal point. The greater the height of the object viewed on the photograph, the greater will be its radial displacement on the photograph.

Thus in Figure 9.3, the top of the television mast is displaced radially from the principal point a greater distance than the top of the tree. It will be observed that similar triangles are formed by the lines subtended from the object on the ground to the optical centre of the lens. As H is large compared with f, it is assumed that H and $H + f$ are the same.

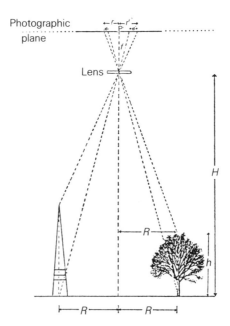

Figure 9.3   Height displacement of object images. The diagram illustrates the radial displacement on an aerial photograph of the image of tall objects, e.g. television mast, tall tree (see text). The radial displacement on the photograph of the tip of the mast is greater than the top of the tree, since, as an object, it is the taller.

Thus

$$r/f = R/H - h \quad \text{or} \quad fR = r(H - h)$$

and

$$(r - d)/f = R/H \quad \text{or} \quad fR = H(r - d)$$

as fR is common

$$r(H - h) = H(r - d)$$

or

$$d/r = h/H \quad \text{or} \quad h = \frac{Hd}{r}$$

This is often known as the basic **displacement formula**. r is the distance from the principal point to the top of the image of the object and is measured radially; d is the displacement or the distance radially from the true base of the object to the top of the object. It is fairly easy on large-scale photographs to measure from the base of a television mast to its tip or similarly an open-grown tall tree; but it is not practical to do so for small trees or a hill on a single photograph. This weakness, combined with the fact that true location of points cannot be determined as accurately as by radial line triangulation and using the parallax formula with stereo-pairs for the height measurement of trees, makes the formula primarily of academic interest.

Tilt displacement

As mentioned in section 5.2, tilt (resultant tilt) is the combined effect of lateral tip of the wings of the aircraft and the dipping of the aircraft fore and aft. Longitudinal tilt or y-tilt is that which is due to the nose of the aircraft being lowered or raised and displacing the nadir point in the $+x$ direction or $-x$ direction along the x-axis, x-tilt or lateral tilt or list causes the nadir point to be displaced along the y-axis. In Europe, omega (ω) is used for lateral tilt and phi (ϕ) for longitudinal tilt. The effect of lateral tilt in displacing the images radially inwards or outwards from the principal point of the aerial photograph is shown in Figure 5.3. Lateral tilt is much more a problem than longitudinal tilt, since during flight the aircraft is more stable fore and aft.

During film processing, photographs incurring major tilt are often **rectified**. This is the process of projecting the tilted photographic negative on to a horizontal reference plane, when making the photographic print or positive transparency (**diapositive**). The establishment of the horizontal reference plane involves determining the x-, y- and z-co-ordinates of ground control points or photogrammetric control points, which, as mentioned earlier, can be expensive unless suitable topographic maps are available. Consequently photographs with minor tilt (3–5°) are not usually rectified during film processing; but the correction is made when preparing the map from the aerial photographs. For further details, including methods used in the correction of inherent tilt in aerial photographs, a standard text on photogrammetry should be consulted. When planimetrically mapping from aerial photographs, which are in theory vertical, simple instruments are available for making the corrections (section 11.5). To transfer forest detail from a slightly tilted photograph on to a horizontal plane such as a map base is relatively easy, and may be carried out using a sketchmaster or overhead projector (Chapter 11).

Other image displacements

The effect of image displacements on aerial photographs, due to aberrations in the camera lens, is usually insignificant in forest resource studies. Lens displacement of images is termed **lens distortion**. Lens distortion may be radial or tangential and reaches a maximum at some distance from the principal point. On a 23 cm × 23 cm photograph the maximum lens distortion might be 0.012 mm at about 10 cm from the principal point.

Displacements developed in films and prints during processing are small, erratic and not radial from the centre of the photograph. The greater distortion usually occurs in the print paper, but polyester as a base is remarkably stable, particularly as compared with cellulose. Film shrinkage probably varies between 0.03% and 0.1% and print shrinkage is about 0.3%. Distortions in stored map sheets can be far greater, and, on old sheets, distortion may need to be taken into consideration when extracting horizontal distances (Howard and Kosmer, 1967).

9.3 PARALLAX MEASUREMENT

The faculty of stereo-vision and the qualitative use of stereoscopes have been discussed in Chapter 8. It is now important to consider how stereoscopy and parallax can be used quantitatively to obtain measurements of trees and forest stands from stereoscopic pairs of aerial photographs. **Stereoscopy** can be defined as the science and art of viewing two different perspectives of an object, recorded on photographs taken from nearby camera stations, so as to obtain the mental impression of a three-dimensional model of the object. Two overlapping photographs taken on the ground, or taken in quick succession from an aircraft, will provide three-dimensional models of objects, when viewed through a stereoscope.

Parallax

This is the apparent displacement of the position of an object (e.g. the top of a tree) in relation to a reference point. In Figure 9.4 it is assumed that the

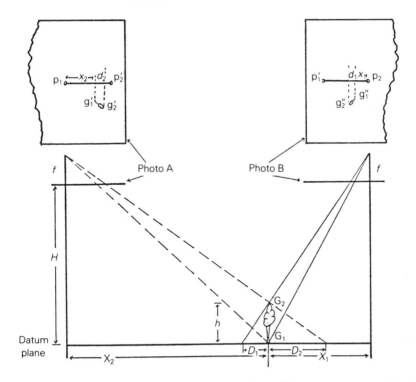

Figure 9.4 Parallax displacement of a tree recorded as an image on a stereoscopic pair of aerial photographs in relation to the reference point (i.e. principal points) (see text). The air bases showing the flight line of the aircraft are $p_1 p_2'$ and $p_1' p_2$.

reference point is the centre of each photograph (i.e. principal point), the photograph having been taken with the camera axis in a truly vertical position, as occurs when the film is parallel to the ground.

On a pair of stereoscopic photographs, the position of the recorded object in relation to each principal point (p_1, p_2) will vary due to a shift in the point of observation or camera station in the aircraft. The distance between the principal points $(p_1, p_2'$ or $p_2, p_1')$, as measured on the photographs, is termed the air base; and the parallax (P) of a point common to the pair of stereoscopic photographs (i.e. the **absolute stereoscopic parallax**) is the algebraic difference, measured parallel to the air base, of the distance of the two images of the same object from their respective principal points.

The object being measured may be a tree, as shown in Figure 9.4. If the tip of the tree is considered as point G_2 and the base of the tree point G_1, then it will be expected that in stereoscopic vision point g_2 will float above point g_1, g_2' to g_2'' being less than g_1' to g_1''. Examination of two photographs stereoscopically will confirm this as being correct. It should be noted that capital letters (G_1, G_2) are used for ground points and lower case letters (g_1, g_2) are used for photographic points.

Parallax difference

Instead of measuring these distances direct, the distances when using the absolute parallax formula are measured in respect of two known controls for all measurements, namely the principal points and the air base. The difference measured in this manner for the top of the tree and the bottom of the tree is termed parallax difference or differential parallax. The parallax difference (Δp) of an object on a stereoscopic pair of vertical photographs is defined as the sum of the radial displacements of the top of the object from its base measured parallel to the air base on each photograph.

In Figure 9.4, the location of a tree on the ground is shown in relation to the photographic station of the aircraft, when the pair of photographs were taken. If it is assumed that the tilt of the camera in flight is negligible and that the photographic scale is constant, then it may be shown that $h = H\Delta p/P + \Delta p$ where h is the height of the tree. This is the parallax formula which forms the basis of all stereoscopic measurement, and is derived as follows:

1. The parallax (P) by definition is: $p = x_1 + x_2 = $ air base $(X_1 + X_2)$, or if algebraic signs are considered

$$P = x_1 - (-x_2)$$

2. The parallax difference (Δp) by definition is:

$$\Delta p = d_1 + d_2$$

3. From similar triangles:

$$h/H = \frac{D_1}{X_1 + D_1} \quad [H + f \approx H]$$

or

$$h/H = \frac{d_1}{x_1 + d_1}$$

$$d_1 = \frac{h(x_1 + d_1)}{H} = h/H(x_1 + d_1)$$

4. Similarly

$$d_2 = \frac{h(x_2 + d_2)}{H} = h/H(x_2 + d_2)$$

5. Substituting in (2)

$$\Delta p = h/H (x_1 + d_1) + h/H(x_2 + d_2)$$
$$= h/H (x_1 + d_1 + x_2 + d_2)$$
$$= h/H [(x_1 + x_2) + (d_1 + d_2)]$$

or

$$h = H(\Delta p)/(x_1 + x_2) + (d_1 + d_2) = H(\Delta p)/P + \Delta p$$

For many purposes the formula may be used as $h = H(\Delta p)/P$, if $h/H < 2\%$. Frequently the formula is written as $h = H(\Delta P)/(b + \Delta p)$ or $h = H(\Delta p)/b$, where the photographic air base or mean of the two air bases (b) is substituted for the absolute parallax (P). It will be observed that, ideal conditions excepted, $P \neq b$; and also that $P + \Delta p$ is the parallax of the top of the object. Often in the USA h and H are expressed in feet and P and Δp in millimetres. Although these are different units of measurement, they can be applied directly in the formula.

The absolute parallax is increased by longer air bases and parallax difference by shorter focal lengths. By increasing the ratio: flying height/air base the intersection of the rays on a pair of stereo-photographs is at a greater angle, thus increasing the accuracy of the stereo-model in preparation of a map. The model is then termed a 'harder model'. To provide a scale model for use in advanced instruments the eye base/space image ratio is made the same as the air base/flying height ratio.

Johnson (1957) provided a mathematical proof that parallax differences smaller than 0.3 mm cannot be detected by average photointerpreters. He suggested logically that instruments graduated in excess of 0.3 mm do not necessarily increase the accuracy of the measurement of tree heights. This provides weight to the argument that simple instruments measuring to this degree of accuracy in forestry may be just as satisfactory as more complex instruments having a greater precision. Thus a parallax wedge reading to 0.3 mm should be as satisfactory as a parallax bar reading to 0.1 mm and first order/second order stereo-plotting instruments with finer readings.

Tree and stand heights

Most frequently the heights of individual trees are measured by the forester in the field with the aim of determining the average height of the forest stand; and the approach is similar when determining the apparent tree heights from aerial photographs. Mean stand height in the field is determined from the measurement of a reasonable number of dominant trees or co-dominants or a mixture of both or occasionally all trees including subdominants. On aerial photographs the latter are excluded since they will not be seen or are difficult to distinguish.

If the tree crown is cone-like, then it will not fully resolve on the photographs, and, unless the ground plane can be clearly seen in the stereo-model, it becomes difficult to place the fused floating mark at the base of the tree. In such circumstances it is necessary to determine the systematic error by correlating the photo-measurements with field measurements of the same trees, or to apply regression analysis using the two sets of data.

Example

Assume the height of a single tree is being measured on a stereo-pair of photographs using a parallax bar and mirror stereoscope and that the photographs have been scribed to show their principal points, transferred principal points and air bases. The mean air air base is then measured (92.3 mm). The fused floating-mark of the parallax bar is placed at the top of the crown of the stereo-image and then at the base of the tree of the same image. By subtracting the two parallax readings the parallax difference for the tree is obtained (1.37 mm). From the the caption on the aerial photographs and with reference to a topographic map, the flying height is calculated as 12 000 ft above the mean ground datum. Applying the parallax formula:

$$h = \frac{H(\Delta p)}{b + \Delta p} = \frac{12\,000 \times 1.37}{92.3 + 1.37} = 175\,\text{ft}\,(53.3\,\text{m})$$

Spot heights

The application of the parallax formula to determining spot heights of terrain features within the stereoscopic field of view of the photographs is useful in several ways. It may be necessary to know the ground elevation of key terrain features (e.g. hilltops), for localities where there are no topographic maps; to provide an input to a geographic information system under similar circumstances; to know the ground elevation in mountainous terrain in order to correct the 'parallax' calculated tree heights, or the calculated scale of small areas measured on the photographs.

In order to determine the approximate ground elevation of a terrain feature, it is necessary to identify on the photographs a point of known ground elevation and to reference the parallax difference measurement to this point. This usually necessitates mounting the parallax bar on a scanning arm, but this is tedious if many feature points are being measured. In such circumstances, it is preferable to use an instrument which has a movable table on which the photographs are mounted and the parallax bar is fixed. A further improvement is to interface the instrument with a personal computer incorporating suitable programmes.

Small areas

On an aerial photograph, an area can be considered as small if it represents only a very minor part of the total area. Small areas constitute sample plots or patches of regeneration or gallery forest surrounded by woodland etc. For each of these small areas, it is necessary to know the local scale on the photograph if the terrain is hilly, and not to rely on the nominal scale in the vicinity of the principal point in order to calculate the scale and hence the area of each. To ignore the local scale of each small area can lead to serious errors, as for example in estimating the number of trees per unit area.

As illustrated in Figure 9.5 for stereoscopic profile of the terrain, there are three photo-plots of equal size at ground elevations of 6500, 5000 and 3500 ft and located at g_1, g_2 and g_3 respectively. Only plot g_2 is at the mean ground datum in the vicinity of the principal point, and hence can have its scale calculated accurately using the nominal scale of the photograph. Assuming a

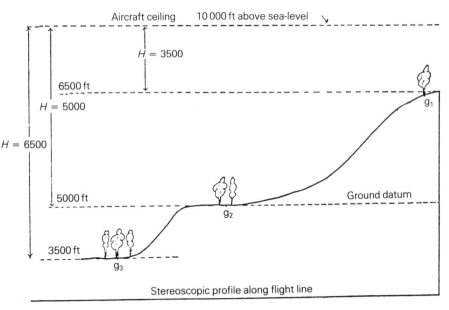

Figure 9.5 Diagram illustrating the effect of ground elevation differences in the stereoscopic model. If these differences affecting local photographic scale are ignored, then sampling on a single photograph with a transparent grid having fixed areas scribed on it, will incorrectly estimate the ground area and hence the number of trees per unit land area. Using the nominal photographic scale, this is shown as providing tree counts of 1 (g_1), 2 (g_2) and 3 (g_3) within the grid cell, although the correct count is two at each ground point.

flying height above the ground of 5000 feet (1640 m) and a focal length of 150 mm (6 inches), the nominal scale will be 1 : 10 000 and can be used to calculate the plot area at g_2. If the nominal scale is used to calculate the ground area of the photo-plots at g_1 and g_3, then their areas will be considerably underestimated and overestimated. Spurr (1960) pointed out that scale differences should be taken into consideration when the ground relief exceeds 3–4% of the flying height. Often in forest inventory, when sampling large numbers of photographs and a large number of plots are used, it is assumed that errors accruing from changes in ground relief are self-cancelling.

When the apparent heights of trees are being determined stereoscopically, moderate changes in the ground elevation in relation to flying height will not affect the estimates too greatly. This is due to the fact that the point elevation on the ground is constant in relation to the two parallax measurements of the top and base of the tree, and that the calculation of the tree height depends on the parallax difference. Under favourable conditions, the error introduced by using the nominal photographic scale is not likely to exceed 10% of the tree height.

Slope

As with determining apparent tree height, two parallax measurements are required to determine the parallax difference (Δp) between the top and bottom of the slope, and hence the vertical interval (Δh); but in addition the horizontal distance (S_h) between the two points is needed. An alternative but less accurate method is to measure the slope distance (S_s) between the two points on one of the photographs, and then to use the sine formula to calculate the slope (b), i.e.

$$\sin b = \frac{\Delta p}{S_s}$$

The horizontal distance between the points at the top and bottom of the slope, as seen in the stereoscopic model, is preferably determined by locating the points on a suitable scale map and measuring the distance between the points; or by determining the difference between the x- and y-co-ordinates of the two points using a photogrammetric plotter or by radial line intersections (Chapter 12). Radial line intersection to determine the horizontal distance (S_h) is simple and provides results of acceptable accuracy for forest resources. The procedure is as follows. A template cut from tracing paper is placed on the first annotated photograph and the principal point, conjugate principal point and flight line are transferred to the template. Next, radial lines are subtended on the template from the principal point to the two points on the photograph located at the top and bottom of the slope. The template is then removed and placed on top of the second photograph of the stereo-pair so that the flight lines coincide. An adjustment may be required for differences in length of the two photographic air bases. Radial lines are again subtended on the template from the principal point of this second photograph to the two 'slope' points. The loci of intersections of the four radial lines on the template now provide the 'true' positions of the points in their horizontal or map positions. The distance is measured between the points on the template in inches or millimetres, converted to feet or metres and inserted in the following formula to calculate the slope in degrees $(\tan b)$:

$$\tan b = \frac{\text{vertical distance}}{\text{horizontal distance between points}} = \frac{h}{S_h}$$

Shadow length

In determining tree height from shadow length, it is necessary to know the angle of the sun to the zenith at the time the photographs were taken. This can be ascertained by reference to nautical or air navigation tables showing the sun's declination $(\cos b)$, latitude $(\cos a)$ and time of the day $(\cos c)$. The angle of the sun's elevation (x) is then calculated:

$$\sin x = \cos a \times \cos b \times \cos c \pm \sin a \times \sin b$$

The angle of the sun's elevation $(\tan x)$ can also be determined approximately by determining the height (h) of a recognizable object on level ground either by field measurement or by parallax difference; and then using this height in conjunction with the photographic shadow length (h_1) to calculate the elevation angle:

$$\tan x = \frac{h}{h_1} \times \frac{f'}{H'}$$

Assuming the elevation of the sun (x) is known, then the height of a tree or similar object, from its measured shadow length (h_1) on the photograph, is given by the formula:

$$h = h_1 \tan x$$

(N.B. It is corrected for scale by multiplying by H'/f'.)

This formula is applicable to level or nearly level ground. When the ground has a noticeable slope then a correction requires to be added, if the slope rises in the direction of the shadow, and subtracted if the slope falls away in the direction of the shadow. As illustrated in Figure 9.6, h is determined by using the formula:

$$h = h_1' \cos b (\tan x - \tan b)$$

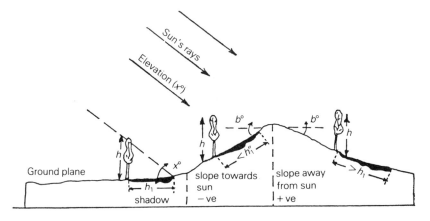

Figure 9.6   The influence of slope on the shadow length of trees of the same height (h) (see text). Shadows are longest for trees on slopes away from the sun.

or

$$h = h_1'' \cos b \,(\tan x + \tan b)$$

where $b°$ is the slope of the ground along the axis of the shadow.

9.4   SATELLITE PHOTOGRAPHY

The first photographs by a manned space mission were taken on board the US Apollo in 1965; was followed by the vertical photography of the US Gemini, which was used in the preparation of a mosaic map at a scale of 1 : 1 million of part of Peru and later in 1975 by Skylab. Nearly all the oblique and vertical photographs of these early missions were taken using either 70 mm Hasselblad or 70 mm Maurer cameras with a format at 60 mm × 60 mm and at scales between 1 : 200 000 and 1 : 2 million.

In recent years, the USSR has been acquiring vertical photographs with 23 cm format Zeiss photogrammetric cameras at an altitude of about 250 km (Oesberg, 1987). A Zeiss RMK 23 cm photogrammetric camera was used in the USA/FRG Spacelab mission (1983) (Konecny, 1985) at about the same altitude to provide strips of photographs of several parts of the world with an endlap of about 60% and a scale of 1 : 840 000. With a ground resolution of about 5 m, planimetric maps are readily prepared at a scale of 1 : 50 000 and thematic (sketch) maps at 1 : 25 000. Photographs having this resolution are suitable for reconstituting the image geometry to ±10 m and for topographic mapping with contours better than 30 m. This type of imagery is urgently needed to meet the world-wide need for up-to-date planimetric and topographic maps, at scales of 1 : 50 000 or 1 : 100 000, which can be prepared using existing photogrammetric plotters. So far computer-processed imagery from data collected by electro-optical sensors on satellites are of lower standard; and there are technical constraints, which may make photogrammetric restitution of the satellite-taken photographs preferable and much less expensive (Konecny, 1985).

Geometric considerations

The geometry of photography from outer space is more complex than for aerial photography. In Chapter 7 the orbiting characteristics of satellites were examined. An on-board altitude control system ensures under normal operating conditions angular stability for pitch (y-tilt), roll (x-tilt) and yaw (swing), which for Landsat have maximums of $\pm 0.7°$, $\pm 1°$ and $\pm 1°$ respectively. These are usually well within the accuracy requirements of thematic mapping for forest resources, but are of concern to the photogrammetrist. Strict altitudinal control in the z-direction or nadir line between the spacecraft and ground is necessary to avoid changes in the photographic scale, and also to control the satellite's velocity, due to geoidal variations resulting from the earth not being a perfect sphere and the tendency of the orbit to be ellipsoidal.

The point on the equator being intersected by the satellite's track on its northwards crossing is termed the ascending node and the southwards crossing the descending node (Figure 9.7; Petrie, 1970). The angle which the orbiting satellite makes with the ascending node is termed the orbital inclination. The

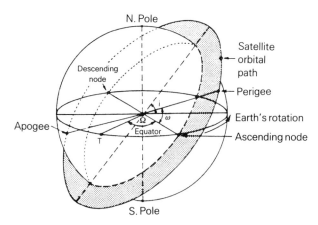

Figure 9.7 Schematic diagram illustrating a satellite's orbit in relation to the earth's surface (see text).

nearer the sub-satellite track is to the polar orbit (inclination = 90°), the greater will be the coverage of the earth. With an inclination angle, for example, of 60° the sub-satellite track along which the satellite will be vertically overhead will not exceed 60° latitude north and south.

Two other factors influencing the geometry are that the earth is not stationary beneath the satellite but rotating from east to west, and that the earth is not flat. The effect of the earth's rotation is that the sub-satellite track will not retrace its track, without correction, along the same great circle on the earth's surface in successive orbits. The ascending and descending nodes will be displaced in longitude by the angle the earth has rotated during the orbital period. On a map plotted on a mercator projection, the sub-track without correction will describe a wave-shaped path. It also means that successive photographs are taken progressively further west in the descending orbit and vice versa in the ascending mode. With data obtained by an electro-optical scanner, the imagery will be sheared (i.e. parallelogram shaped), as shown in Figure 10.3.

Earth's curvature

The errors caused by the earth's curvature are significant. On a single aerial photograph taken at a conventional scale, the values of radial displacement of objects towards the edge of the aerial photograph, due to the earth's curvature,

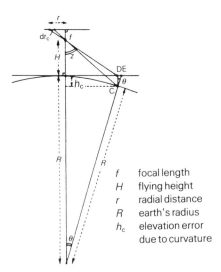

f	focal length
H	flying height
r	radial distance
R	earth's radius
h_c	elevation error due to curvature

Figure 9.8 Radial displacement and the earth's curvature. Errors occur in the taking of measurements from high-altitude aerial photographs and from satellite photographs (see text).

are too small to be graphically plotted (e.g. 0.2–0.4 mm); but these errors must be taken into consideration when stereo-plotting at very small scales from aerial photographs (e.g. 1 : 120 000) and from satellite photographs.

For conventional small-scale aerial photography and high-altitude photography/military reconniassance photography, typical flying heights would be 5000–9000 m, 12 000–18 000 m and above 250 km for satellite photography. Based on photogrammetric experience with high precision stereo-plotters, the standard deviation of single spot heights can be expected to be about ± 0.2% of the flying height and corresponds to a contour interval of about ± 0.67%. With increasing altitudes, relief displacements are reduced. The displacement due to the earth's curvature increases and is easily measurable on wide-angle and super-wide-angle photography at the altitude of Spacelab at about 300 km.

Schematically, relief displacement due to the earth's curvature is provided by Figure 9.8; h_c is the displacement or error in elevation due to the earth's curvature, f is the camera focal length, H is the satellite's orbiting altitude above the earth, R is the earth's radius (approximately 6370 km), r is the radial distance from the principal point on the photograph to the object and its displacement. Based on the radial displacement formula, the elevation error (h_c) caused by the curvature of the earth is provided by the following formula (Smith, 1958). h_c corresponds to h in the displacement formula given in section 9.2, i.e.

$$h_c = \frac{H^2}{f^2} \times \frac{r^2}{2R}$$

Since on a single satellite photograph relief displacement is slight and the tilt of the satellite in orbit is controlled to under 1°, most satellite photographs can be used directly in preparing small-scale planimetric/thematic maps. As errors due to relief and the earth's curvature are radial from the centre of each photograph, the principle of radial line plotting outlined in Chapter 12 for aerial photography can be used, when mapping from contiguous pairs. Further, satellite photography has the advantage that as each photograph covers an extensive area of the earth's surface, relatively few ground control points are needed; and the imagery itself can be used to provide the photogrammetric control points for aerial photographs covering the same or part of the area covered by a pair of satellite photographs. On both a satellite photograph and satellite electro-optical imagery, the tonal variation of images of the same class within a small ground area will be less than occurring across aerial photographs covering the same area. This greater consistency in tone or colour is helpful in image analysis.

10

Computer-assisted image analysis

In Chapter 8, emphasis was placed on the visual interpretation of aerial photographs. In this chapter, the emphasis will be on computer-assisted image processing of satellite-sensed data, since this is the ambit within which has occurred most developments in digital analysis. Consensus of opinion usually reserves the term 'photograph' for images of field objects (e.g. trees) recorded, processed and printed directly on film. A scene recorded digitally using an electro-optical scanner and processed digitally to photographic prints is termed imagery. That is, imagery is the pictorial product (**hard copy**) of image-forming instruments. The term **image** is applicable to both types of product as being the remotely sensed, recorded representation of objects. Sometimes, however, it is convenient to use 'imagery' as a general term to cover (aerial) photographs and (satellite) hard copies.

Digital analysis is equally suited to satellite sensing and some methods of airborne sensing, although most digital systems are specifically designed for handling data of earth resources satellites. If the system is required to handle airborne scanner data or SLAR, additional software will be needed to reformat the computer-compatible tapes and to make corrections to the data in a way similar to that of preprocessing satellite-acquired data (Chapter 7). If the input is to be derived from aerial photographs, additional hardware will be needed to digitize (read) the photographic images and to provide analogue-to-digital conversion (ADC). A special programme is needed to convert the perspective view of rectified aerial photographs into an orthogonal projection unless ortho-photographs are being used. As may be expected, these modifications to a dedicated system for satellite imagery are expensive and will generate large amounts of data. A 23 cm format aerial photograph when digitized using a 25 μm scanning head will produce about 10 million pixels with each containing 8 bits of information.

10.1 IMAGE ANALYSIS SYSTEMS

In the early 1970s, general-purpose mainframe computers were used for batch processing Landsat data. Usually, however, the hard-copy end-products could not be viewed until the analysis process was completed. Sometimes it is not appreciated that, as compared with the later developed dedicated systems, the classification on a general-purpose mainframe computer of the pixels of a single Landsat scene could take many hours and that several man-months of work could be involved in writing a single programme.

The Image 100, one of the first commercially produced dedicated systems in the 1970s, provided interactive viewing on a screen displaying 512 lines by 512 columns of pixels and a reasonable menu of software programmes. The purchase price at that time was over US$500 000, and the equipment did not have the versatility and processing speed of current minicomputer systems at less than a tenth of the price. For forest research and practice, the Canadian Forestry Service was probably the first (1977) to acquire a major dedicated digital image analysis system. This was designed with high processing speed and software specifically for forest mapping. In about 20 minutes the central processing unit (CPU) could complete a maximum likelihood classification of a quarter Landsat scene.

Meanwhile, the trend has been towards lower-cost compact digital imagery analysis systems using micro-computers, faster processing, expanded memories and improved programme menus. Frequently, the software programme capability extends to formatting the imagery to a standard map projection to make it GIS compatible and more suited to field use. In addition, in the 1980s small low-cost systems were marketed (e.g. RIPS) using microprocessors/personal computers and floppy discs, for viewing segments of Landsat MSS (240 lines × 256 pixels within lines) in conjuction with USGS digitized map sheets on diskettes. However, due to the finer sensor resolution of second generation satellites and the consequential increase in the number of pixels per unit land area, the trend is now towards larger viewing screens (i.e. 1024 lines), much greater data storage capacity, faster data processing and being more readily operated by the less inexperienced user (i.e. user friendly). Most systems necessitate a week or longer learning to operate the system and possibly up to a month to achieve full competence.

Mainframe computers

The sharing of general-purpose mainframe computers has not been favoured in the development of interactive digital imagery analysis for a variety of reasons. These include slower processing time, higher cost of user products per unit of ground area, the need often to write special expensive programmes, which are readily available at low cost with dedicated systems, and viewing time, which may block the computer from being used for other purposes. On the other hand, if the programmer's time is available free or at low cost on an existing mainframe computer, its use may provide the means for demonstrating the need for digital equipment in remote sensing analysis and may provide a useful base for some types of research and training.

Historically, computers with a central processing unit of 32 bits or larger were referred to as mainframe computers, 16 bits as minicomputers and 8 bits as micro-computers or personal computers; but, nowadays, this distinction is difficult with the miniaturizing of the central processing unit (CPU). The term **bit** is an abbreviation of 'binary digit', which is the notation in which computers operate. In base 2 (cf. base 10 in logarithms), an 8-digit number permits a range of 2^8 or 256. An 8-bit memory can display up to 256 differences of colour, while 6 bits reduces this to 64 grey levels. A **byte** is 8 bits of digital data and a **megabyte** is 10 bytes.

Dedicated systems

These are sometimes described by manufacturers as stand-alone or turnkey systems, implying misleadingly their readiness for immediate use, like a ready-to-use item purchased off the shelf in a supermarket.

Most dedicated imagery analysis systems are designed around a minicomputer, or more recently a micro-computer. A typical minicomputer, organized around a **central processing unit**, provides the two basic functions of control and mathematical operations and is attached to the memory and peripheries by a multiline connector termed a bus. In the micro-computer, the CPU is contained on a single integrated circuit chip, which is termed a **micro-processor**. The CPU controls the data transfer, initiates processing activities and access to the arithmetical logical unit for binary arithmetic and a set of temporary storage registers. For further information reference should be made to a standard text on computers. A suitable introduction is provided in the *Manual of Remote Sensing* on digital hardware (chapter 20).

Interactive analysis systems

These, which are normally dedicated systems, comprise several major components in addition to the host computer, its supporting software and video-monitor. The peripherals (sections 10.2 and 10.3) include input devices, additional video-display facilities (e.g. mice or cursor controls), mass data storage and output devices and possibly additional keyboard terminals. Figure 10.1(a) shows an interactive system ready-for-use. As the operator/analyst sits

(a)

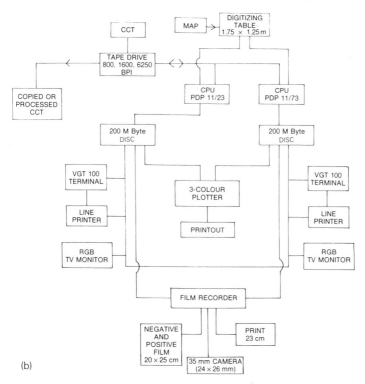

(b)

Figure 10.1 (a) Small digital interactive analysis system (DIAS). (b) Schematic plan of an interactive digital image analysis system with CCT input and map inputs via the digitizing table, two terminals, colour monitor and outputs consisting of copied or processed CCTs, line prints and film products.

at his/her keyboard terminal and views the monitoring/display screen, he requires to have command of software and hardware for accessing data, possibly preprocessing, feature extraction and classification of the identified and displayed data. Figure 10.1(b) provides a schematic plan of this type of system with two terminals, which is well suited to forest and land resources studies. A second terminal increases throughput and avoids delays during the peak season associated with fieldwork.

10.2   INTERACTIVE PROCESSING

Interactive processing requires the operator at the terminal keyboard to provide the computer commands in his/her stepwise analysis. An interactive analysis system is designed to assist the interpreter with the selection of the recorded pixels of ground objects (i.e. **feature selection**) and in the ensuing decision-making process (i.e. **classification procedure**). It is unlikely that fully automated classification systems will become feasible in forestry.

Some of the commands, placed by operating the keyboard terminal, are executed by the system's hardware (i.e. **physical components**) and some by the system's **software**, which are the programmes or instructions that cause the hardware components to process and analyse the data. However, the difference between software and hardware is increasingly less obvious, since the introduction of micro-coding. Sometimes functions are undertaken by the hardware (e.g. vector-to-raster conversion); and sometimes the software undertakes a permanent function from a special programme stored in a read-only memory (ROM) termed **firmware**. Often these special functions are executed by operating a separate digitizer tablet (joystick, trackball). These operations may include zooming the scene to different scales on the screen, splitting the scene, contrast stretching and cursor control in identifying a pixel or groups of pixels of special interest and delineating areas of special interest as sub-samples. The memory display processor must incorporate three separate memory planes for each of the three primary colours, but in practice these may exceed twenty. Memory devices are needed for storing images larger than 512×512 pixels and for arithmetic operations and display other than vector display. Displays for example of entire Landsat MSS scenes are produced by sampling only every nth pixel, which reduces the demand on memory size; but obviously this lowers its image resolution and value for detailed visual interpretation, although providing a synoptic view useful in regional studies.

Normally the colour frame stores are configured to display 512×512 pixels, although, as already mentioned, video monitors displaying 256×240 pixels and 1024×1024 pixels are available. Each scan line is termed a raster line and is read optically in discrete steps. The brightness of each sample corresponds to the grey-scale range (e.g. 0–255) transcribed into the binary notation for storage in the digital memory.

Mass data storage

In comparison with general-purpose computers, dedicated imagery analysis systems often need much greater on-line data storage than initially provided; and this need has rapidly increased with the launching of the second-generation satellites and the growing interest in change monitoring. It is therefore well worthwhile in the planning stage to allow for this factor, even though manufacturers continue to increase the storage capacity of tapes, cartridges and hard (metallic) and floppy (plastic) discs. Until recently, for example, a single 8-inch floppy disc would be filled by a 512×512 pixels sub-scene. Now, Winchester discs have a higher density storage and optical discs have capacities of 1000–2500 Mbytes. Past experience has indicated that disc capacity should allow for a minimum storage of one satellite scene in all bands and off-line storage of all relevant analysed and partly analysed data.

A further consideration, as a precautionary measure, is to provide a duplicate set of analysed data in offline storage. The purchasing of a stand-by tape-drive should also be considered to avoid throughput delays, when major servicing is needed. Nothing is more frustrating than the loss of partly analysed data or analysed data through a disc crash. As mentioned in the *Manual of Remote Sensing* (American Society of Photogrammetry, 1983) a single particle of cigarette smoke resting on the surface of the disc can strike the head with sufficient force to cause the head to sit down on the disc. This is called a **head crash**, which generally destroys the disc pack. The risk is greatly reduced by installing a sealed non-removable hard disc subsystem (e.g. Winchester drive).

Input devices

The commonest peripheral is the tape drive used for transferring the preprocessed CCT data to the interfaced host computer. In the smaller systems, this is replaced by a floppy disc drive or possibly a tape cartridge. If floppy discs are used, then, as a separate operation, part of the satellite sub-scene of interest to the imagery analyst has to be transcribed as a separate operation from the preprocessed CCT to the floppy disc. The main disadvantage of this method is that the sub-scene will cover only a part of the total scene. A Landsat, 4-band, 512 lines $\times$ 512 pixels sub-scene will occupy about 1 million bytes and require in excess of one floppy disc.

Analogue data conversion

If analogue data is being used, as with aerial photographs and some airborne SLAR and scanner data, then analogue-to-digital (A to D) conversion will be necessary. The A to D conversion (ADC) requires the continuous sampling of the analogue signals at predetermined time intervals and their numerical recording on to computer-compatible tapes (CCTs). A common example of A to D conversion is the micro-densiometric line scanning of a film negative in which the densities of the film images (in analogue form) are converted to discrete numbers or digits.

Digitizing

When the input is a photograph, the digitizing is made using either a videcon (TV) camera or electro-optical scanners. A videcon camera, with a spatial resolution of about 550 lines, scans an entire photographic transparency in about 1/30 second. The data is then passed through the A to D converter and stored in the digital memory, which is interfaced with the analysis system. Video-digitizers eliminate human errors associated with manual digitization of maps, are fast and relatively moderately priced.

Scanners have a better spatial resolution (e.g. 10^{-2} mm) and radiometric consistency than videcon digitizing systems; but in price may cost two or more times the price of a dedicated image analysis system. The basic type of scanner samples the photograph line by line, following a geometric pattern which is usually rectalinear. Figure 10.2 shows a film reader–writer. In a film writer, as an output device, the detector head is replaced by a point source of light and the photographic print by unexposed film.

It is also often advantageous to merge remotely sensed data with map data and therefore as a peripheral to medium-sized analysis systems (and larger) the input devices include a co-ordinate digitizer. The most popular type operates

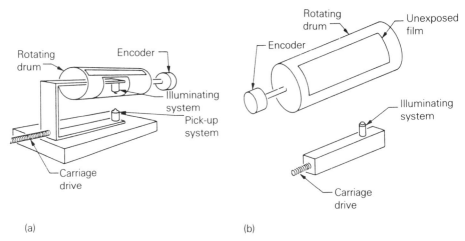

(a)   (b)

Figure 10.2 Illustrated schematically is (a) a drum film reader and (b) the modification of the equipment for film writing.

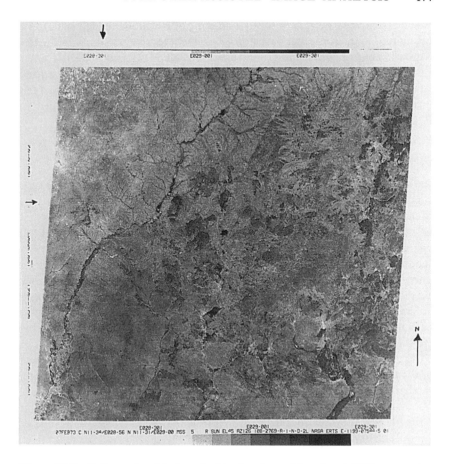

Figure 10.3   Bulk product black-and-white (hard-copy) Landsat-1 MSS scene, taken in the red-sensitive spectral band ($0.5–0.6\,\mu$m). The scene covers approximately $34\,000\,\mathrm{km}^2$ (scale $1:1.8$ million reduced from $1:1$ million); the location of Figures 10.4 (a), (b) and Plate 2 has been indicated by small arrows; a river crosses the area from north-east to south-west; a geological fault parallels the river to the east (top centre); the darker toned natural vegetation (savanna woodland – Sudan) in darker tones to the east of the river contrasts with the lighter toned heavily grazed lands to the west of the river; recently burnt areas have recorded blackish in tone.

electromagnetically. The map is placed on the flat bed of the digitizing table in which a wire mesh is embedded. The flat bed (or tablet) varies in size from A4 to approximately $1\,\mathrm{m} \times 2\,\mathrm{m}$. A cursor is run across the aligned map, and at each feature requiring its co-ordinates to be read a button is pressed, which sends the signal to the host computer. In some co-ordinate digitizers, the cursor assemblage permits a limited amount of additional coded data to be transmitted to the computer data store. This includes relaying the co-ordinates of ground control points and ground elevations taken off a map sheet at larger scale in order to improve the overall geometric accuracy of the remotely sensed data and to improve its information content. The sources of data are merged during processing, when the imagery digital data is geometrically 'warped' to fit the required map projection.

Output devices

The value of any interactive digital analysis will be judged usually by the type and quality of the output pictorial products suited to user requirements. This extends to computer-compatible tapes and floppy discs, which are geocoded for use with low-cost digital analysis sub-systems or to provide inputs to geographically-based land information systems. For example, in the People's Republic of China, a network is being developed for agriculture comprising a centralized dedicated digital analysis system at Beijing Agricultural University and regionally low-cost floppy disc sub-systems. A somewhat similar approach for land resources is being pursued in India with the major computer facility at the National Remote Sensing Centre in Hyderabad and simple digital systems elsewhere.

Output devices range from expensive but accurate flat-bed or drum plotters for thematic mapping (Chapter 11) to the simple use of a Polaroid camera or a standard 35 mm camera with a zoom lens to photograph the display screen. The main disadvantages of this simple procedure are the geometric distortion of the photograph, due to the curvature of the television screen, and a loss of spectral fidelity.

Line printers

Line printers and dot printers are also used in the image-forming process of providing hard copies (Figure 10.4), and have the advantage of being fast

---

Figure 10.4 Comparison of hard-copy devices: (a) part of the Landsat MSS scene (Figure 10.3) obtained in the wet season using a line printer (unsupervised classification) and in the same spectral band; combined with field checking, vegetation types, the railway line, the township and the seasonal water surface have been identified and boundaries can be drawn on the basis of alpha-numeric groupings, as shown in (b). Cr: crops; SWd: dense savanna woodland (e.g. *Combretum glutinosum, Dalbergia melonoxylum, Albizzia amara*); SWs: sparse woodland; RF: riparian forest/riverine woodland (e.g. *Tamarindus indica, Diospyros mespiliformis, Acacia nilotica*); (b) part of the area covered by (a), enlarged to show the alpha-numeric coding used to represent the spectral classes. See also Plate 2 of the colour section.

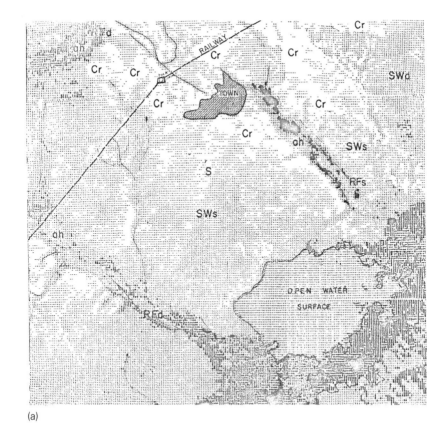

(a)

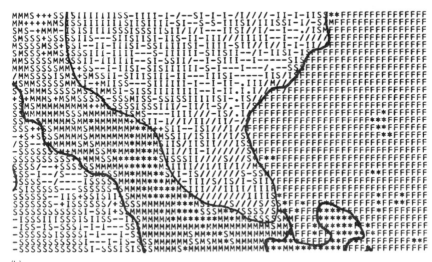

(b)

and cheap. With black-and-white **line printers**, the operator selects a set of alpha-numeric characters, which when typed as representing pixel values and viewed at a distance, span a reasonable range of grey levels and graphically portray the satellite scene (Figure 10.4(b)). Groups of similar alpha-numeric characters are delineated visually, can be hand coloured and a thematic sketch map prepared. However, the end product is not attractive and is preferably replaced by **ink-jet colour printing** or dot printing. In a **dot matrix printer**, dot-forming characters replace the standard characters of the line printer. There are typically 12–25 black dots linearly per centimetre.

The most popular output device is the specially designed film recorder providing spectrally and geometrically controlled photographic negatives. These devices can be grouped as laser film recorders, film writers and CRT film recorders.

Laser film recorders

These provide the highest quality photographic product and are the fastest to operate, but are the most expensive to buy. The laser beam is modulated readily to a fine spot size corresponding to the resolution of the film and has sufficient energy to be split simultaneously into red, blue and green components, which are immediately recombined to provide a colour image.

Film readers/writers

Illustrated by Figure 10.2, these provide a high-quality product, but are slow to operate, more expensive than laser film recorders to maintain and cost more than twice the price of most interactive imagery analysis systems. Less expensive are film writers recording only in black and white and not designed to read (i.e. digitize).

In recording the latent images on the film, which is mounted on the drum of the writer, a variable-intensity light source scans in the x-direction as the drum slowly rotates in the y-direction (Figure 10.2). Each light recording varies with the spectral value (density value) of the respective pixels. Depending on the type of drum writer, the aperture of the light source will vary between 10 and 100 microns.

As a compromise in price and image quality, **CRT recorders** are increasingly popular. A raster display is created on a cathode ray tube (CRT) by successively scanning a regular pattern across the screen – similar to TV display. The electron-gun input is modulated according to the grey-scale values of the pixels held on the satellite CCT input and held and refreshed on the CRT as a display of 512×512 points. The entire screen is 'written' in less than a minute. The recording system is therefore fast and has the advantage that raster graphic prices have declined with the falling prices of memory devices. The permanent record of the TV-type display is made on normal photographic film or at lower quality on Polaroid film. To obtain quality imagery, the film is exposed line by line by focusing the CRT image on to the film. Imagery is obtained in colour by making three successive exposures through red, green and blue filters.

10.3 INTERACTIVE ANALYSIS

Digital imagery analysis can be divided conveniently into three major steps, as supported by the processing system hardware and software. This division, however, should be considered to be arbitrary, since there is often an overlap in employing the techniques. As mentioned in section 7.2, corrections are made to the raw satellite data in primary processing to eliminate radiometric and geometric distortions. This is termed **image restoration**. Image restoration is essential as part of image analysis, if not included in the primary data processing. It forms the first step in image analysis, when, for example, the geometric accuracy of the scene (cf. Figure 10.3) needs to be further improved for thematic mapping, or when the data is to be used multitemporally in monitoring changes of the forest cover.

A distinction is made between image restoration and the second step, **image enhancement**, in that the former uses *a priori* knowledge to reverse the image degradation, while the latter aims only at automated improvement of the images for visual interpretation. Image enhancement involves using well-proven software programmes, which help to ensure that the fieldworker has the photographic products best suited to his/her requirements.

The third step, **interactive image classification**, also uses enhancement techniques. Image classification involves using the computer to identify, evaluate and classify into categories each pixel of the satellite scene. This process is based currently only on the spectral differences between pixels; but it must be borne in mind that while the human eye is better able to discriminate between spatial differences of the images, computer-assisted techniques are superior in discriminating spectral differences; and that in the future use will be made of textural classification.

Image enhancement

The aim of image enhancement is to improve the quality of the imagery for visual interpretation. Digital enhancement techniques are used (a) in image restoration; (b) to improve by changing the grey-scale values of the pixels the interpretative quality of the photographic products in batch processing without recourse to interactive digital analysis; and (c) as the first step in the subjective process of interactive digital classification. Some software programmes are common to enhancement and classification.

Popular enhancement techniques are pass filtering and contrast stretching. Another important technique applied to image enhancement is that of *n*-band spectral data transforms. Essentially the transform reduces the spectral dimension of the pixel data. For example, the two-dimensional spread of the spectral grey-scale values in two bands is converted into new values. Simple examples include the calculation of the spectral differences between paired pixel values and the ratioing of paired pixel values in two bands. This approach, however, becomes cumbersome when the spectral data has been collected in several bands. Therefore other statistical techniques are used. These involve principal component analysis, factor analysis and canonical analysis, which are also used in image classification.

High and low pass filtering

The aim of this **local** operation is to modify the subtle tones and colour contrasts that define the edges and texture of the objects recorded in the imagery. If, in the processing, a known class or classes of pixels have their grey-scale values deliberately changed so that they can be observed more easily in the imagery it is known as **high pass filtering** or **edge enhancement**. The technique is useful for depicting forest boundaries, linear man-made features (e.g. roads), drainage patterns, and in geology surface lineations. One procedure is to compute digitally the average value of surrounding pixels, calculate the difference between this value and the value of the pixel to be enhanced, and then to use the difference, as a weighting, to increase the grey-scale value of the pixel being enhanced (e.g. doubling its grey-scale deviation). When colour coded, these enhanced pixels will record brighter in the processed scene and to the photointerpreter the boundaries will appear sharper and are more easily defined.

The opposite to this local operation is to modify a pixel value by considering the values of the pixels surrounding it. This is termed **low pass filtering**, is used to eliminate 'rogue' pixels or inherent 'noise' among scattered pixels with the result that the imagery loses unnatural speckling and appears smoother and more pleasing to the human eye – hence the term **smoothing**. A common form of smoothing is to use repetitively the computer-calculated average value of a matrix of nine pixels (or sometimes 25) to replace each original central value. A similar technique is used for destriping a Landsat scene, when using average values calculated from nearest neighbour pixels in the two lines adjoining the striping to replace the 'striped' pixels (section 7.2). Image quality can also be improved by reducing or 'chopping' the pixel size from that of the line of pixels as recorded on the computer-compatible tape, and calculating new replacement values for the new smaller pixels, which are based on within-line nearest-neighbour values. A special programme is needed, but the procedure can be time consuming and the cost of the final image product expensive.

Contrast stretching

In addition to local operation enhancement, **general operation** techniques are applied to the data. The most popular is contrast stretching. If the pixels of a forest scene, for example, are displayed in their original form only a small range of the grey-scale values might be used, and can result in imagery with a low tonal contrast. Important recorded characteristics of the forest may not be detectable by the photointerpreter.

To overcome this problem, a histogram of the frequency of the grey-scale values is first plotted. This can be done manually for very small areas, but usually the digital capacity of the computer is used to provide a selected display on the monitor screen and/or a print-out. As illustrated by Figure 10.5, the

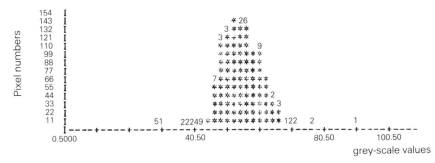

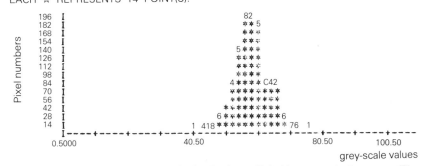

Figure 10.5   Histogram print-out obtained using a digital image analysis system. The frequency within the forested sub-scene of the pixel radiance values has been plotted on a grey-scale axis between 0.5 and 120.5, for Landsat bands red (0.6–0.7 μm) and near IR (0.7–0.8 μm). The full range is 0–256. No grey-scale values occur beyond 70.50; and the curves are not bi-model, indicating that the forest cover (savanna woodland) is fairly uniform.

density values of the pixels are contained entirely in the range of about 45–70. Water in an adjoining sub-scene had a grey-scale value under 5. All other values were not being used. To take advantage of the unused values and to improve the quality of the imagery, new values are calculated for each class using the full density scale. This is termed **linear contrast stretching**.

Linear contrast stretching has the disadvantage that all grey-scale values are treated in the same way, irrespective of their individual frequencies and importance. An improvement is made by weighting all individual grey-scale values according to their frequencies. The technique is then termed **histogram stretching** or histogram equalization stretching. The technique can also be extended to selected features and is then termed **special contrast stretching**.

Classification

Digital image analysis has the major advantage over visual photointerpretation of being able to process spectral and textural feature data in n-dimensional space. What is seen in colour on the viewing screen of the digital analysis system is the compression of the classified data into a two-dimensional colour display.

Feature selection

The basic pattern recognition problem applied in interactive digital analysis is firstly to define the spectral classes of entities of interest within the area covered by the satellite scene (e.g. forest, non-forest, forest types), and then to construct a decision rule, which enables each pixel of the scene to be placed in its previously identified class, and to say what kind of feature or ground object the pixels represent. Choosing the most suitable vector set of measurements to describe the objects of interest is a major task of the digital imagery analyst, and the process is referred to as **feature selection**.

Feature selection can be approached in two ways. Presently, digital image analysis systems in operational use involve statistics throughout the problem-solving process, and this is the approach followed in this section. However, of increasing interest is syntactic pattern recognition, developed for military reconnaisance and particularly useful when using small digital analysis systems having a very limited data storage capacity. The method uses initially a hierarchical structure, similar to using a dichotomous key in plant ecology; and in the process of the pattern recognition those features of the scene not of interest are discarded.

The two approaches use cluster analysis, which, as a form of **unsupervised classification**, provides one of the two major technique groupings used in the classification of the data. The second technique of grouping covers **supervised classification**, which is by far the most important in the 'statistical' approach to image classification. This is mainly due to the wide menu of available operational software programmes.

Unsupervised classification

Basically, unsupervised classification does not require stepwise commands by the image analyst, but relies on a programme that examines the unknown pixels and divides them into logical spectral groupings related to the landscape. In practice there is often limited intervention by the analyst, and it would be more correct to consider these techniques as hybrids.

Irrespective of the intervention by the analyst, the main advantage of unsupervised classification is that it provides a means of obtaining a meaningful pictorial presentation of the uncategorized pixels. This need occurs, for example, when there is no opportunity to obtain field training sets for supervised classification, or, when, due to lack of other information, it is useful to begin the interactive analysis with an unsupervised classification of the pixels. As pointed out by Lillesand and Kiefer (1979), in the unsupervised approach, spectrally separable classes are determined and then their information utility defined (i.e. cluster labelled), while in supervised classification the information categories are first defined and then their spectral separability is examined.

Due to the large amount of spectral data to be handled, there are few examples of entire satellite scenes being directly classified using unsupervised classification. In the 1970s in the USA, the Large Area Crop Inventory Experiment (LACIE) used unsupervised classification involving cluster analysis of entire scenes to estimate automatically parameters for a maximum likelihood classification. In the preparation of the Landsat mosaic of California, 32 Landsat MSS scenes generated initially an impracticable 1200 spectral classes; but after cluster labelling, the number of classes was reduced to sixteen.

Since many of the clustering programmes require extensive computational time and data storage, it is necessary to confine the technique to part of the satellite scene; or, as is more common, to select subjectively subsets or 'unsupervised training areas' with considerable spectral diversity. Using cluster analysis, the **spectral** cover classes of these training areas are determined, and this knowledge is then used in the cluster analysis of the entire scene. Frequently, the image analyst sets the number of spectral classes, and the computer then iteratively clusters the pixels of the training area(s) into classes not exceeding this number around arbitrary centres in n-dimensional space. The latter will vary with the number of spectral bands used.

The pictorial results of the cluster analysis are field checked or compared with existing maps, aerial photo-mosaics, aerial photographs, forest inventory data, etc. This will confirm whether or not the results are meaningful. If necessary, the spectral features are relabelled and the data processed again to improve the results by combining some of the classes or introducing new classes. In supervised classification the training sets are located on the best available information of areas, which are homogeneous on the ground. In unsupervised classification the training areas are selected according to the expected maximum spectral diversity of the grey-scale values of the pixels.

Supervised classification

In the initial approach to supervised classification, features are selected according to the objectives of the study and on the basis of previously acquired knowledge of the ground area and existing related information. The latter includes planimetric and topographic maps, aerial photo-mosaics and stereo-pairs of aerial photographs. It is useful if the image analyst makes at least one visit to the area under investigation to familiarize himself/herself with the local conditions. Also helpful in the selection of the main feature classes is an awareness of their typical spectral reflectance curves, as discussed in Chapter 3, and critical examination of 'bulk'/standard product satellite imagery.

Having identified what are considered to be the main feature classes, it is often advantageous, assuming the programmes are available, to study at least part of the area using image enhancement and possibly cluster analysis, and, accordingly, thereafter to refine the feature classes.

Usually several small training areas are allocated within the sub-scene, and delineated and examined for each feature. The class delineation of each training area, which is assumed to be homogeneous and subject to field and aerial photographic checking, is normally performed using the graphics terminal and viewing screen of the interactive image analysis system. After the training areas have been outlined on the screen with the cursor (joystick), the computer by command accesses the spectral data file, and stores the pixel grey-scale values of each feature class for further examination. Within moderation, the more samples or pixels used in determining each spectral cover class, the more accurate is the training. In practice, between $10n$ and $100n$ pixels can be expected to give satisfactory results (Lillesand and Kiefer, 1979), where n represents the number of spectral bands being used in the study.

Once the spectral cover classes have been satisfactorily determined, which includes examining histograms for each feature and field photo-checking, the analyst is ready to proceed with the classification of the scene under study. For this, using the cover class training sets, there is a choice of software classificatory programmes. These include minimum distance-to-means classification, parallelepiped classification and maximum-likelihood classification, which will now be briefly described.

For simplicity, it will be assumed that only three spectral feature classes are being used as training sets (i.e. water/swamp forest, closed forest, dry grassland/open forest); and that only two spectral bands of Landsat MSS are being analysed (i.e. band 2/red, band 4/near infrared). In normal practice, when initially investigating all spectral bands, the n-dimensional display is processed within the computer. The spectral reflectance curves of the three feature classes would be expected to be similar to those shown in Figure 3.12.

For each feature class shown in Figure 10.6, there are only five grey-scale pixel values. In practice, each feature class would contain about 10–100 pixels. Figure 10.6(a) shows the mean spectral radiance values of the three classes.

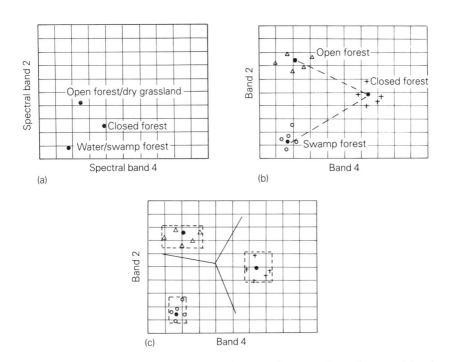

Figure 10.6 (a) Natural surfaces have different reflectances in each spectral band (channel) of the sensor: above is a theoretical plot of the recorded mean pixel values of water/swamp forest, closed forest and scattered tree/dry grassland in each of two channels (red, near IR); no information is provided on the scatter of individual grey-scale values about the means; however, the wide spacing of the means suggests that each of these terrain cover types might be spectrally separated, and this would be further confirmed by a histogram analysis; (b) minimum distance to mean (●) classification (see text); (c) parallelepiped classification (see text).

Minimum distance-to-mean classification

This is the simplest classification technique and is illustrated in Figure 10.6(b). The arithmetical mean spectral values (mean vectors) of the training sets are calculated. The spectral values of all pixels of the scene are then compared with the mean values (+) and are allocated to the feature class closest in value (i.e. at minimum distance). The technique has the major disadvantage of allocating pixels of unknown spectral classes to the predetermined spectral classes, which results in the latter being over-represented in the imagery.

Parallelepiped classification

The problem can be partly overcome, as illustrated in Figure 10.6(c), by introducing a measurement of the range of values likely to occur in each spectral class. By simply using the maximum and minimum values occurring in each training set, a box is established for each set with the mean as its centre point. All pixels within the scene are referenced to the boxes; and when the pixels occur within the arithmetic range of the boxes, they are allocated to that particular 'spectral class' box. The technique is simple, fast and quite efficient; but difficulties occur when the boxes overlap, with the classification of 'unknowns' outside the boxes and the fact that correlations occurring between pixels in two or more bands results in their spatial distribution being elongated (lozenge-shaped).

Maximum-likelihood classification (Gaussian classification)

A further improvement in classification (not shown) is made by replacing the simple range parameter by the statistical parameters of variance and correlation of the pixels of each training set, on the assumption that the sample of the population is normally distributed (i.e. Gaussian distributed). Each training class is described in terms of its arithmetic mean (mean vector) and covariance matrix parameters. All pixels in the scene are statistically compared to determine to which class they belong according to the contours of preset probability levels.

Practice indicates that the spatial distribution and correlation of the pixels conform under most circumstances to the parameters of ellipsoidal equiprobability contours, not being box-like, and the results are improved as the number of spectral bands is increased. Modifications of maximum-likelihood classification include using Bayesian classification and Bhattacharyya distance (Haralick and Fu, 1983). The main disadvantage of maximum-likelihood classification is the large number of computations, and consequently the speed and cost of this type of analysis; but this constraint is likely to diminish with the introduction of new and faster processors.

10.4 ACCURACY

Although this is a short section, it does not imply the lack of importance of testing for accuracy when processing the digital data. As with visually interpreted aerial photographs, the features classified on the imagery should be checked in the field. This field checking should be extended to areas on the imagery shown as unclassified; and, if necessary, the digital image processing must be repeated with due consideration being given to the necessary corrections. The finalized imagery should not be released for general use until the accuracy of the spectral classification has been determined.

Due to the relatively small scale of most satellite imagery, it may be sufficient to check visually in the field the spectral cover classes portrayed on the imagery; or to check the imagery against existing aerial photographs or photo-mosaics; or to supplement the imagery with strips of large-scale photography taken from a light aircraft before reprocessing the spectral digital data.

Frequently existing aerial photographs are used initially to help establish the feature classes; and after the data processing, the photographs are again used to verify the accuracy of the image analysis, but using different sample areas, and if possible including field visits. If available, the usefulness of recent aerial photographs cannot be too strongly stressed. The three-dimensional model provided by stereo-pairs of photographs and the synoptic view provided by aerial photo-mosaics can give an understanding of the forest physiognomy and forest condition only available otherwise by fairly intensive field surveys. In checking the completed spectral classification a study of the aerial photographs will help confirm the accuracy of the image processing, and will help in providing an understanding of any errors and the nature of areas portrayed as unclassified. The digital image analyst will find his/her skills to be greatly enhanced, if he/she is a competent visual forest photo-interpreter (Chapter 8).

For checking the spectral classification of satellite imagery against field observations or the interpretation of aerial photographs and photo-mosaics, several statistical tests are available. These are normally qualitative and not quantitative due to the nature of satellite imagery. The Charlien test can, for example, be used; as is also the determination of correlation coefficients between spectral bands and between image and field data. The most used is the confusion matrix . In this test categorized pixels, excluding the training areas/sets, are compared with the same sites in the field or on photographs, mosaics or up-to-date thematic maps. Statistical sampling is to be preferred to maintain objectivity, but in practice this is difficult, due to difficulties in locating the site of the pixels or group of pixels in the field. Also, access to the forest may be impracticable.

The results of each comparison on the basis of 'right or wrong' are set out in a confusion matrix (Table 10.1; Howard and Lantieri, 1987). If there were no errors all the diagonal figures would be zero. In practice there are errors of omission and commission, which indicate misclassification. If these errors exceed an agreed percentage then further action is needed. This may require, for example, the combination of two of the spectral cover types shown on the imagery, or reprocessing of the data on the basis of acquired additional information.

Table 10.1 Confusion matrix showing the results of comparing selected categorized pixels with the same pixel areas checked in the field and/or on aerial photographs. If there were no right or wrong errors all diagonal figures would be zero (see text)

Digital analysis

		T_1	T_2	T_3	S_1	S_2	S_3	G_1	G_2	BS	Crops
	T_1		$0^{(1)}$	$+$	$+^{(3)}$	$+$	$+$	$+$	$+$	$+$	$+$
	T_2	$0^{(1)}$		$0^{(1)}$	$0/\,^{(1)}$	$+$	$+$	$+^{(1)}$	$+$	$+$	0
	T_3	$+^{(1)}$	$+/0^{(1)}$		$+^{(2)}$	$0/-$	$0/-$	$+^{(1)}$	$+^{(1)}$	$0^{(3)}$	0
Visual	S_1	$0/-$	$+^{(1)}$	$+$		$+/0^{(1)}_{(3)}$	$+$	$+$	$+$	$+$	$0^{(3)}$
interpretation	S_2	$+$	$+^{(1)}$	$+/0$	$0^{(1)}$		$+$	$0^{(1)}$	$+$	$+$	0
of aerial	S_3	$+$	$+$	$+/0$	$+$	$0^{(1)}$		$+^{(1)}$	$+^{(1)}$	$0^{(3)}$	$+$
photographs/	G_1	$+/0$	$+$	$+^{(1)}$	0	$+$	$+^{(1)}$		$+$	$+$	$0/-$
field checking	G_2	$+$	$+$	$+^{(1)}$	$+$	$+$	$+^{(1)}$	$+$		$+$	0
	BS	$+$	$+$	$+$	$+$	$+$	0	$+$	$+$		$+$
	Crops	$+$	$+$	$+$	0	0	$+$	0	$+$	$+$	

Mapping accuracy: $+$, above 80%; 0, between 60 and 80%; $-$, under 60%.

Classes: T, forest or woodland (*) (>5 m height); S, shrub (<5 m height); 1, cover $>80\%$ (closed); 2, cover between 40 and 80% (dense to open); 3, cover between 10 and 40% (open to sparse); BS, bare soils.

Part Four

Remotely Sensed
Data Integration
for Forest Resources

11

Thematic mapping
in forestry

The importance of maps in fieldwork in forestry needs no introduction, since maps are a primary medium of communication in the study of forest resources. They involve the portrayal of visible features on the earth's surface at scales agreeable to cartographic treatment through a suitable base map. A **thematic map** is a graphic representation of a selected number of these features in an already restricted field of study, in which their presentation is related to the surface of the earth. The base map is purposively prepared ahead of representing thematic data on it, or it is developed in combination with existing aerial photographs, or is derived from an existing planimetric map or topographic map.

A **planimetric map** shows the correct horizontal position of features on the ground in terms of their x- and y-co-ordinates, while a **topographic map** also shows elevational differences using contour lines in terms of x-, y- and z-co-ordinates. **Cadastral maps** are planimetric maps showing primarily legal boundaries.

Fieldworkers knowing the magnitude of the errors involved in the survey and management of forest resources are often prepared to accept a lower mapping accuracy than required by photogrammetrists and cartographers in order to save time and to reduce costs. Technically such products should be called thematic sketchmaps, although commonly termed thematic maps.

The recognized international standard of **map accuracy** varies with the scale of the finalized map, since it is accepted that at least 90% of well-defined ground points must be within 2 mm of their correct position when plotted. Thus this tolerance level allows for a greater error in linear distance as the mapping scale decreases. For example, at mapping scales of 1 : 25 000 and 1 : 50 000, mapping accuracy calls for the plotted points to be within about 12 m and 24 m respectively of their true ground positions. In considering elevation accuracy, the **least contour interval** is defined as double the range in elevation within which 90% of the points fall about the mean. Thus if the range is 5 m the least contour interval is 10 m. In the USA the term *C*-**factor** is commonly used to describe machine quality, when referring to the relationship between the least contour interval which can be plotted and the aerial photographic scale (i.e. *C*-factor = flying height above ground/least contour interval). For many stereo-plotters this has a value exceeding 1000.

11.1 MAP PROJECTIONS

If small areas are to be mapped, as is often the case in forestry, then there is usually no need to consider the map projection as the first step in developing the base map, although the survey should be 'tied' into a reference point provided by the national geodetic grid (e.g. 'trig point', UK). On the other hand, for large areas and with the increasing use of satellite data and growing interest in geographic information systems in remote sensing, the map projection is an important consideration.

There are world-wide over 150 different map projections, although few have been widely used. Basically a **map projection** is a system of representing the earth's surface on a plane and taking into consideration that a sphere cannot be projected on to the plane without incurring distortions. In so doing lines, termed parallels of latitude and meridians of longitude, are projected on to the sphere representing the surface of the earth. It is these parallels and meridians that form part of the map base. For map projections in common use, each meridian is a great circle passing through the poles with the prime meridian (0°) passing through Greenwich (UK). Each parallel is at right angles to the earth's axis with the equator at mid-distance between the poles. All of the parallels are parallel to the equator. The several most commonly used projections are conformal in that the shape of small areas (e.g. 1° grid squares is maintained. Since at any plotted point the scale will be the same in every direction, linear measurements are readily and accurately made; but particularly at small map scales (e.g. less than 1 : 2 000 000), the mapped areas do not maintain their correct relationship. Thus using the Mercator projection, Greenland appears, when plotted, to be about the same size as South America, although in actual area it is about a ninth of the size.

The most popular projections include Mercator, Polyconic, Lambert Conformal and Polar Gnomic. The Mercator projection and its derivatives, including the Transverse Mercator and the Universal Transverse Mercator (UTM) projections, are the most widely used. These projections have provided the basis for the making of nautical charts since 1659, and, in the twentieth century, have been used in the preparation of topographical maps in Europe, USA, Australia, Indonesia, etc., in the making of air navigational charts, which provide world-wide coverage at a scale of 1 : 1 million and recently in the geometric correction of satellite imagery. All lines of constant compass bearings can be plotted directly as straight lines on Mercator-based charts, scale changes are slight at lower latitudes and for small areas the relationship of the scale of the meridians and parallels is the same as on a globe.

The **Mercator projection** is illustrated in Figure 11.1. Note that the derived meridians and parallels are projected on to a cylinder, which is tangent to the earth at the equator, and which, when opened and flattened, provides meridians as straight lines. In the Transverse Mercator projection, which is suited to mapping at larger scales (e.g. 1 : 50 000), the cylinder is rotated so that it lies tangent along a meridian of the globe. Other than the equator, the

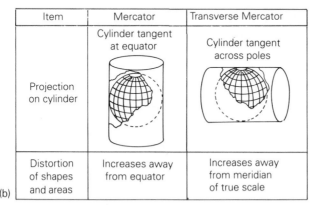

Item	Mercator	Transverse Mercator
Parallels	Parallel straight lines unequally spaced	Curves concave towards pole
Meridians	Parallel straight lines equally spaced	Complex curves concave towards central meridian
Appearance of grid		
Great circle	Curved line (except equator meridians)	Curved line

(a)

Item	Mercator	Transverse Mercator
Projection on cylinder	Cylinder tangent at equator	Cylinder tangent across poles
Distortion of shapes and areas	Increases away from equator	Increases away from meridian of true scale

(b)

Figure 11.1 Two popular map projections: Mercator (left) and Transverse Mercator (right) (see text).

central meridian and meridians at 90° from the central meridian being straight lines, the parallels and all other meridians are curves. In the **Universal Transverse Mercator** projection (not shown), the curves are complex and the globe is mathematically developed into an ellipsoidal form, in which the earth is divided into sixty zones of 6° longitude with each having a common central scale.

For mapping at high latitudes, due to area distortion, a different type of projection is used. These include the **Lambert Conformal Conic** projection with its apex over the pole, and in which points on the earth's surface are projected on to a cone that intersects two parallels of latitude. With this projection the earth's surface has the same shape on the map as on the globe, and hence the projection is well suited to small-scale mapping.

In recent years several projections (e.g. space oblique Mercator) have been proposed to accommodate distortions resulting from satellite sensing, particularly the sinusoidal ground track of the satellite due to the relative motion of the earth. However, as pointed out by Snyder (1981), mathematical formulae are now available to correct for the ellipsoidal shape of the earth and the elliptical ground track of the satellite.

11.2 PHOTOGRAPHS AS MAPS

The use of photographs as maps in forestry has at least three important advantages. A map cannot have a universal function approaching that of remotely sensed imagery, because the presentation of the data depends on the limit of what can be presented physically, and on what human interests and perceptions dictate. On the other hand, to use photographs as maps requires the development and application of special skills. These skills, however, could be introduced into geography courses in schools. Secondly, as maps are expensive to prepare, considerable savings will be incurred by short-cutting this process and confining the end result to photographic products.

Thirdly, as shown by the last UN survey (Brandenberger and Ghosh, 1987) of world coverage by topographic and cadastral maps, this is far from satisfactory. Progress world-wide is even more depressing in consideration of the facts that revision is only about 1% a year, with new map coverage at 1 : 25 000 and 1 : 100 000 being about 0.28%. At this rate, world topographic map coverage at 1 : 25 000 will take about 300 years and at 1 : 100 000 more than 150 years.

As reported by Brandenberger and Ghosh (1985), only about half of the world has been covered in the economically important scales between 1 : 50 000 and 1 : 100 000; and the annual rate in the making of new maps had decreased when comparing the periods 1974–80 and 1968–74. Table 11.1 provides information on the state of topographic mapping world-wide. As might be expected, Europe is the best covered, exceeding 90% at all scales

Table 11.1 State of world cartography: world-wide percentage coverage by topographic maps up to 1980*

Continent	Scales			
	1 : 25 000	*1 : 50 000*	*1 : 100 000*	*1 : 250 000*
Africa	2	24	17	78
Asia without USSR	11	51	62	80
Europe without USSR	91	91	77	95
North and Central America	34	61	7	88
Oceania and Australia	13	42	42	80
South America	10	27	42	50
USSR	5	61	100	100
World	13	42	42	80
Annual progress	0.28	–	0.28	0
Estimated completion period	310 years		170 years	

* Four scale ranges are shown: category I, 1 : 25 000 and larger; category II with an average scale of 1 : 50 000; category III with an average scale of 1 : 100 000; and category IV with an average scale of 1 : 250 000.

excluding 1 : 100 000, and may be compared with the much lower coverage for most regions at a scale of 1 : 50 000, which is in general the minimum scale for renewable resources management. In the case of forest management with the need for maps at scales of 1 : 25 000 or better, the coverage is only 2% in Africa, and most of this is limited to a few countries. It is therefore reasonable to conclude that foresters cannot expect to have up-to-date planimetric and topographic maps for inventory and management. Short cuts must be found with more reliance on photographs as maps covering small areas, on the preparation locally of large-scale sketch maps and the compilation of simple thematic maps either manually or digitally with lowered mapping accuracy, but fully acceptable for forest survey, forest management and monitoring changes in land use.

Aerial photographic products

Aerial photographs are available in the following formats:

Contact prints

At the same scale as the negative these are the cheapest to buy and are the most popular in aerial photointerpretation and simple photogrammetry. Image enlargement is achieved using a stereoscope.

Rectified prints

These are contact-scale single prints carefully corrected for the tilt by re-creating an opposite tilt at the time the print is made.

Ratioed prints

These are enlarged prints or reduced in scale as compared with the original negative. Aerial photographic negatives can be enlarged advantageously at least two to four times and will produce in the print image detail of high pictorial quality. Studies in Japan (Maruyasu and Nishio, 1962) to compare black-and-white photographs and colour prints indicated that the sharpness was similar at about the same scale; but delicate tonal differences became noticeable in the colour enlargements (which were not observed before) and these differences were helpful in the interpretation.

Enlargements are bulky to handle and difficult to study under a stereoscope. However, for study of tropical forests, which often have a great complexity of pattern, enlargements have been found useful both in the office and field, as there is more space for annotations.

Ortho-photographs

Ortho-photographs (section 11.1) are of growing importance in fieldwork as the image displacements in the negative are eliminated during processing: thus they can be used directly as a planimetric or thematic map.

Photo-maps

This term is applied loosely to enlargements of small-scale photographs and often to mosaics of two or more rectified prints. Sometimes the term is used when referring to enlargements of the effective area of a photograph. For example, in New Zealand, by enlarging two to four times the effective areas only of a photograph taken at a scale of about 1 : 60 000, a print is obtained which can be used satisfactorily as a 'map' for many field purposes. In the USA, photographs at 1 : 56 000 have been enlarged 3½ times, gridded and then copied by Ozalid printer to provide field 'maps' for timber assessment in Washington state (Kummer, 1964).

Aerial photographic assemblages

Photographs are used in stereoscopic pairs (or stereoscopic triplets) and in larger groups to form a photographic mosaic. Photo-mosaics are important in mapping forest stands (e.g. Aldred and Blake, 1967). In all cases with the exception of ortho-photo-mosaics, there is the inherent problem of accurate measurement. Each vertical aerial photograph is a perspective view, in which objects are displaced radially outwards due to aircraft tilt and changes in the ground elevation. We have learned from Chapters 5 and 9 that this displacement can be reduced by using only the effective areas of each photograph. In the preparation of photographic mosaics, the gross effect of this displacement is prevented from accumulating across the mosaic, and is retained within the effective area of each photograph used to make the mosaic.

As outlined in Chapters 8 and 9, stereoscopic pairs of aerial photographs are essential in forest photointerpretation and photogrammetry. Occasionally for small areas, a stereoscopic triplet is used with the transfer to the centre photograph, as the permanent record, of the forest-type boundaries etc. Features are delineated on the centre photograph from the two stereo-models provided by the left and centre photographs and the right and centre photographs. This is simple but not recommended, since major radial displacements will exist on the periphery of the centre photograph.

Mosaics are made to a number of quality standards ranging from the simplest index mosaics to the expensive ortho-photo-mosaics, which are as accurate as a carefully prepared planimetric map. For forest survey purposes, mosaics are generally used to give an overall picture of the forest and to act as an inexpensive map for planning management and control of fieldwork. Usually an uncontrolled or semi-controlled mosaic is adequate, though its accuracy is not considered satisfactory for detailed forest-type mapping.

The following types of photographic mosaic are used in forest operations:

Index mosaics

The aerial photographs are assembled so that common images on contiguous photographs overlap and the indices at the bottom of each can be seen. The assembly is then photographed and printed at whatever scale is required. This is primarily used as an index for easy reference to the aerial photographs. In unmapped areas, an index mosaic can be helpful to fieldwork until the photographs or map mosaics are available. Radial displacements will be maximum, since what is seen are the overlapping peripheries of the photographs. In some circumstances the use of satellite imagery is to be preferred.

Uncontrolled map mosaics

A map mosaic comprises the assembled central areas of the aerial photographs. These are fitted together, similarly to pieces of a jigsaw, so that the edges of each central area are carefully matched with the adjoining edges. Normally, the edges are 'feathered' into each other to give the appearance of a single photograph. Single-weight photographs are preferred to the double-weight prints normally used in aerial photointerpretation.

Careful inspection of a mosaic will often show that matching cannot be perfectly attained due to topographic displacement and tilt. There is likely to be considerable difference in scale in different parts of the mosaic, especially if the terrain is rugged. Also, errors are likely to accumulate directionally as there is no means of localizing these.

Semi-controlled and controlled mosaics

These are prepared according to the same general procedure: the aerial photographs (or satellite imagery) are carefully trimmed and feather edged, and successive photographs are assembled on a mounting board by visual matching of the edges of the images (Figure 14.6(a)). Usually the middle photograph of the central line is pasted to the mounting board first and the remainder of the line laid in both directions. The adjacent lines are laid in succession, starting each time with the centre photograph until the photomap sheet is filled.

As implied by the name, each differs mainly in accuracy as affected by the degree of control. Uncontrolled mosaics are assembled without regard to control, semi-controlled mosaics employ degrees of control ranging from rudimentary azimuth lines to that of photogrammetric control points, available from planimetric or topographic maps. Controlled mosaics are based on maximum controls, including that provided by radial line plots, ground control points, topographic maps and the use of rectified prints. When making controlled mosaics, photogrammetrical control points and ground control points are used to provide overall control of the scale of the mosaic. In preparation of controlled mosaics, rectified prints are normally used to remove excess tilt, and ratioed prints are used to provide photographs at the same overall scale. The final scale of the mosaic is provided at the time of printing as for uncontrolled mosaics; but the assembled effective areas prior to printing may comprise contact prints, ratioed prints and rectified prints.

Controlled mosaics are useful in the field for taking horizontal measurements, for the planning of surveys, for the purpose of providing an overall idea of conditions, and, aided by stereoscopic pairs of aerial photographs, for forest stratification/typing. Often a clearer understanding of geomorphic structures and general hydrology can be obtained from the mosaics than from individual and stereo-pairs of photographs. Mosaics are one of the most useful photographic aids in regional geographic studies. In several countries, including Sweden, controlled mosaics are printed with contour lines.

Working in relatively level terrain of the forests in eastern Canada ($\pm$ 30 m per photograph), Aldred and Blake (1967) found that the average error in the location of points on semi-controlled mosaics was about 2½ times greater than on the controlled mosaics. They concluded that the controlled mosaic was nearly as accurate as the topographic map, since 1 : 50 000 maps are only expected to be accurate within 0.05 cm or about 27 m on the ground. The cost of preparing the controlled mosaic was only about 45% of that of a conventional semi-controlled mosaic at the same scale using photographs at larger scale, because the assembly of the former is faster, and production of the mosaic map negative at contact scale is simpler and less expensive than reduction photography.

Ortho-photomaps

The development of ortho-photomaps (ortho-photo-mosaics) follows the development of mosaic maps (aerial photographic mosaics). In 1903 Scheimphlug rectified photographs taken from a balloon. About 20 years later aerial photographic mosaic maps using rectified prints became commercially available in Germany; and in 1966 SOM in France and Zeiss in Germany marketed ortho-photoscopes, by which the entire aerial photograph could be **differentially rectified** and printed.

An **ortho-photograph** is a photographic copy of the area covered by the same aerial photograph, in which the effects of the lens aberrations and the radial displacement due to tilt and topography have been removed. It contains the same detail as would be recorded in an orthographic aerial photograph, if the latter were possible. An **ortho-photomap** is a mosaic of uniform scale and can be contoured by convential photogrammetric techniques to provide a topographic ortho-photomap. The ortho-photomap is no longer a cheap substitute for a planimetric map, but a product in its own right, to which thematic data can be readily added to give a thematic map with a wide range of uses in inventory, management and planning of natural resources. For example in Germany, Sweden and Austria, ortho-photo-mosaics at scales of 1 : 10 000 to 1 : 200 000 are now frequently used for mapping forest cover types and the allocation of field sample units.

As summarized by Kalensky and Nielsen in Canada (1978), the main advantages of resource ortho-photomaps in comparison with the line maps are:

1. Information recorded by a photogrammetric camera is displayed in a geometrically accurate position. The ortho-photomap thus combines the advantages of a photographic display with the geometric accuracy of a topographic map.
2. There is greater ease and higher accuracy of plotting the resource data on the ortho-photograph rather than on a line map. Thematic information can either be derived directly from the ortho-photograph or transferred from aerial photographs. Because of the similarity between the ortho-photograph and the aerial photograph, such transfer is simple even in terrain with few or no prominent features.

3. While on a line map resource data can be expressed only with lines and symbols, the ortho-photomap retains in the background the photographic image, providing the user with more flexibility and enabling him/her to make a more detailed and accurate presentation of resource data.

However, in spite of the above advantages, the acceptance of ortho-photomaps by resource managers has been slower than expected. Some of the reasons were unfamiliarity with the new technique, reluctance to change established procedures, low image quality of input aerial photographs, loss of image definition during printing by older ortho-photographic systems, inadequate cartographic completion, costly continuous tone reproduction of large map formats and, last but not least, the new skills required for map reading. Most of these problems can be solved, but the lack of the three-dimensional observation capability has remained a serious handicap of the wider use of ortho-photographs.

Satellite imagery

Fieldwork is usually confined to areas within a single scene, since each scene covers an extensive area. Obviously, a mosaicking of two scenes or possibly three or four scenes will be necessary when the area of investigation overlaps adjoining scenes. The mosaicking technique is the same as described for uncontrolled, semi-controlled and controlled aerial photographic mosaics, but rectification is not required. Landsat MSS is usually mosaicked at scales between 1 : 200 000 and 1 : 2 million, and SPOT at scales of 1 : 50 000 or smaller.

Satellite mosaic maps have been prepared for entire countries, particularly as an aid to national planning. These include semi-controlled mosaics of Jordan in black-and-white in 1978 (Mitchell and Howard, 1978) and the Lebanon in colour using Landsat MSS and TM, a controlled mosaic of Chile and various digitally derived satellite mosaics in North America, Europe, Australia, Saudi Arabia, etc.

Since satellite mosaics are normally prepared from data collected on different dates due to cloud cover, the tone and colour balance of the different scenes can be expected to differ considerably, and this causes problems, particularly in mosaicking directly from colour prints. Due to seasonal changes in the plant cover and in spectral reflectivity of the vegetation, it may be impossible to match contiguous scenes using the techniques employed in aerial photo-mosaicking. The alternative giving improved results is to use digital matching techniques, but this is expensive on account of the large amount of pixel data to be corrected and processed.

As with aerial photographs, the mapping accuracy of satellite imagery and satellite mosaics will be influenced by the number of ground control points established by field survey prior to or after coverage or by using control points obtained from existing maps. The density of control points per unit of ground area will be far fewer than the number required to cover the same area using aerial photographs. The mapping accuracy is further influenced by the on-board flight recording and the size of the pixels. Difficulty is often incurred in positioning a pixel on the ground from the imagery; and even when ground control beacons are established ahead of satellite coverage, the exact positioning is no better than the location of the pixel within which the GCP is somewhere located.

11.3 DEVELOPING THE FOREST BASE MAP FROM AERIAL PHOTOGRAPHS

Often the forester is concerned with relatively small areas and will use simple low-cost techniques in the course of preparing locally planimetric maps and stockmaps for forest inventory, working plans, etc. This is the approach which will be pursued. For large areas and the preparation of topographic maps, the support and services will be needed of a photogrammetrist and a photogrammetric/cartographic system having a comparator point marker, analytical plotter(s) and possibly digital equipment. The major steps involved in the preparation of a planimetric or topographic map are shown in Figure 11.2. For further information, reference should be made to a modern textbook on aerial photogrammetry covering analytical plotting, block adjustment, rectification and ortho-photomapping. As a compromise between the two approaches, it is often convenient to have the map base prepared by a photogrammetrist on contract and then to transfer the forest detail locally to the base map.

The several steps involved in the preparation of a small forest stockmap are illustrated in Figure 11.3. A forest stockmap shows the distribution of different stand types, which have a bearing on management, and will include information on tree species composition, age or height classes, crown cover, etc. When the primary control points are obtained from existing planimetric or topographic maps, and not by field survey, then these maps should be at the same or larger scales than the forest base map being prepared. Otherwise the new map will not have the required mapping accuracy for its scale. However, as mentioned earlier, fieldworkers are usually prepared to accept thematic maps of lower accuracy, in which the primary control has been obtained from existing somewhat smaller-scale planimetric or topographic maps. In many developing countries, these may be the only maps available (cf. Table 11.1). Thus a topographic map at, say, 1 : 50 000 will be used to provide the basic control points for the forest stockmap at 1 : 25 000. The ground or map control points provide the overall accuracy of the map, while the supplementary

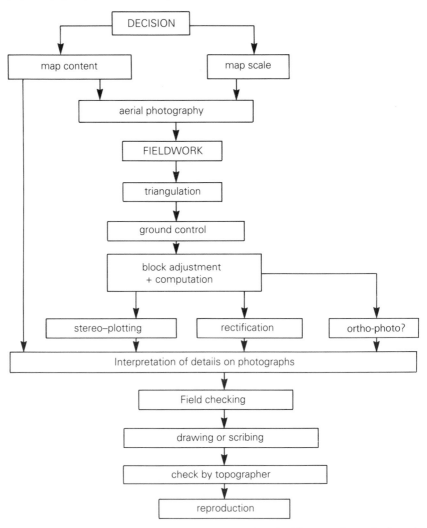

Figure 11.2   Flow chart indicating the major steps involved in preparing a planimetric or topographic map.

photographic control points (Figure 11.3(b)) will prevent the accumulation of errors across the base, when the planimetric details are transferred.

Radial line triangulation

This principle forms the basis of several simple techniques used in preparing the base map, and includes the overlay method outlined below. These techniques require only standard drafting materials, aerial photographs and a

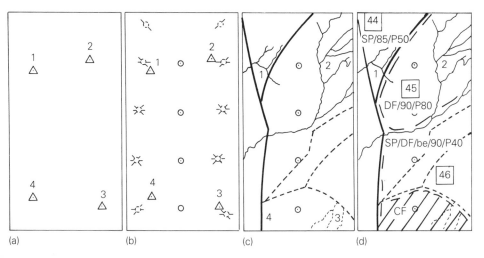

Figure 11.3 Stepwise preparation of a small forest stockmap (thematic map) commencing with plotting on the map base, at the predetermined scale, the ground control points (a). Using the principle of radial line triangulation (section 11.3), photographic control points are established to complete the base map (b). This is followed, preferably in two stages, by the transfer of thematic detail from the individual aerial photographs (section 11.4) comprising (c) principal man-made features (roads, logging tracks, etc.) and major natural features (streams, etc.) and (d) information on the forest stock useful to management. Shown on the completed stock map (d) are the compartment boundaries, the compartment numbers (44, etc.), the species, stocking and planting year (e.g. Scots pine, 85%, planted 1950) and the sub-compartment resulting from clear felling (CF – hatched).

ruler, pencil, school protractor and rubber. Slotted templates, formerly used widely in planimetric mapping, need the use of a special slotted template cutter and centre point punch. As an alternative to radial line plotting, when suitable scale maps exist, the thematic details can be transferred directly from the aerial photographs to the map as described in section 11.4, but again this requires special equipment. If a small-scale map is required, then a satellite scene may be suitable as a base; but the error resulting primarily from the pixel size needs determining.

In radial line triangulation, the photographic/photogrammetric control points are established on the map base using two-dimensional radial line triangulation. This depends on the principle that the centre of each photograph serves as a station from which radial lines, at constant and true angles, can be subtended to ground control points imaged in the periphery of the photograph, irrespective of changes in general scale. It provides a method of fixing control points in space by using overlapping photographs, on which the ground points are imaged. The flight line between the two adjoining

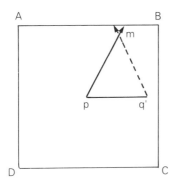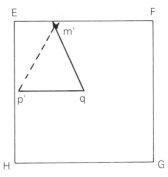

Figure 11.4   Principle of radial line triangulation used to determine the true position of an object recorded as an image (m') on two adjoining photographs (see text).

principal points of overlapping photographs is used similarly to a base line in plane tabling, rays being subtended from each principal point to the images of the control points being fixed.

Figure 11.4 illustrates the method of two-dimensional radial line triangulation. Photographs ABCD and EFGH have centre points p and q and transferred centre points p' and q'. The base lines are formed by pq' and p'q. Bearings of a suitable image (m) are taken from the two station points p and q. Thus the true position of the control point is determined by the intersection at m or m' of the two bearings from p and q. On either photograph, this is achieved by transferring the angle of the ray from the other photograph. The transferred ray is shown as a broken line. The true location of the control point can also be obtained by superimposing contiguous single-weight photographs on top of each other on a light table so that the flight lines coincide. The true position of the control point will again be located at the intersection of the rays. Other points can be located similarly to the location of m. These include all six wing points around the periphery of a photograph.

Overlay method

The method, involving radial line triangulation, is suitable for small areas up to five flight lines and having several photographs in each flight line. The map scale will vary as the scale within each photograph, but the map will be accurate overall provided the terrain is not rugged, since error accumulation is avoided between photographs. However, there is no correction for radial displacement (section 9.2).

Each photograph is scribed as outlined previously to include the principal point, transferred principal point and six wing points (section 9.1), and in addition important objects (e.g. mountain peak, road intersection, ground control points) are also marked on the photographs. Each photograph is then placed separately under the map base consisting of a sheet of transparent or semi-transparent tracing material (e.g. matt acetate, ethylon, Astralon or good-quality tracing paper), and the principal points and flight lines transferred (cf. Figure 11.3(b)). When the flight lines common to two photographs do not coincide, a mean of their lengths is used. Each photograph is separately reinserted under the tracing material and aligned on its flight line segment according to the location of the principal points. Rays are subtended on the overlay from the principal point to the wing points (cf. Figure 11.4). The intersection of the two rays subtended to a common wing point is the correct location of that wing point. After the base map has been prepared by marking the correct positions of the wing points etc., the photographs are once again reinserted under the tracing material and thematic details are traced following the steps shown in Figure 11.3. Normally only the details contained within the effective area of each photograph are transferred. If there is a second flight line, then the corresponding sector of its flight line is positioned on the overlay by resection of radial lines through the already plotted wing points. The same procedure is repeated after completion of the second flight line for the third flight line.

11.4 FINALIZING THE FOREST MAP

In finalizing the map for printing, its scale, the true and magnetic bearing and the legend must all be incorporated in the map. For printing, these and the map lettering are often prepared on separate sheets. The final map should be at a conventional scale, which for forest purposes is usually in the range of 1 : 5000 to 1 : 50 000, but with a preference for metric scales of 1 : 10 000 and 1 : 25 000. It must also be borne in mind that in the printing process the scale may be reduced two times. This fact should be considered ahead of field survey and starting to compile the draft map, in order to avoid recording unnecessary detail in fieldwork and its presentation wastefully on the map base. It is often difficult and adds considerably to cost to have to reduce the scale excessively for printing, since mapped details on the classes and subclasses of the forest resource may have to be amalgamated and new boundaries unsatisfactorily drawn without additional fieldwork/field checking.

Map legend

Important in the compilation of the forest/thematic map is the organization of the **map legend**, which is an integral part of the map and the key to the map content. It should show how the map content is organized, and its terminology requires to be logical, clear and concise, and should be prepared in draft before beginning the transfer of thematic detail to the map base and preferably at the time of field survey. The number of classes and subclasses in the draft legend should take into account the objectives for which the map is being prepared, the expected diversity of the forest or its resources, previous experience and information on what can be reasonably achieved according to available money, time and staff experience and what is feasible in printing in terms of colour, shade and symbols. The initial draft legend will require field checking and updating as work progresses.

Forest vegetation is usually so complex that few of its characteristics can be represented on the map. Often the mapping class is represented by the vernacular or botanical name of the commonest or dominant tree species or co-dominant tree species, and from this it is assumed that the physiognomic or floristic characteristics of the community can be construed. Foresters often represent the vegetation units by the name of the most economically valuable tree species, which may not be the commonest or dominant one. Woody vegetation, when represented by the dominant tree or bush species, is frequently subdivided by naming the commonest understorey species or vegetation cover classes or height/age classes.

Most legends have traditionally been of **hierarchical** type in which the smaller mapping units are grouped into two or more higher categoric levels. Often symbols are used to indicate dominant tree species by capital letters (e.g. SP, Scots Pine), followed in mixtures by the second species (e.g. DF, Douglas Fir), understorey species in lower case letters (e.g. be, beech), stand size or height or site quality (e.g. 25, 25 metres), stocking density (80, 80%) and plantations by the age (e.g. P50, 1950).

Alternatively, colour may represent the forest types etc. and, if necessary, these are subdivided into areas by intermediate colours of lighter or darker shading (e.g. wooded grassland, grassland with scattered shrubs). Hatching or numbering can be introduced to show (a) the height classes of forest stands, (b) the density or crown cover and (c) sub-compartments. Often on forest maps, the stands are divided into age classes by shading or intermediate colours, including those that are mature or approaching maturity and regrowth classes (saplings, poles and sawlog size). Symbols are often again added to indicate the dominant species. Sometimes a concise note on supplementary data is added to the margin of the map, which clearly describes each of the vegetation types of the legend and avoids the need to refer to reports (e.g. Gaussen, 1948; Braun-Blanquet, 1947).

The main, little used alternatives to hierarchical legends are co-ordinate and multivariate schemes using numerical taxonomy. **Co-ordinate systems** are parametric and use a limited number of environmental characters to produce a closed legend. **Multivariate systems** assume that the subdivisions of the environment are polythetic individuals in the form of sites that possess a number of properties. Each property is viewed as a dimension, and the individual is placed in multidimensional space according to its grading in each property. When this has been done for all the individuals, they will form clusters that can be isolated and used as indicators of the groupings to be used as the mapping units.

Sources of map error

When using a single photograph as a map substitute, the governing error source is uneven terrain which produces an error of $\Delta z/z$, i.e. terrain elevation differences/flying height above terrain. In operational work this can commonly be as high as 10%. Less significant error sources in this case are tilt and paper shrinkage (section 9.2). On large-scale aerial photographs, the radial displacement of the tree crowns can result in several metres error in positioning boundaries and sample plots.

With stereoscopes and parallax bars, the governing source errors are linear measurements, tilt and shadow. The latter may result in errors of 5% or more. These affect the calculated distance between points, area measurements and the height of an object such as a tree. The smallest mapping unit is influenced by the scale and the minimum thickness of drawn lines. The latter is usually about 0.2–0.5 mm and the smallest mapping unit is a square of about 3 mm on the side, which at a scale of 1 : 100 000 is about 9 ha.

With stereo-plotters, the governing error source is measuring limitations and lack of focal-plane flatness, which can cause errors of up to 0.01% and corresponds to the operational limits of accuracy of the stereo-plotter (cf. *C*-factor). Analytical photogrammetry with a computer is comparable to graphical plotting with an analogue plotter. The principles that apply to errors on the analogue plotter also apply to analytical photogrammetry. The errors due to focal-plane flatness are in analytical photogrammetry about 0.01%. This allows for a probable error in calculated points to be as low as 1/10 000 of flying height and in analytical photogrammetry about 1/50 000 of the flying height.

11.5 SIMPLE MAPPING INSTRUMENTS

With the exception of the simplest sketch mapping (section 11.3), specially designed (analogue) instruments are essential for the completion of a thematic map. The instruments used in forestry do not usually have the high degree of precision demanded in photogrammetric mapping (i.e. first, second order analytical plotters), but are satisfactory in providing, as the end product, a thematic map of acceptable standard, and are far less expensive to purchase.

Conveniently the instruments can be grouped as being non-stereoscopic or stereoscopic and the latter divided on the basis of being entirely optical, employing the radial line principle or employing a parallax-formed floating mark. The type of equipment will be determined by the nature of the work, the conditions under which the equipment will be operated, and if the equipment is already locally available, servicing and replacement of parts, skills and training of staff, available funds and the purchase price of instruments.

Non-stereoscopic instruments

Sketchmasters and reflecting overhead projectors are used. The latter are the more popular, when mapping from satellite imagery. The sketchmaster employs the principle of the camera lucida, in which the photographic image is seen monocularly superimposed on the map base through a semi-transparent mirror surface of a prism (Figure 11.5). The semi-transparent surface is provided by coating one surface of the prism with a very thin layer of silver. Simple, cheap, home-made instruments require only the purchase of a camera lucida lens (Meyer, 1961) or a front surfaced reflecting mirror, a transparent viewing surface and 35 mm projector (Figure 11.6; Howard and Kosmer, 1967).

A reflecting projector, as the name suggests, uses a light source of adequate brightness to reflect an image of the photograph from the back of the projector

Figure 11.5 Sketchmaster (Zeiss): the photographic images are seen monocularly superimposed on the map base through a semi-transparent mirror mounted on the instrument. The map is plotted on the base from the images observed within triangles formed by the photographic principal point, transferred principal points and the wing points. 1: eye-piece and semi-transparent lens; 2: map; 3: aerial photograph mounted on platen.

Figure 11.6   Simple 'home-made' projector suitable for transferring thematic details to a transparent/semi-transparent map base using a 35 mm projector and front-surfaced silver mirror. A 35 mm 'slide' of the aerial photograph is used with the projector.

through the lens on to the map base. Most reflecting projectors have the disadvantages of being bulky and requiring the use of a semi-darkened room. Several designs are available. Probably the best known resembles a photographic enlarger in which the photographic image is projected on to the map base mounted on a platen that can be adjusted to allow for tilt and photographic scale differences (Figure 11.7). The Canadian-manufactured Procom is popular in mapping from satellite imagery. In a third type of instrument, following the design of the Multiscope, which was popular in the 1950s, viewing is through a pair of binoculars but one eye sees the map base and the other eye the photograph. A current example is the compact Bausch and Lomb Zoom Transferscope.

Stereoscopic mapping instruments

Camera lucida instruments and reflecting projectors have two disadvantages. Firstly, the photographic detail transferred to the map base, within each of the triangles formed by the principal point, transferred principal points and the six wing points, will not be corrected for radial displacement. Secondly, it is often necessary to examine each photograph separately with a stereoscope before monocularly transferring the detail to the map or map base, in order to ensure that the details being seen monocularly are interpreted correctly. These

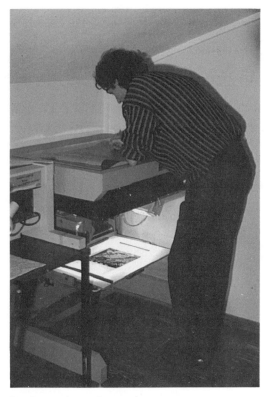

Figure 11.7   Reflecting projector (see text).

disadvantages can be overcome by using an instrument in which a stereo-model is seen as the details are transferred to the map base.

Simple instruments for the preparation of thematic maps from a three-dimensional model may conveniently be grouped into three classes. The first type of instrument uses a mirror stereoscope combined with a semi-transparent lens. The operator sees with both eyes the stereoscopic model formed from the two photographs, and views the map base as he/she did when using a sketchmaster. In so doing, he/she is able to transfer the details of the three-dimensional model on to the map base (Figure 11.8).

A second group, provided by older designed instruments, uses the principle of radial line plotting. Radial line plotters are not popular nowadays due to operating constraints and map error. The instrument uses two straight edges pivoted at the principal points of a stereo-pair of photographs. The true location of objects is given by the intersection of the two subtended lines.

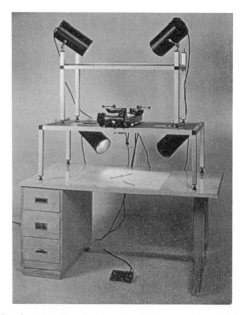

Figure 11.8   A simple stereoscopic mapping instrument by which the operator views the aerial photographs stereoscopically and sees the map base, as when using a sketchmaster (OMI Stereofacet Plotter).

The third group of instruments involves the floating-mark principle introduced with the parallax formula in Chapter 9. A stereo-pair of photographs are viewed under a mirror stereoscope and at the same time a floating mark is introduced. The mark is made to float on the apparent surface of the stereo-model, and is used in conjunction with a suitable drafting mechanism to trace the outlines of the images on to the map base.

In its simplest or earliest form, the floating-dot-cum-tracer comprised a parallax bar having a centrally mounted pencil (e.g. Tracing Stereometer); but the scale of the tracing is the same as the scale of the photographs. This problem was overcome in the Stereopret. Movement of the floating mark and coverage of the stereo-model was achieved by a moving table on which the photographs are mounted. Transfer of the photographic detail to the map base is carried out by a pantograph arm attached to the table. The scale of the pantograph could be adjusted from 1/5 up to 3.0/1.

In the further development of the floating mark principle, image analysis/photointerpretation is combined with a high standard of thematic mapping and involves instrument design having an accuracy between those described and second order photogrammetric plotters. The relatively expensive

Figure 11.9 Stereocord (Zeiss, Oberkocken): this advanced thematic plotter, replacing the Stereotop, provides a digital solution, has a high mapping accuracy and can be used simultaneously for estimating the height of trees, hill slopes, etc.

Stereotop (third order plotter) employed an optical–mechanical solution to the absolute or relative orientation of the stereo-model. The instrument provided not only planimetric maps, on which using the same stereo-model forest data could be superimposed, but also with its capability for contouring, topographic maps could be produced and tree heights estimated. This instrument has now been replaced by instruments, such as the Stereocord (Figure 11.9) and APY System, which provide interfaced computer-assisted digital solutions. In this class of instrument (Ladouceur, Trotter and Allard, 1982), the x-, y- and z-co-ordinates of images in the stereo-model, including tree heights, slope and stand areas, are digitally encoded and stored using a personal computer. The graphical high quality output in map format is by a drum writer or flat-bed plotter.

Mapping in forestry provided by these instruments involves a relative and/or absolute orientation of the stereoscopic model, a scanning and zoom capability, x-, y- and z-measurements, a grid-referenced encoder for the interpreted data and photogrammetric data and the facilities provided by an interfaced personal computer and its appropriate specialized software.

11.6 DIGITAL MAPPING

Before completing this chapter brief comment will be made on digital (computer) mapping due to the likelihood of its wider use in the future in the preparation of highly accurate thematic maps. Since automated mapping equipment is expensive to buy, the preparation digitally of thematic maps of forest resources etc. can be expected to be acquired on contract or equivalent. Publications covering the development of digital thematic mapping include the US Pacific North West Regional Commission (Todd, George and Bryant, 1979), the Image Based Information System at the Californian Institute of Technology (Marble and Pequet, 1983) and Pennsylvania State University (Myers *et al.*, 1989).

Digital thematic mapping is fast and economic in operating costs, and combines the accuracy of computer cartography with the data output of digital imagery analysis or a geographic information system (section 12.5). Fully automated thematic mapping, however, is unlikely to be achieved, since the human element is essential and adaptable in visually interpreting variations in the local landscape (e.g. tree shadows). Automated pattern recognition is parametric, relying on means, variances, etc., has difficulty in handling many interpretative functions; and when successful will provide differing results for the same algorithms according to the threshold values of classes etc., which have been chosen by the different analysts.

A digitally prepared topographic or planimetric map represents the spatial compression of large amounts of data obtained from stereo-pairs of aerial photographs and topographic maps, which are processed and stored by computer. Existing maps are digitized using, for example, a drum reader; and planimetric and topographic digital data is obtained from the stereo-pairs of photographs using an analytical plotter or a modified analogue plotter. For example, the UK Ordnance Survey modified in 1984–85 three analogue plotters (Wild A-10, Kern PG-2) to analytical plotters using microcomputer software and hardware. Equipment for automatic contouring, however, has been available since about 1968 (e.g. Zeiss Autocart); but the initial concept was to plot the contours on the base map in pencil and then to digitize them manually. The modern photogrammetric concept, using for example a Wild

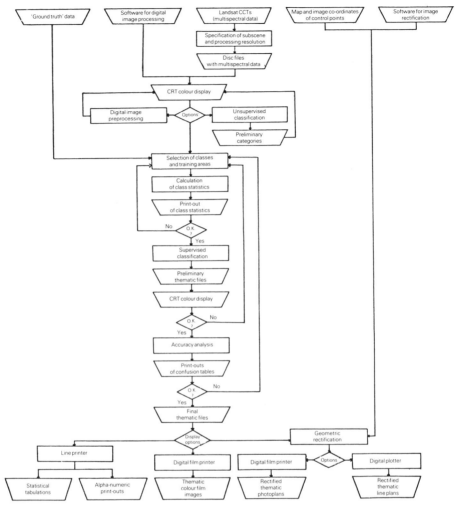

Figure 11.10 Flow chart illustrating the procedure of digital mapping from satellite imagery.

B-8S analytical stereo-plotter, is for the map data to be stored directly on magnetic tape or disc and not to be firstly recorded in pencil as a line map.

For a stereoscopic pair of aerial photographs separate files (magnetic tapes) are maintained for the restitution/rectification of the stereo-model using the analytical stereo-plotter. The true positioning of photogrammetric control points is obtained by field survey or from suitable quality maps and by photogrammetric bridging across strips of the aerial photographs and also computer stored on magnetic tape. Relevant map data in the digital form of points, lines and polygons is acquired from the stereo-model using the analytical stereo-plotter and an attached encoder. The x-, y- and z-co-ordinates of these digital data are then edited and stored on the true position magnetic tape. This is followed by field checking of the data and its further editing, with addition of place names and feature codes, before storage on the 'mapping' magnetic tape and the final print-out as a planimetric or topographic line map. If the 'true position' data is to be used to prepare a thematic map or to provide inputs to a geographic information system, the 'true position' magnetic tape will require to be edited/structured to accommodate the information obtained visually and stored in additional computer files (e.g. road grades, forest types, stand heights, etc.), or if the structures are considerable a separate 'structure' magnetic tape would be added. Figure 11.10 shows the steps involved in digital mapping from satellite imagery (Z. Kalensky, pers. comm., 1989).

Digital mapping is being influenced by recent developments in GIS technology in not only providing line maps showing the location of major physical and man-made features and contours, but also attributes of other thematic features of the landscape, including forest cover. A digital data base is first created with no coding attributes for vector or raster sets. This null attribute data set provides the base map or template for assembling the thematic data. The thematic data classes are cast separately as a raster layer with the cell size within each raster set determined by the scale/resolution of the thematic map. Under very favourable conditions, a few thematic classes (e.g. water surfaces, green vegetation cover) could probably be transferred directly to the cartographic data base from scanned imagery.

With respect to map revision, computerized mapping is advantageous in cost and speed, if it uses an already established digital map base or an existing map that can be readily digitized. The digitizing consists in converting the line and point data to the computer-readable format. Usually the hardware and software employed in mapping in space is readily adapted to mapping in time, since basically the addition is one or more historic layers of vector data to an existing system. The initial true map data and the revision map data are digitally combined to provide new line, point and polygon features and possibly tabular outputs.

12

Spatial information systems for forest resources

Due to increasing interest in geographic information systems, some attention needs to be given to spatial information systems before proceeding to consider remote sensing in forest practice. Information systems can be grouped as being management or geographically based and further subdivided as being manually or computer operated. Management information systems (MIS) using computers were developed rapidly in the 1960s for the purpose of providing decision-makers with computer-organized information. For example, MIS is used widely by large industrial, wholesale and retail companies to improve their efficiency and competitiveness.

Computer-assisted spatial information systems emerged later in the 1960s, being strongly influenced by MIS and concerned mainly with the handling for management of large land-use data sets. The term land information system is often used for spatial systems handling landscape data at larger mapping scales. Technology, as a management tool, in the storing, retrieving, analysing and updating of large quantities of map and forest survey data, has been motivated by budget constraints, the need for quick decision-making and the rapid decrease in computer costs.

A **geographic information system** (GIS) involves the computer-organized grouping of activities and procedures covering the input, storage and manipulation, retrieval and presentation of spatially based and related data. Each bit of data is spatially referenced and processed digitally within the computer. In the broadest context, a GIS is tailored digitally to the user's data needs and channels the locations of spatial data at the intelligence level of the decision-maker. A geographic information system handles spatial data in terms of their x-, y- and z-co-ordinates and non-spatial attributes of that data. It is much more than a digital mapping system; but, up to the present time, the information output of remote sensing analysis, whether digital or visual, has contributed little as a primary source of the multilayer data stored in a GIS. In most geographic information systems, existing maps have provided the primary inputs (cf. Marble and Pequet, 1983; van Roessel, 1986).

Geographic information systems help to reduce human errors and to eliminate many mapping and draughting tasks, and they are quick and efficient in providing spatial information, including multilayered maps, point and line proximity calculations and the interactive transfer of data to and from management information systems and digital imagery analysis systems. Although cost effective in operation, a geographic information system has the serious drawback of being expensive in establishing its data base; and there is usually a lack of readily available spatial data, other than that obtainable from topographic maps, planimetric maps and the occasional thematic map (e.g. soil map, geological map, forest stockmap). Large amounts of existing natural resource data are seldom in a form that can be quickly and automatically computer coded and geographically referenced. The usefulness of any GIS will depend primarily on the type, accuracy and detail of the input data it contains.

12.1 MANUAL SPATIAL INFORMATION SYSTEMS

Low-cost manually operated methods of storing and retrieving geographically related information have existed for many years. The presentation of systematized spatial information for decision-makers has existed since the compilation of the first maps. Today, the simplest manual methods used in forestry comprise map overlays, or a single map and referenced punched cards recording attributes of field-collected data. An extension of these methods at small scale is to transfer each class of attributes on the basis of geocoded grid cells to separate base maps and for general use to provide a microfiche of the derived maps.

Overlays

The simple concept of using map overlays on a light table unfortunately requires each map to be available on transparent material for viewing on a light table. The manual transfer of existing map detail to a new transparent base map at a common scale can be laborious and costly; and the viewing and visual synthesis of more than two or three map overlays on a light table is impracticable. It is more difficult than with computer-assisted systems to reconcile type boundaries recorded on the separate overlays. If a new thematic map has to be drawn from the reconciled line data, this again will be laborious and probably will require the assistance of a competent draughtsman/woman.

These problems help to explain the reason for the increasing popularity of computer-assisted techniques.

Consider, for example, the thematic map needs of a forester, who is concerned with soil erosion assessment, and the preparation of a map showing by location the categories of erosion in subtle grey shades or (colour) hues. Information on soil types in most developed countries is available from existing soil maps, and slope can be obtained from topographic maps, being not usually available as a slope percentage map. Plant cover, however, will involve the analysis of recent aerial photographs. It will be necessary to bring the existing maps to a common scale, which requires re-drawing or photographic reduction or enlargement; and new slope and plant cover maps, if prepared manually, will be costly. Reconciling the maps on a light table will be difficult as soon as several layers are involved.

Punched cards

The alternative to using overlap maps of transparent material on a light table is a **punched card** system with a single map as the visual spatial reference and preferably supported by aerial photographs and if appropriate satellite imagery. A forest stockmap can be simply kept up to date by transferring to it, using only a pencil, ruler, protractor and rubber, information obtained from field collected data, current aerial photographs or possibly satellite imagery. Additional information can be stored on two separate sets of cards. Punched feature or master cards are prepared for each geocoded or grid square or compartment number. As each important feature or event occurs, it is reduced to a single coded number, which is punched on the compartment card. For each event a separate descriptor card is prepared showing details of the event such as planting, thinning, clear felling. If the cards are few in number they can be filed alphabetically for retrieval. Otherwise the relevant cards are retrieved from the pack using a strong needle inserted into the punched marginal holes of the cards.

As data is accumulated, it is often useful to sort the cards for fast or slow retrieval. The slow retrieval can be developed as a repository of all material unsuitable for reduction to simple classification. This includes maps, aerial photographs, satellite imagery and bibliographic references. Intermediate between this manual approach and employing a geographic information system is to computerize the punched card data, but still to rely on a master map as the spatial reference.

12.2 GEOGRAPHIC INFORMATION SYSTEMS AND GIS INPUT

The main advantages of using computer-assisted spatial information systems over traditional methods of preparing maps and manually filing data or information, are the reduction in human error; the ability to recall quickly map overlays from the GIS data files; to combine these overlays – but reconciling boundaries can be difficult; and to revise, in view of environmental changes, statistics and boundaries and areas shown on the maps. The labelling of change is readily checked on the video-screen of the system. The print-outs in map format to scale can be obtained in shades of blue, green and red (or their combination), showing the changes in the forest types, stocking, silvicultural treatment, volume classes, etc.

The technology involved in geographic information systems extends the use of maps, cartographic modelling and spatial statistics by providing analytical capabilities not otherwise available for developing complex terrain models and examining landscape/land use problems. Currently, the commonest use of GIS is for the production of urban thematic maps and to provide for their revision. In comparison, digital mapping (section 11.6) has been concerned mostly with the production of planimetric and topographic maps conforming to international mapping standards.

A geographic information system is designed to accept inputs in different formats (e.g. maps, tabulations), to store the data on magnetic tape, magnetic disc or floppy disc, to process the data, and to present processed data on video displays and as computer maps and tabulations. The purchase price of a geographic information system, suitable for integrating with a digital image analysis system, will vary widely, depending considerably on the data-handling features. A modest GIS component of an integrated system is available at less than half the cost of currently marketed DIAS; but a large geographic information system can cost several times the price of a turnkey digital imagery analysis system.

In a comprehensive geographic information system, several major subsystems are identifiable (cf. UNESCO, 1976). Management planning of the system needs to cover long-term staffing, fixed and on-going costs including maintenance, replacement of parts and materials, training of the users of the system, publicity on outputs and innovations resulting from feedback by the users.

The **data acquisition subsystem** addresses the problem of acquiring all the data elements required to analyse particular problems and the physical data inputs into the system. The digital rendering of the input data, including information derived from remote sensing, is frequently termed **data capture**. This function should not be confused with the primary capture of remotely sensed data by aircraft or satellites. The capture of data by a geographic information system also refers to the digitizing of existing maps and satellite imagery. Frequently the capture (encoding) of input data is slow and costly. Devine and Field (1986) estimated that about 80% of the time spent on GIS application involves digitizing the input data, and pointed out that even with the fastest system about five minutes is needed for each data entry of area (polygon) from a map.

All input data are always geographically referenced to points on the earth's surface. The registration of the data can be achieved photogrammetrically, but it is achieved usually more cheaply and less accurately by establishing control positions extracted from existing maps. Control after entry of other data into the host computer uses programmes which fit the data to the map base. This is termed **rubber sheeting** and conjectures the stretching of a sheet of data to fit the base.

The input hardware devices are those as mentioned in Chapter 10. As illustrated in Figure 12.1, which is taken from Figure 11.3, the data organization is referenced by grid cells, vectors or strips (rasters). Grid cells are the simplest to use and require the least storage capacity. The disadvantage is

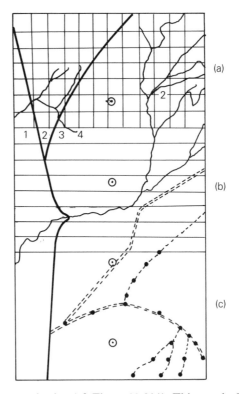

Figure 12.1 Data organization (cf. Figure 11.3(c)). This may be by grid cells (a), by rasters (b) or by vectors (polygons) (c). Sometimes no disticion is made between (a) and (b). In the polygon approach, a new set of vectors is provided at each point coinciding with a change of direction. In (a), the data is recorded at each intersection of the grid lines, which constitutes systematic point sampling or alternatively the contents of each grid cell may be interpreted, coded and geographically referenced.

that the contents of each cell are generalized, and hence the difficulty and inaccuracy in relating point data to each other and to the geographic reference co-ordinates. The larger the grid cells, to conserve storage, the more inaccurate is the method.

The preferred approach in geographic information systems is to use vectors and polygons (Figure 12.1(c)). Points pose few problems, since they are represented by a co-ordinate (i.e. a pair of vectors). A single line is represented by two pairs of vectors, a polygon by a pair of vectors each time there is a change in direction and a circle or arc by the sides of the polygon representing the circumference. With small vector systems, hand tracing using a cursor and a crosswire digitizing table is popular, due to its low cost and simplicity. The digitizing table should be large enough to accommodate the largest documents (e.g. 60 × 80 cm) and have a precision of 0.002–0.03 cm for registering the co-ordinates of map sheets and line data respectively. Commonly the digitizing tablet contains a menu for adding attributes/descriptors to the polygons, lines and points.

When automated techniques are used, re-drafting manually of the input maps etc. is often necessary as the first step. This is to eliminate colour coding, which appear as shades of grey to the machine reader, and extraneous detail (e.g. fine drainage networks) as occurs on some maps. Automated line following alleviates the operator of some tedious work, since he/she needs only to place the tracking spot on the line of interest on the map; but the operator needs to enter the attribute data and has to ensure that line junctions are corrrectly followed (Marble and Pequet, 1983).

Raster representation of input data favours automated scanning, is fast, and, as numbers and not vectors are used for referencing the raster lines, the storage space per pixel is reduced. The data is recorded sequentially (Figure 12.1(b)). The concept of the raster matrix was introduced in Chapter 7, when considering the line form of pixels in a Landsat scene. Each pixel of the line is numbered and referenced to a common starting point (i.e. pixel 1). The size of the cells will depend on the requirements of the GIS user and the final map scale. The disadvantage of raster representation of data is the considerable increase in total data storage when recording lines and polygons. Usually modern geographic information systems have the ability to convert data in vector form to data in raster form (or vice versa) using either hardware or software.

12.3 STRUCTURE OF THE GIS DATA BASE

The data base consists of interrelated digitized data, which is stored together with control of redundancies in order to facilitate analysis and retrieval and to serve optimally one or more types of application. The captured data is a matrix of numbers, which encode specific information about a particular area in two- or three-dimensional space. Once the data has been stored, there are various analysis options depending on the software and hardware of the system. Among the most common are recoding, updating, indexing, searching, displaying, digitizing and overlaying. The display programme is used to show the results of a GIS analysis on the (CRT) video screen.

In a comprehensive geographic information system at least two subsystems are identifiable related to the data base. These are the data storage subsystem, which overlaps the data acquisition subsystem, and the retrieval and analysis subsystem. FAO (1986) gives guidelines on information processing for developing countries, which includes examples using remotely sensed data (e.g. Brazil, India).

The significant steps in the **storage subsystem** are the partitioning of the data into workable units, the control techniques for processing the workable units, the encoding of the descriptive data corresponding to the graphic inputs, verification and correction, data reduction in which the encoded data is simplified and compacted, the development in the computer of data files, file organization, address and access methods, and the linking of the graphic data with descriptive attributes.

The **retrieval and analysis subsystem** covers extracting the data from computer storage and performing the necessary analytical operations to meet the objectives of the exercise. These involve the retrieval of the data, measurement of areas, calculation of linear distances, the referencing of point data, the overlaying of one data set on another, the comparison of multiple data sets, and statistical analyses of spatial data.

Labelling and edge-matching are both necessary functions of the subsystems. **Labelling** describes the attributes (i.e. contents) of the polygons and the characteristics of points and lines, and calls for the development and testing of a suitable map legend prior to the major processing of the data. This helps to ensure that the labels are topologically related to the graphic entities. Labelling of the input data is made at a convenient time in the input process. **Edge-matching** is required to ensure the joining of common boundaries, as occurs between map sheets and the lines of adjoining polygons. The join needs to be topological as well as graphic. That is, the polygons so joined become a single polygon and lines so joined become single line segments in the data base. A single attribute will describe each new data element. The edge-matching software needs to cover the smoothing of small gaps in the processed data, double lines which should be one, missed lines and other associated minor discrepancies.

An **overlay programme** is used to combine two or more GIS files and to produce a new file holding either the maximum of each cell's values across all the input files or the minimum of values. The stored data will primarily represent a series of maps according to the type of available input data and the

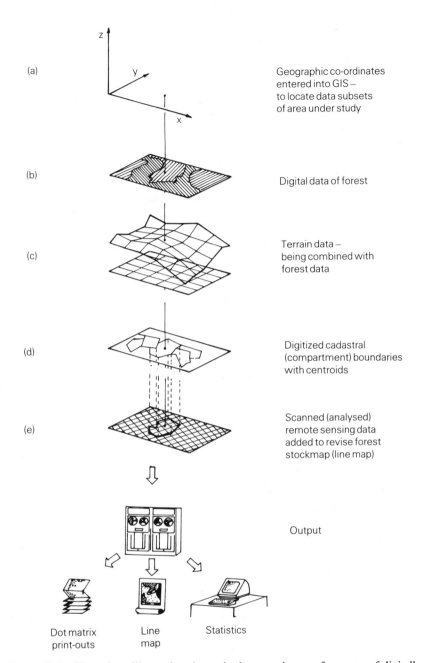

(a)   Geographic co-ordinates
entered into GIS –
to locate data subsets
of area under study

(b)   Digital data of forest

(c)   Terrain data –
being combined with
forest data

(d)   Digitized cadastral
(compartment) boundaries
with centroids

(e)   Scanned (analysed)
remote sensing data
added to revise forest
stockmap (line map)

Output

Dot matrix   Line   Statistics
print-outs   map

Figure 12.2 Flow chart illustrating the revised output in map form etc. of digitally stored layers of file map data. Hard-copy output devices include not only dot matrix printers and drum and flat-bed line plotters, but also film writers, inkjet printers, matrix camera recording in 23 cm format the raster video display and simple photography of the video screen using a 35 mm reflex camera or Polaroid camera.

purpose of setting up the geographic information system. Each map layer (Figure 12.2) will comprise a set of vectors or grid cells or rasters that correspond to their spatial locations, and which are entered into the system one at a time. Each map in the data base will have the same grid cell, vector or raster reference, and will usually contain only one kind of information (e.g. soil types, forest cover types). Two or more of these layers can be combined digitally for the purpose of analysis.

On specifying the number of files for overlaying, the programme identifies the name of each file, provides on the display screen the variable name and the class name and provides the opportunity to recode its values. In combining boundaries on maps originating from several disciplines (e.g. geology, forestry, soils), it may be preferable to rank them hierarchically according to the phytogeomorphic sequence, as will be discussed in Chapter 14, in order to avoid the problem of boundary matching. For this purpose, it is recommended that landforms take precedence over plant communities, since the latter tend much more to 'fragment' the landscape; and both take precedence over soil boundaries, since these are not usually observed directly on imagery.

12.4 GIS OUTPUT

The information output subsystem, after data analysis, provides graphic displays in map form of entire data sets (and subsets); and provides the result of overlaying data sets, including their revision and the tabular listings of measurements, comparisons or statisticial analyses. The subsystem incorporates verification procedures to check that the data retrieved from storage is accurate or within predetermined statistical limits, that the analytical procedures are valid in terms of data accuracy, and that the graphic displays of the finalized data conform to specified mapping accuracy. The range and quality of the output data will obviously be limited by the nature of the inputs.

The outputs of geographic information systems are increasingly oriented towards both statistical and mapping functions involving carefully chosen algorithms suited to the data structure. Once the working data base is established and provision is made for information retrieval and output, management output functions can be undertaken, involving spatial and geometric calculations. These include, for example, point referencing on print-out maps of data within the search radius to lines and polygons, including district forest boundaries, compartment boundaries and extraction routes. Using cross-tabulation techniques, data aggregated for one area or forest compartment can be aggregated for use in thematic mapping of other areas. An example of mathematical modelling is the creation of a solar radiation map combining the attributes of latitude, slope, aspect and cloud cover with fundamental insolation equations (e.g. de Steigeur and Giles, 1981).

The output in map form can be visualized as stored digitally on file in the computer, which is adjusted to the required output scale and over which is 'stretched' the overlay of merged data from two or more digitized input thematic maps. This is illustrated in Figure 12.2. At step (a) the geographic co-ordinates (x, y, z) are entered into the system to locate the subsets of data covering the area of interest. At step (e), the map data, the cadastral map data (e.g. forest compartment map) and the terrain/topographic data are merged with new forest data obtained from image analysis. The output could be either a thematic map or a thematic map showing monitored changes; or in combination with mathematical modelling, a line map showing logging potential or present or potential forest productivity, and supported by statistical tabular print-outs.

During output, display, editing and symbolizing facilities are needed. The GIS display functions are similar or the same as used during the input of data and the management of the data base for points, lines and polygon areas. A wide range of symbols is required for points, lines, alpha-numeric strings and polygon areas. Double lines are used for roads, when registering the map. Editing facilities are needed not only for these symbols, but also to delete and move symbols, to shape names and to adjust functions including scale and 'rubber banded' lines etc.

The hardware of the GIS information output subsystem will be similar to that already commented on in Chapters 10 and 11. These are a CRT video display for monitoring and 'quick look' and preferably with a large screen (1024×1024 pixels), a line printer for tabulations etc., a dot matrix plotter (or equivalent) for quick graphic inspections in a range of grey-scale values or multicolours; and preferably a drum or flat-bed line plotter for registering the finalized thematic map on good-quality paper or mylar or an already printed map sheet or a photogrammetrically prepared base map. A plotting precision of 0.01–0.03 cm is usually acceptable. It should be borne in mind that the mapping accuracy of the output map will be only as good as the accuracy of the input maps and quality of the equipment. Thus, if the scales of input maps vary between 1 : 20 000 and 1 : 50 000, the output map cannot have an accuracy approaching 1 : 10 000.

Since maps have always been an important tool in the practice of forestry, it is readily appreciated that the graphic outputs of a geographic information system will be increasingly important to forest management. GIS as a separate concept of the more general ideals of MIS provides quickly maps in various forms, which previously were not feasible economically. The significance of data presented in map form, and not as a tabulation, will be readily appreciated by comparing published tabular data and related maps.

12.5 LINKING GIS AND REMOTE SENSING

In this chapter several references have been made to the linking of GIS and RS. The information derived by analysis from the remotely sensed images is made to mesh with data stored in the GIS data bank. Examples of integrated systems include those operated by the Pennsylvania State University (Myers *et al.*, 1989) and the Image Based Information System (IBIS) developed by the Jet Propulsion Laboratory of the California Institute of Technology (Marble and Pequet, 1983).

The main interest in integrating remote sensing and GIS has come from remote sensing specialists; and this interest is reflected in the growing numbers of small digital imagery analysis systems with a GIS capability. Usually, the potential inputs of remotely sensed data (captured data) to the GIS must be accompanied, as remote specialists well know, by human intervention in the analysis. This latter point is often not fully appreciated and understood by the GIS specialist, whose work is highly automated using computers; and who does not make maximum use of remote sensing in updating GIS data banks, particularly if imagery analysis does not appear to have the required mapping accuracy. Frequently the classification and positional accuracy of remote sensing analysis is coarser than the classification requirements of the GIS specialist. In fact, the pixel size of remotely sensed satellite data can be expected to remain in the near future much coarser than the needs of many geographic information systems, including those designed to support large-scale urban studies and civil engineering. As pointed out by Myers *et al.* (1989), although computerized pattern recognition handles a few thematic categories well, the basic difficulty for automation of integrated systems lies in the differences existing between the less understood spectral context of images suited to computer-assisted techniques and the spatial context of images requiring visual interpretation. For the former, algorithmic pattern recognition is derived primarily from multivariate statistics.

The need for human intervention in GIS and integrated systems must be fully recognized, as a key requirement to achieve stepwise separability of thematic classes using initially the computer-assisted techniques. When a class has been located within specified limits of accuracy, which is not amenable to digital pattern recognition, it must be delineated and identified using conventional visual analysis (photointerpretation); and the null attribute code is displaced by the identified code. This type of editing is in effect a two-stage classification technique. The second stage might for example necessitate simple visual reconnaissance by flying the forest area or taking oblique or vertical aerial photographs of it, using a small-format camera, and followed by their analysis.

In the process, it would often be cost effective to register firstly the identifiable data classes on the base map at the required scale and to omit (suppress) any class areas which will not be of interest or cannot be adequately identified. This is followed by labelling these as 'other', and generalizing or incorporating categories likely to incur substantial errors of commission. It necessitates at the earliest possible time an adequate map code or legend and its modification as work progresses. For example, the retention of the map code riverine (gallery) forest may not be justified on area and it would be incorporated in the category 'lowland forest'.

Finally, it should be appreciated that a common oversight in the development of integrated remote sensing/geographic information systems is the assumption that the flow of data is unidirectional. That is, from the remote sensing analysis system to the geographic information system. The reverse flow is not only desirable, but is also realistic and needed by the remote sensing analyst. The major constraint to this approach is that of cost in constructing the GIS digital data base; but these costs can be expected to fall with improvement and increased availability of hardware and software and of maps already in digital format. There is little doubt that the ready availability of a GIS/land information system to the imagery analyst helps improve efficiency, as is well supported by the older practice of referring visually to existing thematic maps in the course of aerial photointerpretation.

When a forest imagery analyst is studying a point or area on aerial photographs or satellite imagery, it would be advantageous to have map data accessible quickly on a range of themes covering geology, landform, altitude, slope, aspect and forest vegetation types and their permutations, as can be provided by a GIS/land information system. Similarly, it is useful to be able to overlay digitally at the corrected scale topographic, planimetric and cadastral GIS data on digitized satellite imagery. The overlaying of roads, streams, forest boundaries, etc. on video-displayed satellite imagery helps in the location of pixels in advance of fieldwork and makes the CRT video display more meaningful.

The ready access to a geographic information system by the photointerpreter/image analyst might also give the desirable stimulation for the modernization of conventional forest aerial photointerpretation, which remains much as it was immediately following the Second World War. This is particularly needed with the increasing quantities of image data available, in an increasing choice of formats and scales, the decline in the availability of skilled services provided by the traditional photointerpreter and the tendency to reduce fieldwork and to concentrate management at more distant points of control.

Part Five

Remotely Sensed Data in Forest Practice

13

Planning for remote sensing applications

As remote sensing is far reaching in its applications, it calls particularly for a critical examination of the purpose and use of the analysis system ahead of acquiring equipment. The system is likely to involve airborne and satellite data/imagery and visual analysis and, if permanent, digital analysis. Depending on the purpose, permanency and complexity of the envisaged remote sensing operations, its physical core and managerial structure may be termed a remote sensing field team, a remote sensing unit, remote sensing branch, remote sensing laboratory, remote sensing centre or remote sensing space agency.

The planning of an aerial photographic contract (section 5.6) may require action to be taken (e.g. call for tenders) up to a year ahead of the need to use the photographs in the field; and the time period between the ordering and the supply of satellite imagery or archived aerial photographs may be several months. The decision to use existing satellite data or archived aerial photographs does not involve flight planning, but does require careful evaluation of the tasks ahead in relation to the different types of available remotely sensed data, and their ordering well ahead of field needs (section 7.4).

When the decision is reached to use remote sensing data, consideration needs to be given immediately to whether this will be on a temporary or permanent basis, since it will influence the choice of equipment, staff training and the type of inputs and outputs. If planned on a temporary basis, for example to assist local forest inventory, the programme should be examined to see if the remote sensing activities will have a wider and more permanent function, including direct support to forest management and forest policy. On the other hand, if the imagery analysis system is being established on a permanent basis, which may justify the purchase of expensive and sophisticated equipment and considerable staff training, the wider and interdisciplinary implications of this decision and the identification of all would-be users should be considered. It is likely that any permanent system and its outputs will be used not only in forest inventory, but also in the management of a wide spectrum of forest resources. This may involve land use and forest policy decision-making, and possibly extend to users in other disciplines. The latter may justify establishing an interdisciplinary advisory board or steering committee under the chairmanship of the lead department, and to represent all eventual users, who at some stage will be prepared to contribute financially or in kind to the development.

Point of departure

The idea of establishing a permanent remote sensing imagery analysis system will usually evolve from a set of economic and technical problems for which it is recognized that remote sensing, as an advanced technology, has an important role to play by improving efficiency, reducing costs and providing new types of data/information. Frequently the idea is supplemented by small informal meetings to ascertain the needs for the system; to find out if a system already exists, which could be used; to determine who will be the users of the new system, and what should be the extent of the photogrammetric capability of the system.

This can be considered in planning as the point of departure. It is associated with the **group event** at which the participants create the **group situation** for follow-up activities, including responsibility for developing the project outline and project document, identifying the source(s) of funding and possibly initiating a cost/benefit study. The **point of departure** is characterized by identifying the common problem and recognizing the level of knowledge within the group and the skills, attitudes and political tasks of the individual participants.

It is important for the group leader to become familiar with the skills of the participants in solving the problem. Homogeneity of the group is desirable, but heterogeneity can be helpful, particularly on complex matters. It may be decided at the group event that the group needs the participation of one or more external experts, that the group lacks experience, that there is a significant gap in knowledge and skills and the learning process will involve seminars for senior staff interspersed with future meetings, a training course for technical members of the group and possibly advanced training of some length for one or more staff. Usually in order to maintain the greatest possible support, conducive training and learning situations must be created, as well as an awareness outside the group on progress. The latter involves the wide distribution of the minutes of the meetings, the preparation and distribution of technical notes and the opportunity of those not in the group/planning committee to participate in seminars and training.

13.1 PROGRAMME FORMULATION

Programme formulation usually begins immediately following the successful conclusion of the meeting(s) at the point of departure; and involves a thorough evaluation of the problem to be solved, a brief study of the conditions under which the solution is to be achieved and a careful matching of local capabilities, limitations and identified new inputs. These are brought together with other relevant material in the project document or initially in a project outline. If a large programme including its implementation is envisaged, then the monitoring of progress along the lines of network analysis introduced in section 7.4 will be found useful to management. This approach or equivalent methods used in industry bring together timewise information on schedules, deadlines, costs, etc. For example, **back programming** plans the project from the end point back to its inception; and thereby enables the manager to know deadlines for the starting and ending of each phase, and so ensures completion on time.

Whether immediately following the point of departure the full project document should be prepared detailing the programme, or simply as an outline containing the major pertinent facts, depends very much on personal choice, project size, custom, experience and practicality. There is much to be said for initiating a project with an outline, since the feedback from senior management can be taken directly into consideration, when the time-consuming detailed project document is prepared; and which, for reasons not known at the time, would not be acceptable at the decision-making level without considerable revision. Preparation of a major technical project document, which is approved without serious delay, involves skill, experience, appropriate relevant scientific know-how and clarity in writing. There is no advantage in preparing a document full of technicalities, which will be little understood by the decision-maker(s). A lucid project outline is more likely to be read and appreciated by the busy administrator than a lengthy project document.

Project outline

As an introduction to project formulation, only the format of a project outline will be considered. The outline, whether for a large or small project, should be carefully prepared in consultation with those who will be later involved in implementing the work. It takes into consideration the likely views of the decision-maker(s), who will eventually fund/approve the project. Also it is often not appreciated that the time period between the concept of a project idea and the approval and implementation of the project document is seldom less than 12 months and for a large internationally funded project can be as long as three years (Howard, 1984). Figure 13.1 shows the stages involved in preparing and approving a project request submitted from a developing country for funding by an international organization or donor country.

A project outline, which justifies and summarizes the project programme, usually needs to be not more than a few pages in length, but should set out concisely the key events and budget. There is no generally agreed format; but the sequence of events should be logical and easily read without query by the non-technical decision-maker. It may be compared with an executive summary accompanying a lengthy report; but a project outline always contains several subject headings, the content of which is more specific than in most executive summaries. The preparation of a project outline can also serve as an excellent training exercise for students; and may be arranged to take only a few hours of preparatory work. It will bring together as a test, what the student has learned about remote sensing and its application. A parallel may be drawn with the requirement that final year forestry students prepare, as an exercise, a forest management plan.

The first section of the project outline provides the **background** leading up to why the proposal is being prepared and contains relevant historical and/or geographical facts. This is followed by the second section (**justification**),

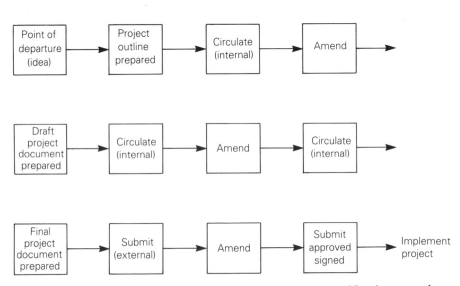

Figure 13.1 Project formulation stepwise. Time scale 1–3 years. Not shown are the steps between the submission to and the approval by the funding body. This may involve political and technical departments of a donor country.

which justifies logically why the project is needed and mentions the major technical, economic and social benefits. The third important section sets out clearly but briefly the **objectives**. These are broken down into long-term goals and the medium or short-term goals. The former are often expressed in economic or social terms, while the latter would be entirely technical and concerned with immediate results. The fourth section, **work plan**, sets out the institutional framework, including the project location, the nature of the input support needed in kind (e.g. office space) and moneywise for implementing the project, the suggested timing and period of the project and the main events timewise in sequence of the implementation of the project. Often the latter is accompanied by a bar chart. The fifth section sets out briefly any **preparatory activities** which will be needed ahead of project implementation (e.g. aerial photography, overseas training); and is followed by a sixth section specifying the physical **outputs** and other benefits and the evaluation operations and criteria for periodically monitoring progress and achieving the targeted objectives. The final section, **budget**, provides in summary the funding, its breakdown into internal and external sources; and the latter (donor contribution inputs) is preferably shown on an annual basis with a further general breakdown into major items. These would separately include the cost of equipment with specific mention of high-cost items, materials, study tours and fellowships, travel, consultancies and needed secretarial/technical support. However, at this stage the budget is not so detailed as the budget of the final project document, but the two should agree on the total expenditure and major items.

13.2 STRUCTURING THE IMAGE ANALYSIS SYSTEM

This section begins with the assumption that programme/project funds have been approved. Irrespective of the complexity of the remote sensing imagery analysis system, its function must be to support the user, whether this involves fieldwork or research or both. The Beijing Agriculture University in the People's Republic of China, for example, has successfully combined the two functions; and, world-wide, many academic institutes and national space agencies are increasingly associating research with the user. In this respect, a distinction needs to be made between the user in applied research and the user as the worker needing the data for field operations. If a committee is being formed for the latter, care needs to be exercised that its members are not primarily researchers. If emphasis is to be placed on structuring the analysis system for fieldwork in forestry, this will emphasize initially the importance of aerial photographic interpretation including stereoscopic viewing. A larger imagery analysis system would handle both analogue and digital data and be suited to multistage methods, involving the use of aerial photographs and imagery acquired by earth resources satellites.

If the analysis structure is on a temporary basis to support forest inventory, then it is likely to be integrated closely in the field with the survey; but this will have the constraint that the equipment must be portable, simple and easy to use. Normally in the assessment of forest resources, the equipment will be structured around the use of aerial photographs, which will be in panchromatic black-and-white or sometimes colour; or when these are not available or when the aerial photographs are several years old, then also higher resolution satellite imagery (e.g. Landsat TM, SPOT).

The minimal equipment for fieldwork would comprise a pocket stereoscope, a school protractor, engineer's linear scale, tubular magnifying scale and/or magnifying glass, colour chinagraph pencils and felt pens, a small bottle of alcohol for removing markings made on the photographs, a clipboard to hold the stereo-pair of photographs, a set of dot grids for measuring areas, a crown density scale, a pocket calculator and possibly a parallax wedge. If forest stock mapping is contemplated on good-quality paper (e.g. cartridge paper) or on an existing map, then an overhead projector or vertical sketchmaster (Chapter 11) would be needed.

In reaching a decision on installing digital analysis for satellite-acquired data, it is important to recognize the difference in needs for satellite remote sensing (SRS) of the developing and industrialized countries. For the former, hard-copy satellite imagery may be well suited to much-needed reconnaissance surveys, to provide a small-scale map substitute where no maps or only old maps exist, to assist in the thematic updating of maps at medium scale (e.g. 1 : 50 000/1 : 100 000), to monitor changes in forest cover and to help in the stratification of large areas ahead of field survey. Radar imagery, if available, may be suitable for some tasks (Chapter 6).

On the other hand, in the industrialized countries, large-scale forest stockmaps and planimetric maps are usually available, and the role of the satellite image analysis system will be to support research into new methods and techniques, and for map revision at medium scales (e.g. 1 : 50 000 and possibly 1 : 25 000). As already mentioned, the derived forest information is not likely to be better than that obtainable from very small-scale high-altitude aerial photography. In advanced countries with extensive land areas and lacking thematic maps and large-scale topographic maps, the needs may be in-between and extended to using satellite data in environmental and change monitoring and small-scale thematic mapping.

In structuring a remote sensing system to accommodate both aerial photointerpretation and satellite digital image analysis, there is obviously a wide choice in methodology, equipment and integration of staff into the system. In addition, the use of a small photographic processing unit may be essential if the grey levels and colour balance of aerial photographs and satellite imagery obtained from negatives are to be maximized for visual interpretation. The cost of a small photographic unit is comparable to that of a small to medium-sized stand-alone digital imagery analysis system; and if extended to cover satellite hard-copy processing, it will need a relatively low-cost diazochrome processor/printer (Chapter 8). Costs will be minimal for handling 35 mm black-and-white film and their enlargement and maximum for 23 cm colour film, including its associated negative-to-print enlarger. Success will largely depend on employing a highly skilled laboratory assistant, experienced in aerial photographic processing and printing, and having a high-quality photographic enlarger, temperature control, clean chemical-clear water supply, a reliable electric supply, fresh chemicals for colour work and a light-proof dark room.

For digital image analysis, as discussed earlier (Chapters 10 and 12), several major decisions will be needed for the processing of the digital data. Care needs to be exercised in choosing a system with a suitable menu of programmes. The system should be able to handle more than the quantity of data envisaged, and after-sales servicing must be effective. For training and some research, a low-cost system with a monitor display of 240 × 256 pixels may be sufficient; but with the advent of the second generation satellites with higher resolution sensors and consequently many more pixels per unit ground area, a CRT display of 1026 × 1026 pixels is to be preferred, albeit at higher system cost. With the supplier trend of higher cost for hard-copy imagery than for CCTs and with the data mainly available on CCTs or disc in the 1990s, it is becoming essential for the user to have access to a digital system providing geometric corrections, image enhancement and registering the images to an acceptable map base. In the UK, the latter requires the incorporation in the system of a programme for referencing selected pixels to the national grid. Additional software/hardware will be needed to provide not only unsupervised classification and image enhancement, but also operator-controlled image analysis for feature selection, training sets, supervised classification, editing following accuracy assessment and the production of hard copies at acceptable mapping accuracy and meeting the operational needs of the user. The source of output devices will need to be identified (i.e. line printer, dot matrix printer, film writer, drum or flat-bed plotter).

The potential user will require to be determined, as being involved in forest inventory, forest management, possibly forest policy, other renewable natural resources, planimetric/thematic mapping, geographic information systems, etc. For planimetric mapping and GIS, consideration needs to be given to interfacing the systems. If stereo-mapping is contemplated, then a plotter (Chapter 11) should be chosen, which is compatible with the digital imagery analysis system and the mapping standards of the user.

13.3 EDUCATION AND TRAINING

The success of any formulated programme involving imagery analysis will depend not only on the careful selection of equipment and techniques, but also on the adequate training and experience of the interpreter/analyst. The inclusion of all aspects of remote sensing in the undergraduate professional education of foresters is becoming increasingly difficult, as the subject expands from aerial photointerpretation and elements of aerial photogrammetry to include satellite digital image analysis and geographic information systems. There is a danger that the future professional forester on graduation will not be adequately trained to apply remote sensing as an art and science to fieldwork in forestry, as has long been the practice in North America. Probably, the graduate will have only a general understanding of digital analysis using satellite data and at the sacrifice of some essential training in the practice of aerial photointerpretation. For expediency, it is inadvisable to integrate remote sensing courses and forest inventory courses, since remote sensing has a wider function. There appears little academic logic, outside the research field, to abandon visual photointerpretation in undergraduate forestry courses in favour of satellite digital analysis. If in the future, aerial photographic analysis becomes 'digital', possibly through the impact of geographic information systems and the decreasing cost of computer equipment, then this rationale may no longer hold.

At the postgraduate level, there is much greater versatility in education and training in remote sensing. Universities are providing course-oriented master degrees in remote sensing applied to forest resources; and for many years some schools of forestry in North America, Europe and Australia have awarded master and doctoral degrees by thesis with specialization in remote sensing applications. In Europe the best-known centre of learning specializing in remote sensing is ITC in the Netherlands; and several national schools, including, for example, two schools of forestry in the Federal Republic of Germany, regularly award doctoral degrees by thesis in the applications of remote sensing.

Course content

The practising forest photointerpreter through training and lectures needs to know how to mark up the aerial photograph, to be able to interpret stereoscopically pairs of aerial photographs and to use a parallax bar, to make accurate photographic measurements of linear distances, areas, tree heights and slope, to correct for photographic scale differences, to relate images on aerial photographs and satellite imagery; and in the field to 'read' aerial photographs and satellite imagery correctly and to position satisfactorily in the field photo-sample plots and the pixels of satellite imagery. An understanding of statistical theory is needed for stratification, random sampling and the correlation of field and photo samples (Chapter 18). Failure to quantify the stereoscopic model exposes aerial photointerpretation to the erroneous statement that it is qualitative, while digital imagery analysis is quantitative.

Both satellite data analysis and aerial photointerpretation are qualitative and quantitative, as previously mentioned.

Although a sound grounding in the principles and techniques of photogrammetry is desirable, it has unfortunately to be kept to the minimum, due to other priorities in the education of students in forest resources. The minimum elements in photogrammetry are probably those presented in this text. The level of awareness of the principles of photogrammetry should be sufficient to discuss technically common problems with the photogrammetrist and to plan an aerial photographic mission.

The minimum training required in the application of digital imagery analysis to forest resources is likely to expand in the next few years as finer resolution satellite data becomes available. The training currently required in day-to-day field activities is very much geographically related. In industrialized Europe, where excellent large-scale maps are available, training is minimal other than for research; and calls for being able to visually interpret satellite imagery, to understand the pixel composition and geometry of the imagery in terms of spectral and radiance values and the relationship of remote sensing application with GIS. An awareness of computer-assisted processing and analysis will be found useful for communicating with the digital image specialist. Of countries large in area, particularly if inadequately mapped, and many developing countries, the use of satellite imagery can be expected, however, to form an essential part of fieldwork, and therefore through education a working knowledge of digital imagery analysis is needed, including an understanding of how/when to use the commoner software programmes. At the research level, hands-on experience of a digital imagery analysis system is essential, as well as an in-depth understanding of its relationship to digital mapping and geographic information systems.

Short-period education

At an international meeting in Toulouse at the Centre National d'Etudes Spatiales (CNES) in 1985, there was consensus that at least 10 000 experts trained in remote sensing applications would be needed in the next decade. Short-period education in remote sensing will help in this matter, as distinct from forming a part of degree and diploma courses. Training is provided through technical tours, on-the-job training, short courses and seminars. Visiting scientist schemes operate in several countries as diverse as Brazil, Canada, France, India, Japan, Thailand and the USA. An outstanding example of on-the-job training in co-operation with industry was the training of Indian scientists and engineers to build and test equipment being supplied by industrialized countries at the time the Indian Space Research Organization (ISRO) was being established.

An example of short-course training in applied remote sensing, including land use and forest resources, is that operated for many years (1974 onwards) through the Food and Agriculture Organization of the United Nations (FAO) in co-operation with other bodies of the UN and several industrialized countries (e.g. Italy, Germany; Howard and Travaglia, 1985). In the period 1974–89, over 1100 participants from more than 100 countries attended the courses, which involved about 250 lecturers from all continents. The courses are held either at the FAO headquarters in Rome or in a developing country. There is considerable merit for developing country-based introductory courses, since these may only require 1–3 lecturers and simple equipment, and larger groups of nationals can be exposed to the courses. In a host industrialized country, the infrastructure base for holding international special-subject courses is usually better developed; and resources and technical skills are more accessible and readily available.

Due to the time constraint, FAO international courses are intensive and narrow in theme concept. The remote sensing subject matter is examined critically in depth in the context of operational use. Each course is designed to equip the selected professional participants, of similar technical background and with several years' field experience, with practical know-how and confidence in applying newly gained skills. The number of participants attending each course out of the many applications needs to be strictly controlled (i.e. 20–25 persons), since experience has shown that when the number approaches 30 per lecturer, the participants cannot receive individual attention and often fail to have their questions/problems adequately addressed. A course length of under two weeks becomes superficial and is much more costly per participant in terms of the content of the transferred technology; courses longer than about four weeks are increasingly difficult for senior officers to attend due to their being absent from their decision-making duties. The core time for a course is probably three to four weeks.

For a three weeks' course, for example, there are usually three main elements. That is, a refresher week, a week emphasizing case studies and a final week aiming at self-analysis, acquiring further experience and achieving self-reliance in the theme work of the course. If the course is four weeks, then the third week can be advantageously mainly devoted to fieldwork, and report writing can be introduced in the fourth week. During the course, the course director ensures that the theme of the course is adequately covered, with sufficient emphasis on application, that each lecture is oriented correctly and merges with the next lecture and related to associated hands-on workshops.

14

Phytogeomorphic and land-use considerations

The two major factors of the natural environment, important to forestry, to be observed directly by remote sensing, are the landforms and the plant/forest cover. In combination, the study of these two factors can be termed **phytogeomorphology**; and this provides the forest image analyst with a powerful interpretation aid in forest practice and a means of integrating studies of the landscape. Whereas the present-day landforms, as related to the forest site, usually predate history; the existing forest cover reflects the interaction of recent and current environmental factors. In combination, the two are more informative on the forest site quality and land-use suitability than when either of the two factors are examined independently.

Soil, a further major factor, is unfortunately not directly observable on remotely sensed imagery of forests; and this restricts, in image analysis, the long-established integrating role of soil survey. As compared with the analysis of remotely sensed images, the digging of soil pits and the taking of auger borings are slow and relatively expensive. It must, however, be appreciated that remote sensing forms an essential part of soil studies in the field; and that information concerning the soil can be inferred from the vegetation and landforms recorded as images. Thus, it may be possible to delineate the boundaries of the soil types of the landscape on the imagery and within the boundaries to describe the soil as sandy, clayey, silty, etc. The term 'land' embraces the landform, soil and vegetation; while the term 'terrain' usually excludes the vegetation.

14.1 THE GEOMORPHIC FACTOR

Through education in the life sciences, most foresters are aware of the important role plants and plant ecology play in fieldwork. Regrettably, the same cannot be said for geomorphology. Few foresters receive formal training in geomorphology and hence tend to ignore or at least to minimize this important factor and the ways in which geomorphic data can assist in forestry and land-use planning. The writer ventures to add that with the increasing importance of remote sensing, geomorphology is an essential element of professional forest curricula. Follow-up references linking geomorphology and remote sensing include Lueder (1960); American Society of Photogrammetry (1960); Mitchell (1973); Townshend (1981); American Society of Photogrammetry (1983); and Howard and Mitchell (1985).

In the context of this book, only the briefest of comments can be made on geomorphology. The parent materials, topography of the landscape and the processes that result from the interaction of these with the climate provide the three principal components of geomorphology. In forestry their influence is seen by changes in site quality, in the physiognomy of the natural forest, by abrupt variations in the forest cover and by the distribution of indicator species. With respect to the first component, parent materials, an example is provided by the unglaciated land division with its maritime climate that comprises part of southern England and north-west France. There is a clear demarcation in land use and forest types between the chalk lands with their calciphilous vegetation and the sand deposits with their heath and conifer-dominated vegetation. The first is best suited to arable farming including the growing of cereals, and the second to forestry and recreation. A skilled photo-interpreter is able to classify a region by the surface expression of its principal rock types, to identify deformation structures (e.g. faults, joints) and to extract information on the degradational and depositional landforms. An example of subdividing a region based on its landforms is illustrated schematically in Figure 14.1.

The second component, namely the topographic form of the land, is best observed using stereoscopic pairs of aerial photographs. Topography is a prime determinant of land use. This is particularly so with the need for mechanization to compensate for the high cost and often scarcity of labour. Altitude, slope, aspect and exposure all influence the distribution of natural and induced vegetation, the water run-off and the accumulation of ground water. At a point where the slope becomes too steep for mechanized agriculture, although the soil may be well suited to arable farming, cultivation of the land may have to be abandoned (e.g. terraces on the Appenines in central Italy), or the land use changed to forestry. Similarly, where the slope is too steep for track vehicles in forestry, the land may have to be used for protection and recreation forestry or logged using overhead cable systems.

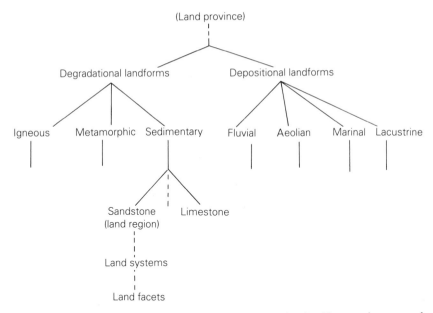

Figure 14.1 Simplified geomorphic approach to mapping landforms using remotely sensed imagery. Its association with land units, which are phytogeomorphic, as discussed in section 14.3, is indicated by the broken lines.

Accompanying slope from its highest to lowest points, there is usually a topo-sequence of land-use activities, which is often also associated with aspect. For example, in northern Britain, the descending topo-sequence to be observed will be hilltops with lithosols, which sustain on open peaty moorlands, sheep grazing, rough shooting, turf planting of conifers and, in a few areas, deer/reindeer ranching. On the hill slopes podsolized soils occur which favour the planting of exotic conifers and native pine. These give way on lower slopes to the brown forest soils with mixed farming and deciduous woodland on the piedmonts; and finally to wet meadowlands, possibly fringed with willows, on the gley soils of the valley bottoms.

The third factor, weathering processes, is categorized initially as depositional and degradational/erosive (Figure 14.1). All have an effect on the land use and land potential. Alluvial deposits in the lowlands often provide freely drained soils with a high mineral content well suited to market gardening, and too highly priced for forestry. In contrast, surfaces exposed to degradational processes, if not controlled by afforestation or by maintaining the forest cover, can experience severe sheet and gulley erosion, flash floods and the deposition of surface materials where it is not wanted. The upper

Table 14.1 Comparison of aerial photographic parameters of three different rock types providing contiguous land systems (black-and-white panchromatic photographs at 1 : 20 000 scale)

Landform	Stream density (metres/ha)	Bifurcation ratio	Altimetric frequency	Stream confluence angle (°)
(i) Folded exposed sediments	27	15	0.09	80
(ii) Regolith (deep weathered layer over sediments (plateau))	6	4.0	0.025	60
(iii) Granodioritic intrusion (plateau)	12	3.5	0.05	45

elevational limits in valleys of periodic flash floods may be evidenced on photographs as occurring at long intervals of time, by bench terraces and, at frequent intervals, by the physiognomy of the vegetation; and provides a useful demarcation line between planned urban development and retaining the land for agriculture and forestry.

As shown in Table 14.1 aerial photointerpretation can provide a parametric approach to identifying adjoining rock types (Mount Disappointment, Australia), and valuable information on stream density, etc. Figure 14.2 shows the stream patterns recorded on aerial photographs at a scale of 1 : 10 000 of the same area. **Stream density** is defined as the total length of all water courses measured on the imagery within a circular area of 1 mile^2 or 1 km^2 and

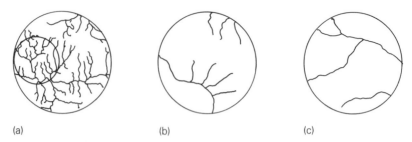

(a)   (b)   (c)

Figure 14.2 Differences in stream patterns recorded on the aerial photographs (Table 14.1) of the contiguous terrain types: (a) exposed sediments; (b) regolith on sediments; (c) granodiorite.

expressed as length in yards or metres per 1 mile^2 or hectare. As described by Strahler (1964), the **bifurcation ratio** is the number of stream confluences per mile or 1 km; **confluence angle** is the angle of the highest order stream with the stream of the next lower order; and **altimetric frequency** (f) is the difference in elevation (e) divided by the distance between the interfluves (d), i.e.

$$f = \frac{e}{d}$$

14.2 THE VEGETAL FACTOR

Foresters usually need no introduction to methods used in the field to identify and to map differences in the plant cover. The intention in this section is to examine briefly the identification and classification of woody vegetation covering extensive areas in the context of remote sensing. The term **plant cover** refers to the proportion of the ground covered by the vertical projection on to it of the overall vegetation canopy and is not to be confused with leaf area index (LAI), discussed in Chapter 3. When the numerical result is expressed as a percentage of the tree cover (e.g. 80%), it is termed **crown cover** or **crown closure**; and if expressed as a decimal coefficient it is termed **crown density** (e.g. 0.8).

Land cover

This relates to the type of cover present on an area of land (Table 14.5); while **land use** relates to the human activity associated with the piece of land. **Land evaluation** is concerned with assessing the **land potential**, as being suitable for different types of management and different crops. What is seen on the imagery are different types of cover and from this the land use and/or land potential are inferred.

Classification of woody vegetation

Classification involves the grouping of objects into classes on the basis of common properties or relationships. The classification of vegetation most frequently embodies floristic characteristics, and occasionally habitat characteristics or the combination of floristics with habitat. However, in remote sensing it is the physiognomy of the plant community or forest stand which is usually most readily observed or can be inferred. The identification of the tree species is often difficult and in the tropics usually impossible. The choice of method also reflects the purpose of the work, training, experience and the natural environment in which the vegetal survey is carried out (cf. Dansereau, 1951). For example, Braun-Blanquet (1964) considered that a physiognomic classification was not precise enough in western Europe; while Poore (1963), working in tropical forests of Malaysia, judged it to embrace faithfully the ecological factors of the habitat. Kückler (1967) treads a middle path by saying 'the vegetation mapper must realize that his work can be a lasting success only if he is thoroughly familiar with both the physiognomic and floristic features of the vegetation'.

In a physiognomic approach to the classification and mapping of vegetation, emphasis is placed on the appearance of the vegetation irrespective of its floristic composition. It is concerned with the shape of the individuals forming distinctive groups and the internal and external shape of plant communities. Schemes used on an international basis in recent years include UNESCO (1973), Holdridge (1966) and Blasco (1984). The UNESCO scheme, although globally tested, requires revision for use with remotely sensed data. Holdridge uses climatic factors for identifying major units (i.e. plant formations/sub-formations), which are then mapped using when required aerial photographic mosaics.

Blasco's method (or Toulouse method), being eco-floristically based and using satellite imagery, is well suited for regional mapping at small scale within defined floristic zones; but is not equally suited to global mapping, due to being floristically based. In each eco-floristic zone, the presentation of the formations is in the order of the woody types that are the most dense to the types that are the most open. This is an expression of the different degradational stages or progressive evolution (Howard, Kalensky and Blasco, 1985).

The approach adopted by the FAO Remote Sensing Centre in Rome (Howard and Schade, 1982; Howard, Kalensky and Blasco, 1985; Howard and Lantieri, 1987; Howard, 1990) makes maximum use of multiphase/multistage remote sensing data and employs a physiognomic classification, since this is usable world-wide and enables comparisons to be made between comparable vegetation types of different floristic regions. The method involves hierarchical subdivisive classification of vegetation based on its physiognomy as observed on satellite imagery, photo-mosaics and sample stereo-pairs of aerial photographs.

Stepwise, the revised approach assumes that the most appropriate method of classifying vegetation is (a) to identify the plant formations in terms of the macro-climatic zone in which they occur (cf. Holdridge). Each zone is based on mean annual temperature, annual total precipitation and humidity province (i.e. arid, semi-arid, sub-humid, humid); (b) to delineate on the remote sensing imagery each plant formation in terms of its physiognomy, including lifeforms (but not floristics); (c) to divide the plant formations into land systems (section 14.3) or into plant sub-formations according to the stand structure of the vegetation, including the sub-dominant structure. The vegetation is then classified as natural or induced and preferably followed by

Table 14.2   A classification suited to the global remote sensing of forest vegetation and forest change (see text)

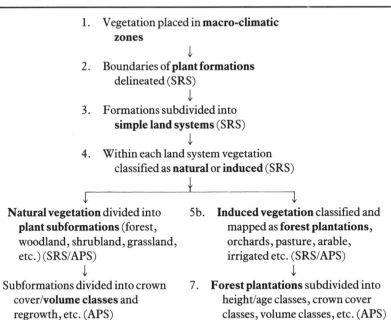

1. Vegetation placed in **macro-climatic zones**
    ↓
2. Boundaries of **plant formations** delineated (SRS)
    ↓
3. Formations subdivided into **simple land systems** (SRS)
    ↓
4. Within each land system vegetation classified as **natural** or **induced** (SRS)

5a. **Natural vegetation** divided into **plant subformations** (forest, woodland, shrubland, grassland, etc.) (SRS/APS)
    ↓
6. Subformations divided into crown cover/**volume classes** and regrowth, etc. (APS)

5b. **Induced vegetation** classified and mapped as **forest plantations**, orchards, pasture, arable, irrigated etc. (SRS/APS)
    ↓
7. **Forest plantations** subdivided into height/age classes, crown cover classes, volume classes, etc. (APS)

SRS: satellite remote sensing. APS: aerial photographic sensing.

multistage sampling, including stereoscopic examination of contiguous pairs of aerial photographs; (d) to ensure that the final classification is open-ended, so that, if required, further attention can be given to seral and induced vegetation, the floristics of plant communities, dominant timber species, plant succession, regrowth and habitat. Reference should be made to Table 14.2 (Howard, 1990), which at level 4 becomes phytogeomorphic with the introduction of land systems.

The term formation describes a group of many plant communities having a recognizable common lifeform. A plant formation is the major vegetation unit within a geographic region, to which are referred all climax plant communities exhibiting similar lifeforms. Its sub-division on the basis of stand structure is the sub-formation. Each sub-formation has a distinctive and easily recognizable structure. Based on macro-climatic criteria, two or more contiguous plant formations may be grouped into a larger geographic unit termed a **pan-formation** (e.g. boreal forest in northern Europe). In western Tanzania, for example, the savanna woodland formation, 'miombo', is hierarchically divided into subformations (e.g. closed woodland, open woodland, scattered tree savanna). The severity of the die-back of forest types now being observed in Europe and North America seems often to be associated with the tree species growing within a recognizable plant formation.

Within a plant formation or plant subformation or land system, **regrowth** replacing mature or over-mature woody vegetation can be divided on the imagery into economic classes or size-classes: seedlings, saplings, poles, timber. Forest **succession** can be described using imagery as 'progressive' or 'regressive'; and fire damage (**pyric**) as current (this year), recent (i.e. within the last few years) or old.

Physiognomic characteristics of woody vegetation images

The three main physical characteristics of woody vegetation obtained from the stereoscopic images and used in physiognomic classification are their qualitatively recognizable lifeforms, and the mathematically definable parameters of stand height and crown cover. To these are added the correlated parameters of crown diameter and height/crown width ratio. Typical lifeforms are coniferous forest, deciduous broad-leaved forest, sclerophyllous forest (hard leaved) and rainforest (hygrophilous).

There is growing recognition of height when choosing between crown cover and height as the first criterion of vegetal classification, since it is not so strongly influenced by human activities (Howard, 1970a, d). Figure 14.3 provides examples of the stereoscopic profiles of forest types based on stand height and crown cover; and Table 14.3 provides an example of the stereoscopic classification of forest subformations based on lifeform, stand height and crown cover.

As already mentioned, stand height is obtained from the parallax measurement of a reasonable number of dominant or co-dominant trees, as the trees would be selected on the ground. Height based on parallax measurements can also be used as the criterion for identifying the lifeforms of mature vegetation, as described for field observations by Raunkiaer (1934).

Following height, as the first parameter, although there are no generally accepted limits for crown cover classes, it is useful to class woody vegetation as having over 80% ground cover (closed forest), 40–80% (fairly dense/dense), 10–40% (open), 10–40% (sparse) and under 2% (trees rare or absent). Often for timber stands 10% classes are used. For comparison, Table 14.4 shows the albedos obtained for woody vegetation types along a south-to-north flight transect of approximately 150 km in central Victoria, Australia and extending from the high rainfall wet sclerophyllous forest to semi-arid saltbush (Howard and Barton, 1973; Howard, 1974). The albedos were distinctive for stands of different physiognomy; and for mature natural vegetation, there is a conspicuous trend in the albedo values with increasing/decreasing stand height, crown closure and observable biomass.

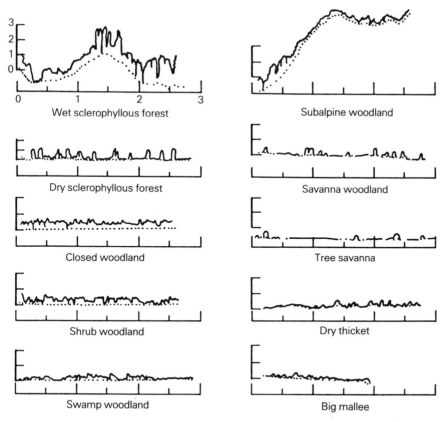

Figure 14.3 From the stereoscopic images of woody vegetation, profiles can be drawn as shown above, which helps in classifying the forest types (south-eastern Australia). The x-axis shows the horizontal ground distance in 100 m and the y-axis woody vegetation heights in 30 m steps. The dotted line indicates the ground plane.

Land cover classification

This extends beyond the classification of vegetation, since it includes urban cover, barren lands, etc. The most widely used classificatory key is that developed by the US Geological Survey (Anderson *et al.*, 1976). The classification is designed for application at four levels, and satisfies the three basic criteria of giving common names to objects or groups of objects, transmitting information about those objects and allowing by inductive reasoning generalizations to be made. Level I is assumed to be obtainable by Landsat MSS, level II by high-altitude aircraft using a 150 mm camera lens (e.g. scales exceeding about 1 : 120 000), level III by photography at about

Table 14.3 Principal forest subformations and their photogrammetric parameters in south-eastern Australia

Lifeform (formation)	Subformation	Stand height (metres)		Crown cover (%)		MS crown width	H:MSCW ratio	Photographic tonal range	
		Main storey	Under-storey	Total	Main storey			MS	US
Forest									
Rainforest	Temperate	30 ± 8	11 ± 6	~100	35 ± 10	13 ± 3	2.5	0.2–0.7	>0.7
Sclerophyllous forest	Wet scler.*	55 ± 16	16 ± 10	90 ± 10	75 ± 15	15 ± 5	>3.2	0.3–0.1	>0.5
	Dry scler.	35 ± 8	<5	65 ± 20	60 ± 25	13 ± 5	2.4	0.3–0.0	<0.5
Woodland	Dense (closed)	18 ± 6		45 ± 15	45 ± 15	10 ± 3	1.8	0.3–0.5	<0.3
	Layered	18 ± 5	10 ± 2	50 ± 10	30 ± 10	11 ± 2	1.6	0.3–0.7	>0.3
	Shrub	16 ± 3	<5	50 ± 10	35 ± 10	11 ± 3	1.5	0.2–0.5	>0.4
	Swamp	16 ± 5	<5	35 ± 10	35 ± 10	11 ± 3	1.6	0.3–0.7	<0.3
	Open (savanna)	15 ± 3		25 ± 5	25 ± 5	13 ± 2	1.1	0.5–0.7	<0.2
	Heath	12 ± 3	<2	35 ± 10	35 ± 10	10 ± 2	1.2	0.5–1.0	>0.5
	Tree savanna	10 ± 3		30 ± 15	30 ± 15	8 ± 2	1.2	0.3–1.0	<0.5
	Subalpine	10 ± 2		15 ± 10	15 ± 10	8 ± 2	1.2	0.5–0.7	<0.3
Tall shrubland (scrub)	Thicket (wet)	8 ± 3		60 ± 15	60 ± 15			0.7–0.1	>0.7
	Thicket (dry)	6 ± 3		45 ± 15	45 ± 15			0.6–0.8	<0.7
	Big mallee	8 ± 2	<2	50 ± 10	50 ± 10	<5	<1	0.7–1.0	<0.3
	Mallee	6 ± 2		30 ± 10	30 ± 10	6 ± 2	<1	0.5–0.7	<0.1
	Savanna	6 ± 2		25 ± 5	25 ± 5	5	<1		<0.2
	Mallee heath								
Low shrubland	Shrub-steppe	<2		<40	<40	<5		<0.3	<0.1
	Heath (wet)							0.7–0.3	
	Heath (dry)							0.3–0.7	

Table 14.4 Results of radiometric measurements from low-level light aircraft showing the relationship between albedos (in brackets) and different woody vegetation types according to height class and gross cover of the woody vegetation

	Lifeform			
	Megaphanerophytes	*Mesophanerophytes*	*Microphanerophytes*	*1 Nanophanerophytes* *2 Chamaephytes*
Site class – Raunkiaer (Height – metres)	>30	8–30	2–8	1 0.25–2.0 2 0.25
Gross cover (dominant woody strata) c. 100%				
Dense e.g. 75%	Wet sclerophyllous forest (0.079) Dry sclerophyllous forest (0.090)	Tall woodland (0.098) Cypress pine woodland (0.116)		
e.g. 10–50%	Dry sclerophyllous forest (0.105)	Shrub woodland (0.125)	Shrubland Big mallee (0.096) Savanna mallee (0.100)	Steppe Saltbush (0.140) Blue bush (0.155)
Open e.g. 10%		Tree savanna (0.165)		Grassland (0.155)

Table 14.5 Land cover classification for use with remotely sensed data – levels 1 and 2 (USGS)

Level I	Level II
1 Urban or built-up land	11 Residential
	12 Commercial and services
	13 Industrial
	14 Transportation, communications and utilities
	15 Industrial and commercial complexes
	16 Mixed urban or built-up land
	17 Other urban or built-up land
2 Argicultural land	21 Cropland and pasture
	22 Orchards, groves, vineyards, nurseries and ornamental horticultural areas
	23 Confined feeding operations
	24 Other agricultural land
3 Rangeland	31 Herbaceous rangeland
	32 Shrub and brush rangeland
	33 Mixed rangeland
4 Forest land	41 Deciduous forest land
	42 Evergreen forest land
	43 Mixed forest land
5 Water	51 Streams and canals
	52 Lakes
	53 Reservoirs
	54 Bays and estuaries
6 Wetland	61 Forested wetland
	62 Non-forested wetland
7 Barren land	71 Dry salt flats
	72 Beaches
	73 Sandy areas other than beaches
	74 Bare exposed rock
	75 Strip mines, quarries and gravel pits
	76 Transitional areas
	77 Mixed barren land
8 Tundra	81 Shrub and brush tundra
	82 Herbaceous tundra
	83 Bare ground tundra
	84 Wet tundra
	85 Mixed tundra
9 Perennial snow or ice	91 Perennial snowfields
	92 Glaciers

3000 m to 12 000 m flying heights and level IV photography below 3000 m (i.e. scales larger than about 1 : 20 000).

At each level, it is assumed that an interpretation accuracy is attainable at about 85% for the categories of land cover (Table 14.5) and involving more than one interpreter. Level I and level II are principally of interest to users of satellite data at national, regional and international levels. In the context of forestry, level I calls for the division of the land into forested and non-forested. At level II, the forested lands are divided into coniferous, deciduous and mixed forest and rangelands. Shrublands (e.g. chaparral, mesquite), due to their spectral response, are grouped with rangeland. The ecological concept separating rangeland from pasture is that the former supports climax vegetation (i.e. shrubs, forbs, grasses), while pasture is improved grassland.

Levels III and IV were designed for local use using aerial photography. In the context of forestry, this would imply at level III the division of the forest lifeforms into at least forest types or tree species groupings. That is, the classification is at genus level (e.g. spruce, pine); but, for reasons given in the next chapter, levels III and IV are not well suited to the natural forests of the tropics and much of the southern hemisphere. Level IV in forestry would include recognizing site quality classes by species and age classes.

There is no constraint on the system, which precludes its use with the subdivisive hierarchical classification of vegetation outlined in the last section and with land unit classification as will now be discussed. Level III often equates with the plant subformation and land facet and level IV with the land element. When commencing image analysis, as observed by the writer, it is often useful to divide levels I and II into two groups (i.e. hygrophylous, non-hygrophylous). This takes into account the very low spectral reflectivity of water (cf. Chapter 3).

14.3 COMBINING LANDFORMS AND VEGETATION – A PHYTOGEOMORPHIC APPROACH TO FOREST LANDS

The concept of a phytogeomorphic approach to land use and land potential calls for the study, simultaneously or in sequence, of landforms and vegetation and their interdependent relationships (Howard and Mitchell, 1980, 1985). Vegetation and landforms are complementary in providing forest information, whether this is based on field collected data or image analysis. In the context of applying aerial photointerpretation to the classification of the landscape, the initial work was that of the forester Ray Bourne in the late 1920s and early 1930s, who worked in Zambia (1928) and at the University of Oxford. Bourne (1931) recognized land units of three different magnitudes. This can be viewed as the forerunner of **hierarchical** subdivisive land unit **classification** (cf. Brink et al., 1966). Bourne recognized the need to divide the surface of the earth into natural regions of uniform character, and suggested that aerial photographs should be used to identify distinctive unit regions within larger physiographic regions. As provided by geological maps, he proceeded to map unit regions (i.e. the forerunner of land subprovinces/land regions) and unit sites. He viewed an association of sites as constituting a distinct region (i.e. **land system**) and that a site (cf. **land facet**) for all practical purposes has similar physiography, geology, soil and edaphic factors.

Whether emphasis is placed on landform or vegetation in recognizing a land unit in a specific locality will depend primarily on the magnitude of the unit, but also on the variations in the landscape and the training and experience of the worker. As pointed out earlier, what is important is the simultaneous examination of the two factors, since they synthesize important land characteristics that are readily observed on the imagery. In the hierarchy of units, the geomorphic factor is considered more significant earlier in the classification of the forested landscape than the vegetal factor, but after macro-climate, macro-physiography and geology. It is more expressive and unifying of the larger units of the landscape than vegetation and plant communities, although the unit boundaries on the imagery will be enhanced by the natural vegetation and the vegetal land-use patterns it contains. At the level of the land province and land system (Table 14.6), emphasis on vegetation is likely to result for mapping purposes in the fragmentation of the area into undesirable small units, which are difficult to map or to recombine for management/planning purposes. At the level of the land system or land facet, it is preferable to distinguish between natural and induced vegetation types. For the smaller land unit (i.e. land facet), the physiognomy or external structure of the woody vegetation on aerial photographs, and not floristic groupings (e.g. plant associations), is sufficient for the mapping of these land units.

Classificatory land units

The term **land unit** is favoured as a general term to be used when referring to a landscape unit of any magnitude in the same way as **plant community** is applied to distinctive plant groupings of any size. Information on land units, based on subdivision of the landscape, as observed on aerial photographs and satellite imagery, is provided in Table 14.6 (Howard and Mitchell, 1980).

A subdivisive hierarchy of several terms, although appearing complicated, is justifiable in that it gives robustness to the classification, rationalizes the landscape for planning and management and enables the survey to be terminated at any level. It may then be pursued later in more detail at a lower level, without having to commence the work again. For GIS, the ranking of disciplines facilitates reconciling map boundaries. The disadvantage of a classification based on synthesis and not subdivision would be that classificatory errors made with the smaller units are magnified in synthesizing larger land units, but it is quite justifiable for management to synthesize larger management units from the identified and mapped smaller units.

Plate 1 Film types used in the study of forest resources. Note the differences in image contrast, sharpness and texture which vary with the film type: 1: bracken and grasses; 2: water tank; 3 and 4: forest; 5: dirt road but partly sealed (scale about 1:15 000). In the infrared colour the green vegetation records reddish while other materials are bluish or straw-coloured (see also p. 63).

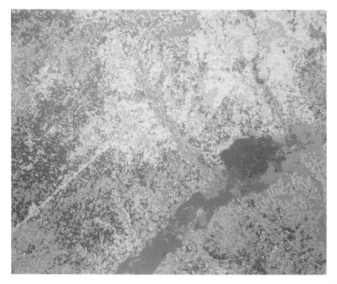

Plate 2 For comparison with Figure 10.4 (a) and (b), colour composite in three spectral bands (green, red, infrared sensitive) obtained by supervised digital image classification of early dry season Landsat data. Nine spectral classes are represented at a scale of 1:250 000. The finalised thematic map (not shown) was comparable to a thematic map produced at the same scale using 1:40 000 black-and-white aerial photographs.

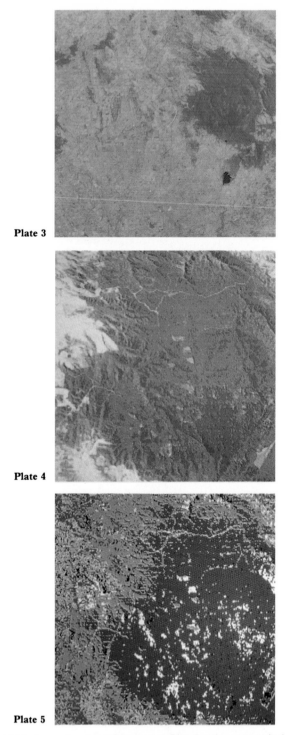

Plate 3

Plate 4

Plate 5

Plates 3, 4 and 5 Hard-copy Landsat MSS products. Plate 3: colour composite bulk product 70 mm transparencies and additive viewing compared with hard-copy products of the forested area obtained by digital analysis. Plate 4: unsupervised classification; Plate 5: principal component analysis. The land sub-province can be divided into two land systems, principally by the differences in the hue and texture of the forest cover and drainage patterns; but the recognition of smaller land units is erratic. Figures 14.6 (a) and (b) show the same area using (analogue) aerial photographic products (see p. 263–4).

Plate 6   **Plate 7**

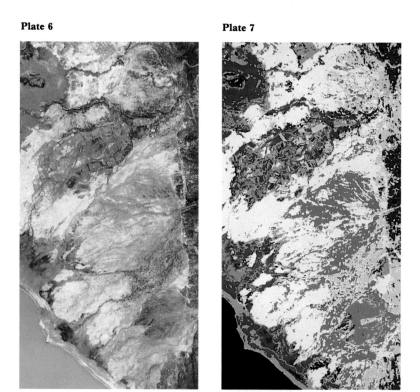

Plates 6 and 7 A comparison of hard-copy imagery (SPOT MS Lake Baringo, Kenya). Plate 6: maximum-likelihood classification of the vegetation without mapping the land-system boundaries; Plate 7: maximum-likelihood classification after delineating the land-system boundaries. Note the improved information content in Plate 7.

Plate 8 High altitude colour infrared photograph (part, 10 x) taken at a nominal scale of about 1:75 000. Shown are remnant rainforest (bottom right, coarse texture, deep red), regrowth forest of several age classes (intermediate textures and hues), areas returning to forest after agricultural crops (smooth texture, light orange – red), and areas cleared and used for temporary agriculture (smooth texture, greyish). Note the distinctive dendritic drainage lines with an absence of forest.

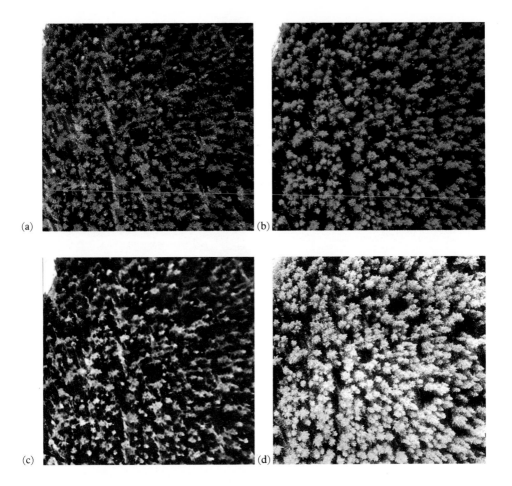

(a) (b)

(c) (d)

Plate 9 Comparison of four film-filter combinations taken with a 35 mm cluster camera at a flying height of 240 m showing the die-back, due to drought, of *Pinus radiata* (1:5000 scale) (Victoria, Australia). About 10% of the 16-year-old trees affected by drought were not identified on the normal colour photographs, as compared with 8% omitted by ground survey. As compared with the count of infected trees on the normal colour photographs (Ektachrome-X with UV filter), only 56% were counted on the colour IR photographs (Ektachrome Infrared Aero type 8443 with Wratten 12 filter) and 38% on panchromatic black-and-white prints (Plus-X with Wratten 12 filter). Infrared black-and-white photography (HSIR film with Wratten 89b filter) was found unsuitable. With colour IR, unhealthy crowns often registered bluish and thus, were difficult to separate from some shadows and some dead ground vegetation. (a) Normal colour; (b) colour IR; (c) panchromatic black-and-white; (d) infrared black-and-white (see p. 286).

Plate 11 (opposite) Computer-supervised classification of Landsat TM data (1984) showing pollution damage of forest cover in the vicinity of the German-Czech border. Deep red: heavy damage; magenta/mauve: moderate damage; and green: slightly damaged/healthy (conifers — mainly Norway spruce). Deciduous forest has recorded in yellow. Note the lake: light-blue (see p. 289).

Plate 10 Part of a 70 mm normal colour photograph (Ektachrome-X) at a scale of 1:1300 showing die-back of *Pinus radiata*. Five crown condition classes were identified (see text, p. 286). Note the ground control markers in yellow plastic.

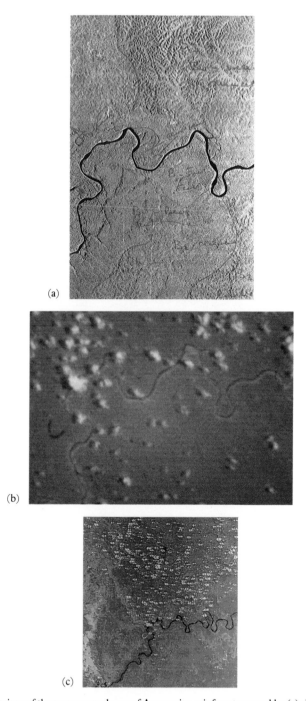

(a)

(b)

(c)

Plate 12 Comparison of the same ground area of Amazonian rainforest covered by (a) side-looking airborne radar (SLAR) — 40–60 m ground resolution; (b) 'bulk product' Landsat MSS colour composite (green, red, IR sensitive bands), 79 m ground resolution; and (c) digitally-analysed Landsat MSS scene (supervised classification) comprising approximately 800 000 pixels. The SLAR scene has been stratified using a chinagraph pencil into forest types prior to field checking. Note that the SLAR provides considerable information on the terrain and shows forest clearings along the river. The bulk product provides negligible information, but the digital imagery provides some information on type boundaries with changes in ground elevation (to the left). The speckling is caused by clouds, which with SLAR are absent due to its cloud penetrating capability. Yellowish: savannah woodland; reddish: scrub/shrubland; green: rainforest; black: river; blue: (human) induced vegetation; light-blue circles: clouds.

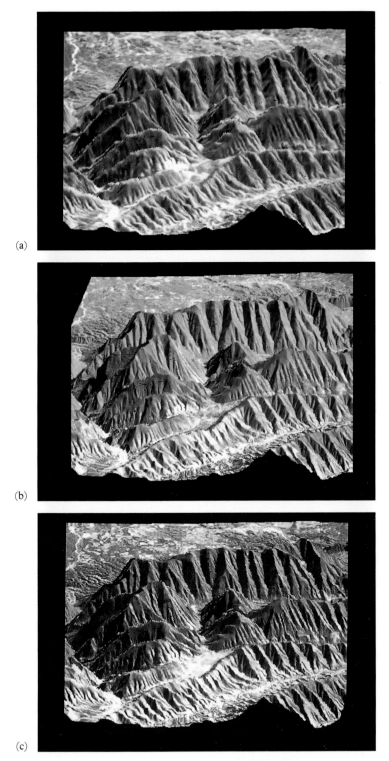

(a)

(b)

(c)

Plate 13 Digital elevation models of the same area using (a) Landsat MSS (channels 4-2-1) (green, red, near-IR sensitive bands), (b) Landsat TM (channels 7-5-2) (red, near-IR, mid-IR sensitive) and (c) SPOT MS (channels 3-2-1) (blue, green, red sensitive). (See also text, p. 325.)

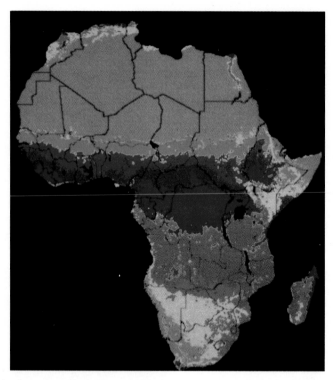

Plate 14 Environmental satellite image composite in vegetation index format covering eight periods of 21 days and showing plant cover of Africa in the period April 1982 to February 1983. The feature space used to label the land-cover classes is shown in Figure 17.4.

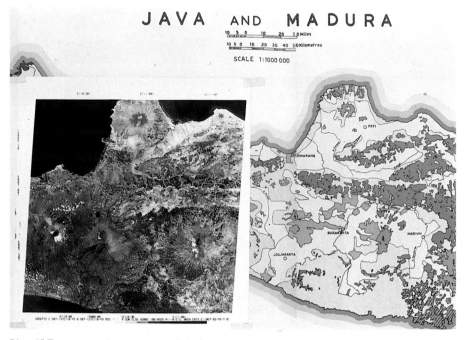

Plate 15 Forest cover change over a period of about 20 years provided by a comparison of Landsat MSS imagery and the existing map. The forest cover has declined by about 40% (central Java).

The **land system** (or equivalent) is judged to be the most widely used land unit at the present time associated with remote sensing, although on occasions the term system may be misunderstood. It is useful, for example, in defining strata boundaries for forest inventory. It is primarily a geomorphic unit having a predominantly uniform geology, climate (past and present) and a characteristic soil association. A first approach to land system classification was devised in 1946 by the Australian CSIRO Division of Land Research (Christian and Stewart, 1947). These authors, working in reconnaissance land surveys in northern Australia, described a simple land system as a group of closely related topographic units, usually small in number, that have arisen as a product of common geomorphic phenomena and are appropriate for mapping at scales between 1 : 250 000 and 1 : 1 000 000. They also recognized that simple land systems with affinities in common can be synthesized into larger complex/compound land systems. The equivalent subdivisive unit is the land subprovince (Howard and Mitchell, 1985).

An intermediate unit between the land system and facet is necessary to provide 'robustness' to the classification as otherwise the land system and land facet will vary greatly in meaning, size and composition from place to place. This unit, the **land catena** (Howard, 1970b), consists of a chain of geographically related land facets, and forms the major repetitive component of the land system. The **land facet** is the smallest topographic unit. It is usually easily mapped from aerial photographs, and is distinctive in soil series and vegetation. The land facet/facet corresponds to Bourne's sites and represents the basic practical subdivision of the landscape for forest management.

Figure 14.4 provides a block diagram of part of a land province divided into land systems on the basis of its landforms. Figure 14.5 shows stereoscopic profiles of two land systems within which the land catena and land facets are identifiable. A smaller unit, the land element (Brink *et al.*, 1966), can sometimes be identified on aerial photographs, and in image analysis is considered to be the smallest observable subdivision of the landscape. It is indivisible on the basis of landform/topography, is characterized by its soil type (soil phase) and is identifiable on the imagery only by its plant subformation/plant association or the pattern of land use. Two adjoining land elements may contain vegetation of similar structure but the floristic associations will differ. In Figure 14.4, land systems I and II are characterized by their land facets, while land elements are likely to predominate in land system III. In Figure 14.5 (Howard, 1970c), land facet (valley bottom) is likely to contain two or more land elements.

Land-unit classification in forestry

When applying subdivisive land-unit classification to the forested landscape using remotely sensed imagery, the most important units are those shown with an asterisk (*) in Table 14.6. These are the land system, land catena and land

Table 14.6 Subdivisive categories of land units observable on imagery**

Land unit	Paramount discipline used to identify unit	Comments and description
Macro units Land division	Geography	The synoptic view provided by environmental satellite imagery is valuable for the identification of these extensive land units having a gross landform expressive of continental structure in which the climatic zones are evidenced by the fit of the natural vegetation (panformation) to its continental landforms.
Land province	Physical geography	Major physiographic unit. Recognizable as a distinctive extensive assemblage of landforms expressive of a second-order structure and the fit of the natural vegetation (plant formations). Regional forest and land-use patterns recognizable, which tend to fit the landforms on Landsat MSS.
Land subprovince	Physical geography	A major subdivision of a land province evidenced by the fit of vegetation and land-use patterns within an identifiable grouping of land systems (compound land system). Can map using Landsat MSS, TM, SPOT, etc. Probably useful in global and continental monitoring.
Land region	Geology	A land unit, usually of considerable magnitude, which is identifiable mainly through the image characteristics of its simple or compound land system(s). The land region has surface properties of a geological unit with a small range of surface forms. Vegetation and land-use pattern may fit. Can map using Landsat MSS imagery.

Micro units (useful in forest survey and management)	Discipline	Description
Land system* (simple land system)	Geomorphology	A recurrent landform pattern of geographically and geomorphologically related smaller land units (e.g. land facets). Its image drainage/topographic pattern is distinctive and provides boundaries coinciding with major geomorphic features. These patterns are enhanced by the vegetation. Vegetation and local land-use patterns usually fit. Has characteristic forest types/site qualites. Can map using Landsat TM, SPOT, Salyut/Soyuz imagery and AP mosaics.
Land catena*	Geomorphology/botany	Each land catena contains a recurrent pattern of geographically related smaller land facets to which land use and forest types fit. Stereo-pairs of aerial photographs useful.
Land facet*	Botany/geomorphology	Normally cannot be mapped on Landsat MSS. Use aerial photographs stereoscopically. A land facet comprises a distinctive unit of topography with which is associated an equally distinctive vegetation structure at the level of the plant subformation/forest type. Usually, climatic uniformity can be inferred from the vegetal structure. Local land use usually fits. Has a characteristic soil series/soil association and forest site quality.
Land element	Botany/soil science	Sometimes too small for forest management, but on flood plains can be extensive. For practicable purposes uniform in vegetation (e.g. plant association), climate, lithology, landform, hydrology and soil (i.e. characteristic soil phase/soil type). Uniform in topography. Field mapping usually essential.

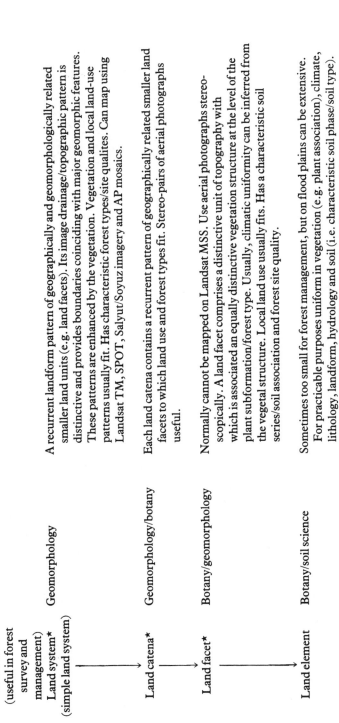

** *Note:* Not shown as the first macro-unit is the 'land zone', which is a climatic unit representing a very extensive part of the earth's surface (e.g. cool temperate zone).

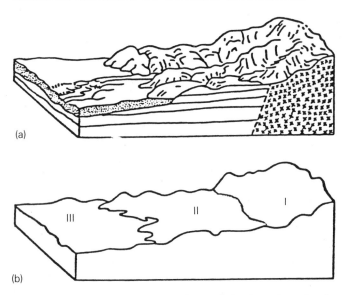

(a)

(b)

Figure 14.4   Simplified block diagram (a) of part of a land province showing two main rock types/land regions (igneous, sedimentary) and (b) the same area divided into three land systems, each of which can be expected to have distinctive forest types/tree species distributions.

facet. Both the land system and land facet provide units within which the tree species are progressively reduced in number. This therefore assists in the photointerpretation of the species, and the stratification of the forest into more homogeneous units. It also reduces the number of ground samples required to achieve an acceptable sampling error, since the strata are more homogeneous. In the tropics, since tree species tend to be abundant per unit ground area and often impossible to identify on aerial photographs, the most effective approach to stratification as a part of forest inventory can be expected to be geomorphic/ phytogeomorphic and not based only on the forest cover. The forest photointerpreter must develop skills in being able to examine the terrain beneath the tropical forest canopy in the stereoscopic model provided by pairs of aerial photographs; and may consider, in this part of the interpretation, that the forest canopy is a nuisance factor or 'noise factor'.

At the level of the land facet, forest site units are obtained which can be expected to be uniform for forest management in terms of the site quality of existing crops and for the choice of species for planting. Frequently, the same results are obtained in the field by using the plant/tree cover types as the indicator of site quality; but in situations where the plant ecology becomes complex due to plant succession etc., as caused by man or natural events, the phytogeomorphic approach is again to be preferred.

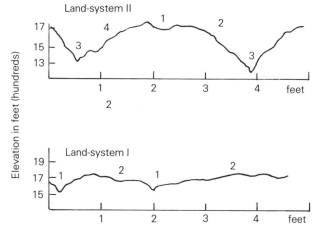

Figure 14.5   From the stereoscopic model, profiles of the terrain can be drawn, which are useful in studying the soil series, the land catena and the distribution of tree species. Land-system II is on granodioritic intrusion forming part of a plateau merging with deeply weathered sediments (regolith) over Silurian sediments (Land-system I). The facets have been numbered 1–4 and 1–2. A land catena is represented by the numbering 3, 4, 1, 2, 3 (Mount Disappointment, Victoria, Australia).

Varying degrees of detail on the land division, land province and land subprovince are observable from the higher resolution environmental satellite imagery (e.g. NOAA AVHRR). Land subprovinces are readily mapped from Landsat MSS imagery, while aerial photographs are not suitable, due to the limited ground area covered by each exposure. If satellite imagery is not available, then an attempt can be made using small-scale aerial photographic mosaics. The (simple) land system is readily identified and mapped using SPOT and Landsat TM data; but is often difficult to map from Landsat MSS imagery. Frequently in publications, what has been termed a land system using Landsat MSS, is in effect a landscape unit usually corresponding to the land subprovince or occasionally land provinces.

For accurate and detailed interpretation of the land catena and land facets, aerial photographs are essential and these should be used stereoscopically. A superficial impression of these units can be obtained from aerial photo-mosaics, sometimes SPOT and Landsat TM and, if available, ortho-photomaps and topographic maps. Usually the contour intervals on topographic maps are insufficient to provide the required geomorphic detail. Forest mapping is only occasionally at the level of the land element, unless the land elements are large. An example is provided by the extensive mangrove forests in the coastal tropics. In coastal Malaysia, there are up to five

(a)

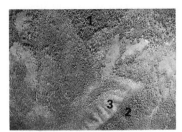

(b)

Table 14.7   Grouping of tree species by land systems (Mount Disappointment State Forest, Victoria, Australia)*

Land system	*regnans*	*obliqua*	*gomiocalyx*	*gomiocalyx hybrid*	*rubida*	*radiata/dives*	*macrorrhynca*	*elaeophora*	*polyanthemos*	*melliodora*	*viminalis*	*ovata*	*aramaphloia*
							Eucalypt species						
I	19	8	6	–	–	–	–	–	–	–	–	–	–
II	–	10	–	3	4	12	8	–	–	–	–	–	–
III	–	7	–	–	–	10	7	9	1	1	1	–	–

* The numbers show the species frequency as occurring in 40 sample plots per land system.

associations of mangroves within the land system controlled by tidal inundation which can be delineated as land elements on aerial photographs or as two elements on Landsat MSS.

These comments will now be illustrated by examples from south-eastern Australia and East Africa (Kenya Highlands). In south-eastern Australia, land-unit classification was applied by the author to the Mount Disappointment State Forest (Victoria) using both visual and digital image analysis (unsupervised, supervised classifications). The forest physiognomy is provided mainly by (evergreen) wet sclerophyllous forest, (evergreen) dry sclerophyllous forest and (evergreen) sclerophyllous woodland; but adjoins to the west and south pasture (grassland), which can be broken down also into land systems, land catena, land facets and land elements. Plates 3, 4 and 5 and Figure 14.6(a) and (b)show the forest recorded digitally by Landsat MSS, the main forest area on an aerial photographic mosaic and smaller land units on an

Figure 14.6   (a) Semi-controlled aerial photographic mosaic of the same area as shown in Plates 3, 4 and 5, at a scale of 1 : 200 000; it is doubtful if the land sub-province/land region would be identified without the information obtained from the satellite imagery; the three simple land systems can be recognized and mapped and land catena and land facets also identified on the folded sediments (left side of photograph); (b) aerial photograph (part) at a nominal scale of 1 : 15 840 (lower right side of mosaic) showing a land catena and land facets dominated by wet and dry sclerophyll forest and sclerophyll woodland. The information to be extracted from the stereoscopic model provides much more information on the terrain including slope and drainage patterns than on satellite imagery. The forest types, forest cover and stand densities can be mapped and stand heights estimated directly from parallax measurements. 1: wet sclerophyllous forest; 2: dry sclerophyllous forest; 3: woodland.

aerial photograph. The two plant formations comprising sclerophyllous forest and sclerophyllous woodland were identifiable on the satellite imagery; but the breakdown of the sclerophyllous forest into its two subformations is difficult without using the geomorphic factors of drainage patterns and topography; and only the land sub-province and two of its land systems are identifiable. On the aerial photograph (part), the wet sclerophyllous forest, the dry sclerophyllous forest and the sclerophyllous woodland can be identified. As shown by Table 14.7, the mapping of the land systems aggregates the eucalypt species into groups. The most important timber species, ash (*E. regnans*), is confined to a single land system. This is not only useful in stratifying the forest ahead of forest inventory, but also, with the reduced number of tree species in each land system, it assists in the photointerpretation of the species and contributes to the design of multistage sampling (Chapter 19).

The second example covers an area of central Kenya (near Lake Baringo). The employed classificatory method is that given in Table 14.2. SPOT data was digitally analysed using unsupervised classification and field checked to determine the degree to which plant formations and plant subformations had been identified and successfully mapped (Lantieri, 1986; Howard and Lantieri, 1987). This was followed by supervised classification (maximum likelihood) without and with the land system boundaries used to stratify the digital data. Land system boundaries were readily identified on the SPOT imagery (20 m, 10 m resolution). The improvement in the mapping of the plant cover is illustrated in Plates 6 and 7. The conclusion was reached that SPOT data with its spatial resolution of 20 m, is well suited to the mapping of (simple) land systems and that this, as a form of stratification, improves the identification of forest types/land-use categories. In finalizing the study using remote sensing data, it was found that dichotomizing the vegetation into natural and induced is most opportune after mapping the (simple) land systems.

15

Airborne remote
sensing application

In operational airborne remote sensing, attention needs to be given to defining carefully the objectives of the work, the composition of the forest under examination, the relationship of the remote sensing component to other sources of the information to be obtained from the work, the available field support, the training and skills of the image analyst/photointerpreter, the available image analysis equipment and the type(s) of image data to be used including the scale/resolution of the imagery. As work progresses, the image analyst/photointerpreter must field check each successive stage.

Up to the present time, there has been no substitute for the human brain of the experienced forest photointerpreter for extracting the most appropriate quantitative and qualitative usable information from aerial photographs. In the application of airborne remote sensing to operational forestry, computer-assisted techniques have been little used, except in the processing of SAR imagery and the occasional thermal data collected by electro-optical scanner. However, the practical use of digital imagery analysis is likely to expand in the future. This is due to several major factors including the influence of digital analysis developed for satellite remote sensing, the progress made in digital mapping, the growing importance of geographic information systems as a management tool, the need in monitoring to merge aerial photographic and satellite data and the possibility of digitally handling texture in addition to spectral radiance values.

15.1 CONSTRAINTS OF THE FOREST COMPOSITION

It is frequently not appreciated that the influence of the composition of the natural forest on image analysis commences at the regional level. Probably when applying remote sensing image analysis techniques, the forest cover of the world should be typified as comprising at least three zones (Howard, 1970a; American Society of Photogrammetry, 1983). These are the tropics, the boreal forests of the northern hemisphere and the evergreen (sclerophyllous) forests of the southern hemisphere. Further zonation would separate other regions of the northern hemisphere, and, on the basis of management, separate many of the temperate forests of North America from the temperate forests of Europe with their long history of silviculture.

Tropical forests

Nearly half of the world's forests are in the tropics, and within most of these forests, there is a multitude of tree species of many families. Many of the species seem to compete at regeneration for the same site. By chance and in the same locality, a single tree species may be at different phenological stages and develop polymorphic crown shapes. Consequently, the direct identification of tree species on the aerial photographs is usually impossible. Several hundred tree species may occur within $1\,km^2$, and within tropical rainforest up to 100 species have been recorded per hectare of land.

Over extensive areas of tropical Africa and tropical South America, where the annual rainfall is less than $1000-12\,000\,mm$, the high forest is often replaced by savanna woodland; but even these forests are rich in tree species. The author, for example, when establishing and inventorying the Kigwa-Rabugwa Forest Reserve in western Tanzania recorded on the ground over 80 tree species in an area of about $100\,km^2$, and had to confine the use of the aerial photographs to stratifying the forest into types based on land systems and land facets (Howard, 1958). Special inventory forms were designed to accommodate recording the large number of species, with emphasis on the commonest species, their size classes and their economic importance.

Tropical rainforest, unlike tropical woodland which is usually single storeyed, can be characterized as having three tree storeys, a well-developed shrub layer, a sparse herb layer and interlaced crowns. Richards (1952), quoting Schimper (1903), defined rainforest as evergreen, hygrophilous in character, at least 30 m high, but usually much taller, rich in thick-stemmed lianes and in woody as well as herbaceous epiphytes. Occasionally, where environmental factors are limiting to the development of rainforest, the forest may have locally a single dominant tree species.

The (tropical) montane forests differ structurally and floristically from the tropical rainforests of the lowlands by a reduction in the number of tree species, decreasing tree-and-stand heights with increasing elevations, the disappearance of emergent tree species above the top or A-storey, less interlacing of the crowns, fewer or an absence of lianes and a reduction in tree storeys from three to one or two. However, the large number of tree species continues to make their photographic identification very difficult; and it is only at highest elevations with the occurrence of consociations that this may be practicable (e.g. *Juniperus procera* in East Africa).

Boreal/cool temperate forests

Probably the best results on using remote sensing have been achieved in the boreal forests of Canada, northern USA, northern Europe and Siberia. The forests of north-eastern China (formerly Manchuria) also resemble these forests in structure and genera. As pointed out by Isaev (1986), the forests in Siberia are vast, remote and relatively homogeneous in physiognomy and species (i.e. pine, spruce, larch, aspen and birch). In boreal forests, there are few tree species per unit area of land, the standing trees are fairly widely spaced, permitting the ground to be readily seen on aerial photographs, crowns do not interlace, the tree shadows can be observed and measured with snow on the ground, the heights of individual trees measured using the parallax formula, their crown diameters measured and, if available, stand volumes estimated using aerial volume tables (Section 16.4).

A perusal of literature relating to the interpretation of tree species, forest types, etc. using remote sensing will show that many of the present-day techniques were developed in the boreal/cool temperate forests of North America and to a lesser degree in northern Europe. When the various techniques have been applied without further testing to the warm temperate forest of the northern hemisphere and the southern hemisphere, the results have varied greatly and have sometimes been disappointing. Often the crowns of the different tree species are much more similar than for species of the cool temperate northern hemisphere.

Evergreen (sclerophyllous) forests of the southern hemisphere

In Australia there are over 500 species of the single sclerophyllous evergreen genus, *Eucalyptus*, and few gymnospermous species. Thereby, the role of image analysis in forest inventory tends to be intermediate between that of the tropical forests and cool temperate forests.

Referring back to Chapter 3, it will be recalled that since tree foliage is not a uniform diffuse reflector, it is not surprising that the erectophilous eucalypt leaves, although angiospermous, provide radiance values markedly different from those of many angiospermous planophilic tree species of the northern hemisphere. In general, the spectral reflectivity of eucalypts appears to be similar to that of many northern hemisphere conifers. Failure to recognize this fact resulted in the 1960s in infrared black-and-white film being used in the national forest inventory of Portugal in expectation of being able to separate two important species, namely the angiospermous *Eucalyptus globulus* and the gymnospermous *Pinus pinaster*. The results were so disappointing that it led to the abandoning of IR photography and to relying on the standard panchromatic black-and-white photography. Again in the 1960s, IR and modified IR black-and-white photography were used experimentally without success in Australia in an attempt to separate the gymnosperm *Callitris glauca*

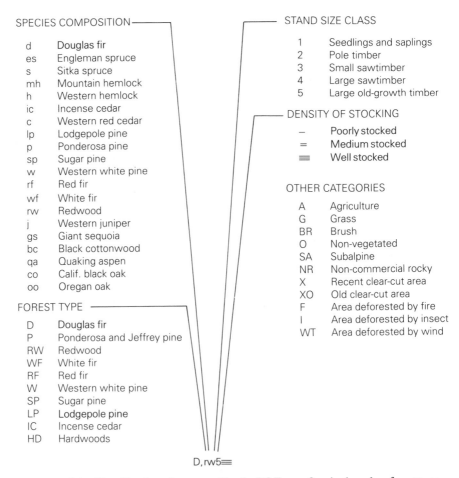

SPECIES COMPOSITION

d	Douglas fir
es	Engleman spruce
s	Sitka spruce
mh	Mountain hemlock
h	Western hemlock
ic	Incense cedar
c	Western red cedar
lp	Lodgepole pine
p	Ponderosa pine
sp	Sugar pine
w	Western white pine
rf	Red fir
wf	White fir
rw	Redwood
j	Western juniper
gs	Giant sequoia
bc	Black cottonwood
qa	Quaking aspen
co	Calif. black oak
oo	Oregan oak

FOREST TYPE

D	Douglas fir
P	Ponderosa and Jeffrey pine
RW	Redwood
WF	White fir
RF	Red fir
W	Western white pine
SP	Sugar pine
LP	Lodgepole pine
IC	Incense cedar
HD	Hardwoods

STAND SIZE CLASS

1	Seedlings and saplings
2	Pole timber
3	Small sawtimber
4	Large sawtimber
5	Large old-growth timber

DENSITY OF STOCKING

–	Poorly stocked
=	Medium stocked
≡	Well stocked

OTHER CATEGORIES

A	Agriculture
G	Grass
BR	Brush
O	Non-vegetated
SA	Subalpine
NR	Non-commercial rocky
X	Recent clear-cut area
XO	Old clear-cut area
F	Area deforested by fire
I	Area deforested by insect
WT	Area deforested by wind

D, rw5≡

Figure 15.1   Classification scheme used by the US Forest Service based on forest type, species composition, size classes and stocking density. The condensation of the information into a small group of symbols (e.g. Drw5,≡) facilitates annotating the aerial photographs and forest stockmap.

from eucalypt species and in New Zealand to separate *Podocarpus* from *Nothofagus*.

In general, the natural forests of this category are delineated on the aerial photographs by size classes of species groupings using stand height and crown cover as the criteria, and no attempt is made to interpret the species unless growing as consociations. This is followed by the field inventory, when the tree species are identified separately. As in the tropics, age classes are not readily determined, since the trees do not have annular rings. It is therefore incorrect under such conditions to refer to forest stands by age classes, and classification should be by height classes or size classes.

15.2   AERIAL PHOTOINTERPRETATION OF FOREST TYPES

Probably in the initial stages of any forest survey, an attempt will be made to identify directly from the photographs the objects of interest, whether these are individual tree species, groupings of tree species, forest stands differing in crown density/volume or dead, dying, diseased and damaged trees and stands. Failure to attain acceptable results by direct identification will lead to classification of the images into recognizable groups within which sampling must be carried out in the field. The overall operation is often referred to as typing and coding or simply **forest typing**. An example of the symbols used in forest typing by the US Forest Service is shown in Figure 15.1. This is based on forest stand types, species composition, size classes and stocking.

When classifying the data contained in the photographs, relatively few criteria should be chosen, as the combinations provided by even a few important criteria can easily result in a map legend with a hundred or more different classes! Also, a minimum area is usually fixed for each type delineation, below which areas are not broken down. For example, Seely (1960) pointed out that in forest inventories in Canada, the minimum area of stand delineation varies between 5 and 40 ha. On finalized maps, as the smallest recorded area is usually $2\,mm \times 2\,mm$, this influences the smallest areas to be interpreted and delineated on the aerial photographs.

If forest succession is an important factor due to previous logging operations, shifting cultivation, damage caused by natural factors, etc., these must be taken into consideration in the photo-typing. At an early stage of the work in natural forest, the photographs should be examined in conjunction with field checking to ascertain if any of the tree species are gregarious and if the stands are homogeneous in log classes.

15.3   AERIAL PHOTOGRAPHIC SCALE

In the practice of forest photointerpretation several information thresholds can be detected according to the photographic scale or image resolution. These thresholds, based on personal experience (Howard, 1990), are summarized in

Table 15.1   Forest information thresholds over a range of aerial photographic scales

Photographic scale	Forest characteristics	Diagnostics/remarks
1. 160 000/120 000 (micro-scale)	Forest/non-forest; evergreen/deciduous/mixed; some size classes; occasional forest types	Tone; colour contrast; textural contrast; site
2. 90 000/75 000 (micro-scale)	As 1; stand size classes; stand physiognomy; some forest types	As 1; stand height; crown cover; crown sizes (qualitative)
3. 30 000/25 000 (medium scale)	As 2; more detailed	As 2; API easier
4. 20 000/10 000 (large scale)	As 2; tree sizes; forest types; stand volumes; tree volumes; some tree species; understorey data	As 2 but quantitative; photogrammetric parameters (e.g. tree height)
5. 10 000/8000 (large scale)	As 4; stand composition by tree species	As 4; colour/tone contrast within tree crowns
6. >5000/4000 (macro-scale)	As 5; individual tree species	As 5; branching habit
7. >2000 (macro-scale)	As 6; tree species identification most reliable tree volumes; improved shrub characteristics	As 6; crown details

Table 15.1. Frequently only standard photogrammetric photography is available at scales between about 1 : 30 000 and 1 : 80 000. Photographs at scales of 1 : 120 000 to 1 : 160 000 provide little information on the structural composition of the forest stand, and probably can be equated in usable information content with the fine resolution satellite imagery that is now available. In the range of scales between about 1 : 75 000 and 1 : 90 000, taken at high altitude and using CIR film, details of the stand structure can be studied in the stereo-model; and the usable information content extracted from the model may be comparable with that obtainable from old panchromatic black-and-white photography taken at a scale of about 1 : 25 000. To obtain photogrammetric parameters of the stand at an acceptable level of accuracy for fieldwork and for aerial volume tables, the well-established scale range of about 1 : 10 000 to 1 : 15 000 is needed, although under favourable conditions a scale of 1 : 20 000 may be acceptable. Tree species identification seems to

be little improved over a wide range of scales, but a marked improvement with colour photographs is often observed in the scale range of 1 : 10 000 to 1 : 8000 and at larger scales. For the same reason, scales larger than about 1 : 4000 are preferred in studies of disease and insect damage; and also at these macro-scales shrubs can be adequately studied. Except for cost, scales larger than 1 : 2000 would often be preferred for identifying forest species, provided they can be combined with photography at small scale and/or possibly remotely sensed satellite data.

Tropical forests

The largest scale photographs available may be at 1 : 30 000, but are more likely to be as small as 1 : 50 000/1 : 60 000, which have been taken for mapping. Aerial photography at smaller scales is particularly prevalent in the tropics, due to economic factors, and this adds to the problem of having forests rich in tree species and of which few species are economically exploitable. In such circumstances, consideration should be given to using recent good-quality satellite imagery, since the forest information content of the aerial photographs may be very limited and no better than satellite data.

Some workers in tropical forestry have favoured the use of larger scale photographs for photointerpretation, particularly in intensive surveys (Hannibal, 1952; Loetsch, 1957; Swellengrebel, 1959). For example, Swellengrebel, using photographs at a scale of 1 : 10 000 in Guyana, was able to identify groups of *Mora excelsa*, *M. gongrrijpii* and *Ocotea rodiaei* (greenheart). In Sri Lanka, de Rosayro (1958) concluded that a scale of 1 : 20 000 was to be preferred for larger scale forest mapping and interpretation, but is possibly not justified economically due to the paucity of readily marketable species. Francis (1957) commented that for tree species identification and the examination of the tree crowns a minimum scale of 1 : 20 000 is needed and 1 : 10 000 may be preferable. However, in Tanzania, panchromatic black-and-white photographs at 1 : 10 000 and 1 : 15 000 were tried in 'inventory' studies of tropical deciduous woodland species, but were found to be of little help in the identification of tree species (Howard, 1959, 1970a).

Consociations are normally readily recognizable on aerial photographs at all scales, and often also on satellite imagery provided the area is represented by several pixels. For example, in Brazil consociations of *Mora excelsa* and pure upper storeys of *Goupia glabra* were identified on 1 : 40 000 photographs (Heinsdijk, 1957). Occasionally, a genus or a species is photographically distinctive in rainforest, although growing in association with many others. *Dipterocarpus* as a genus is often recognizable by its large crown and characteristic light photographic tone produced by shiny leaves (e.g. Loetsch, 1957). In New Guinea, at high elevations, Boon (1956) was able to identify the conifers *Agathis* and *Araucaria* by their distinctive crown shapes.

Boreal/cool temperate forests

Photographic scales of 1 : 10 000 to 1 : 20 000 have long been the most popular for typing in the cool temperate zone forests of Europe and North America. At these scales, several stand parameters can be measured with acceptable accuracy, as will be discussed in the next chapter, but tree species identification varies greatly. At scales smaller than about 1 : 20 000, tone becomes increasingly important in photo-typing. At scales between 1 : 10 000 and about 1 : 15 000 forest typing is usually based on stand height, crown closure and identifying the species or species mix within the stands delineated on the stereo-pairs of aerial photographs.

For example, in Scandinavia, spruce stands (*Picea excelsa*) can usually be separated from pine stands (*Pinus sylvestris*), provided one or other of the species comprises 75% of the crop and is older than five years (Francis, 1957). In Finland, Nyyssonen, Poso and Keil (1966), using 1 : 10 000 enlargements from 1 : 28 000 panchromatic photographs, found about 80% of the pine and spruce stands were correctly identified, but only 50–60% of the hardwoods. The identification of species in pure stands was noticeably better than for mixed stands; and in the lower volume stands pine was over-estimated to the detriment of spruce. As pointed out by Stellingwerf in Holland (1966), a higher correctly identified percentage (70–90%) can be expected for pure stands than for mixtures by groups of species (60–80%) and single tree mixtures (40–60%).

Sclerophyll forests

Until recently, most photography was black-and-white at small scale (e.g. 1 : 80 000), and consequently, other than serving as a field sketchmap, was of limited use in forestry. An example of photo-typing an extensive temperate evergreen formation at larger scale is provided by the forest assessment of 80 000 ha of the gregarious *Eucalyptus camaldulensis* (red gum) along the Murray River (Australia). Much smaller areas of cypress pine (*Callitris glauca*), box and mixed red gum–box within the formation could be identified and delineated on the black-and-white panchromatic photographs at a scale of 1 : 10 000, so as to leave virtually pure *E. camaldulensis* for further studies. The red gum was divided into classes according to recognizable characteristics, which normally influence the criteria used in a ground-volume assessment (i.e. height, stocking, crown diameter). In Tasmania, a photographic scale of 1 : 25 000 has for many years been found adequate for typing mature and regrowth forests (e.g. *E. delegatensis*). Studies, carried out in Victoria by the writer using natural colour film, indicated that species identification within delineated land systems is considerably improved at scales between about 1 : 8000 and 1 : 10 000 by using the characteristics of crown texture, crown shape and colour contrast within and between the crowns of the several different tree species.

Man-made forests

Man, as a biotic factor, has greatly influenced the development of forests, particularly in western and central Europe, the eastern United States, South Africa, Chile, Brazil, Australia and New Zealand. The image analyst of forest plantations can usually rely on stockmaps at scales between 1 : 10 000 and 1 : 25 000, showing the species planted, the year of planting and the boundaries of the subcompartments and documentation giving the frequency and dates of thinning and volumes extracted. It could be reasoned therefore that aerial photography is not required. Practice has shown, however, that this is not the case, that remote sensing is increasingly used in the forest management, that photointerpretation is economically sound and that valuable information can be obtained. Hildebrandt (1960) estimated that, in the Federal Republic of Germany, the use of aerial photographs had resulted in a 40% reduction of time and expense as compared with former field practice.

Studies carried out by the author in west Wales (1960–61) highlighted the value of standard mapping scale black-and-white aerial photographs in the preparation of working plans and in general forest management. The immediate advantage is that the boundaries of the planted species, usually in pure stands and recorded on the stockmap, can be verified and the stocking of the stands assessed. Stocking of young stands may be irregular due to weed growth, frost damage, disease, etc. and beating up may have been irregular. In addition, the photographs were used by the writer for mapping the land systems and land facets, and for locating degraded agricultural land (e.g. bracken slopes, old coppice). The latter were suitable to acquire for planting as coniferous plantations or suitable for amelioration treatment as mixed or hardwood forest. The photographs were also used for locating hedgerows with large trees suitable for felling or retaining under a preservation order. In the Scottish border counties using superior quality black-and-white photographs at a scale of 1 : 7500 (f: 152 mm), an attempt was made (1966) to identify the five planted species directly from the photographs after preparing a photographic key with reference to the stockmap. Although this type of study is usually unnecessary due to the information being available from the stockmap, it can be useful in verifying species boundaries, subcompartments and species composition, particularly some years after heavy beating up (i.e. restocking after planting).

Specially planned aerial photography of man-made forests should be scheduled at large scale in the period between the trees being large enough to resolve on the photographs and canopy closure, or in the period immediately following thinning, since otherwise the stocking is too dense and remains so until after (further) thinning. A dense canopy prevents a parametric approach using image analysis to volume estimates etc. (Chapter 16). In eastern Victoria (Australia), the canopy of *Pinus radiata* is so dense at about eight years that the ground is only seen along rides; little tree detail is visible between the crowns, except in occasional gaps, and necessitates planning the photography for the sixth or seventh year after planting (Spencer and Howard, 1971). For many plantations in Australia and New Zealand, the boundaries and areas of newly planted *P. radiata* are verified and mapped using specially commissioned small-format aerial photographs.

15.4 HIGH-ALTITUDE AERIAL PHOTOGRAPHY

Unfortunately, in agriculture and forestry, high-altitude aerial photography (HAAP), with its image resolution better than 2 m, has received little attention in recent years. The method was used intensively in airborne 'run up' experiments prior to the launch of Apollo-9 in 1969 and of Landsat in 1972. It warrants being carefully considered whenever small-scale photography is planned for land use and forest cover studies, when coverage by SLAR is favoured or when a national reference base is required for monitoring by satellites.

HAAP provides rapid aerial photographic coverage, its image resolution at larger scales is superior to that available from second-generation satellites, and for aerial photography it is highly cost effective. Ortho-photo-mosaics or topographic maps can be prepared in the scale range of 1 : 25 000 to 1 : 100 000 at relatively low cost, since ground and photogrammetric control is reduced – possibly two stereo-models per map sheet.

As conventional photography for mapping is taken with an unpressurized aircraft at maximum ceiling of about 9000 m with an 88 mm focal length lens and panchromatic black-and-white film, this provides a scale limit below 1 : 85 000. As mentioned in Chapter 5, an 88 mm lens is unsuitable for use with colour IR film; and therefore for this type of small-scale colour photography a longer focal length (e.g. 152 mm, 205 mm) and a high-altitude aircraft is needed.

The usefulness of high-altitude aerial photography for civilian purposes was highlighted over 20 years ago by Bock (1968), using a Lear-Jet at 13 500 m. Test strips of aerial photographs were obtained in Arizona, Minnesota and California, with a Wild RC-8 camera and Kodak Plus-X, Double-X and Ektachrome MS Aerographic films. The study showed that a very favourable C-factor was obtained in mapping and the film resolution was as fine as 0.0032 mm for high contrast detail (e.g. white lines 0.3 m in width on an aircraft runway). From the study of Gut and Hohle (1977) with improved cameras further geometric information emerged. They reported very small errors during flight due to atmospheric refraction, change of focal length with temperature and tip and tilt. However, it was found necessary to correct the stereoscopic model for warp (about 0.02% of flying height), due mainly to curvature of the earth; but the accuracy of elevation (vertical) measurements at about 0.004% of the flying height was a new achievement in analogue photogrammetry. The error in the ground resolution (with panel targets) and a flying height of 13 400 m was better than 2.5 m. With increased flying height and a wide-angle lens (150 mm focal length) distances of about 60 km between ground control points were bridged across the stereoscopic model. They concluded that planimetric maps and ortho-photomaps can be produced at scales of 1 : 25 000 and 1 : 50 000 from photographs at a scale of 1 : 100 000.

In interpretation, results have indicated that colour IR film with a yellow filter is superior for land use, agriculture and forest studies to other film types, since it has the qualities of both providing the imagery in colour and better resolution due to the better haze penetration. Analysis of the data of Arizona showed that Landsat MSS imagery and HAAP at a scale of 1 : 120 000 could both be used to classify land cover (section 14.2) into about ten broad types (i.e. category I: urban, rangeland, agricultural, forest, water, etc.), and subsequently using mainly HAAP into finer classes. That is, category II land-use types including deciduous, evergreen and mixed forests; grass, savanna, chaparral and desert as grazing classes, and pasture, arable, croplands and orchards as agricultural classes. In California (Lauer and Benson, 1973), it was concluded, using black-and-white panchromatic photographs at 1 : 15 840 and IR colour photographs at 1 : 120 000, that the latter enabled forest-type boundaries to be properly mapped, that forest-type identification with the two film–filter combinations had a high correlation and that cost was greatly reduced using HAAP. It was judged that the cost of forest aerial photography would be reduced up to six times and photointerpretation by half or more, as compared with using conventional aerial photography. However, Hudson, Amsterburg and Meyers (1976), working in Michigan with colour infrared photographs at scales of 1 : 60 000 and 1 : 120 000, concluded that only four tree species groups were distinguishable at those scales as compared with six species groups using CIR at 1 : 36 000.

A comparison by the author of high-altitude aerial photographs from several countries (USA, Canada, Australia, Sierra Leone, Cameroon, Indonesia) at a range of scales between 1 : 70 000 and 1 : 150 000 suggests that there is a recognizable reduction for forestry in the information content of the photo-graphs at contact scales between about 1 : 90 000 and 1 : 120 000 (cf. Table 15.1). At scales below about 1 : 90 000 (e.g. 1 : 74 000 in Sierra Leone), the data content usable for forestry may equate with that of older black-and-white aerial photographs at scales as large as 1 : 25 000; but at scales of 1 : 120 000 and smaller the data content for forestry and land use may be no better than that obtainable from recent satellite imagery. Also, as the scale is reduced, the smallest areas that can be delineated in the aerial photographs increase, although enlargements can be used. M. P. Meyer (pers. comm., 1977) pointed out that the field forester delineates forest cover types to a minimum of 1–1½ ha, while on photographs at a scale of 1 : 80 000 the minimum area is about 5–8 ha. Further, for fieldwork, all colour photographs are more difficult to use, requiring a transparent overlay (e.g. matt acetate) and felt pens, as compared with simply a chinagraph pencil for typing black-and-white photographs.

Extensive coverage with high-altitude aerial photography has now been completed in several tropical countries (e.g. Indonesia, Liberia, Sierra Leone). The high-altitude aerial photographic coverage of heavily forested Sierra Leone (72 000 km^2), initiated by the writer in 1975, was probably the first national coverage by this technique, and provided coverage within two months, as compared with previous national coverage by conventional black-and-white panchromatic photography in eight missions over a period of 14 years (Schwaar, 1978). In 15 flying days, 99.3% of the country was covered by the CIR photography at a flying height of 11 140 m using a 152 mm focal length lens. Simultaneous panchromatic black-and-white photography was taken using a super wide-angle lens (88 mm), as a precaution against CIR mishaps and to reduce the number of stereo-models for mapping.

Visual image analysis (cf. Howard and Schwaar, 1978) showed that (a) broad forest stand classes and palm oil plantations could be identified and delineated; (b) individual tree species could not be identified; (c) many woody plant subformations could be separately delineated; (d) meaningful parallax measurements of stand heights could be obtained for open forest using regression analysis and these stands could be placed in broad height classes; (e) three to five stand density classes could be identified; (f) emergent tree crowns could be counted; (g) control points could be easily established for photogrammetric mapping at a scale of 1 : 50 000; (h) timber extraction routes could be planned; (i) the landscape portrayed on the photographs could be divided into land systems and smaller land units (e.g. land catena, land facets) (cf. Blecher and Birchall, 1977); (j) information is obtainable on forest degradation and shifting cultivation including stages in vegetal succession (Plate 8).

In addition, the hierarchical subdivisive classification outlined in section 14.2 was used in the vicinity of the coast, to identify and classify the woody vegetation. This was based on stand physiognomy using stereo-pairs of the high-altitude aerial photographs.

The identified vegetation classes are shown in Table 15.2 (Howard and Schade, 1982). Numerical codes have been allocated to the classes to facilitate computer processing. On Landsat MSS imagery only forest, mangrove and arable were identified. Elsewhere (Schwaar, 1978), a comparison was made between high-altitude aerial photographic interpretation and digital image analysis of Landsat MSS data for mapping forest cover. This indicated that the satellite imagery is unsatisfactory for mapping forest regrowth (cf. Figure 15.2).

Table 15.2 Identified vegetation classes obtained from high-altitude aerial photographs in Sierra Leone and following the classification outlined in section 14.2, with numerical coding added to aid computer processing

A Formation/subformation	Local classification	Numerical code (indices)				
Humid tropical forest closed, evergreen broad-leaved		54	11	11		
Humid tropical woodland, closed evergreen broad-leaved		54	21	11		
Humid tropical woodland, dense evergreen broad-leaved	Advanced regrowth	54	22	11	3	
Humid tropical woodland, open evergreen broad-leaved		54	23	11		
Humid tropical tall shrubland, dense, evergreen broad-leaved		54	32	11		
Humid tropical tall shrubland, dense, evergreen broad-leaved	Mangrove	54	32	11		
Humid tropical tall shrubland, dense, evergreen broad-leaved	New regrowth	54	32	11	1	
Humid tropical tall shrubland, dense, evergreen broad-leaved	Intermediate regrowth	54	32	11	2	
Humid tropical tall shrubland, open, evergreen broad-leaved	In degradation	54	33	11	2	
Humid tropical low shrubland, open, evergreen broad-leaved	New regrowth	54	43	11	1	
Humid tropical low shrubland, open, evergreen broad-leaved	New regrowth, in degradation	54	43	11	12	
Humid tropical low shrubland, sparse, evergreen broad-leaved	In degradation	54	44	11	2	
Humid tropical low shrubland, sparse, evergreen broad-leaved	New regrowth, in degradation	54	44	11	12	
B Induced						
Induced	Grassland	54	00	00	00	1
Induced	Arable	54	00	00	00	2

Climatic zone ⌐
Height class ⌐
Cover class ⌐
Evergreen/deciduous ⌐
Broad-leaved/needled ⌐
Regrowth ⌐
Succession ⌐
Induced vegetation ⌐

(a)

(b)

Figure 15.2 Comparison of the mapped boundaries of mature rainforest and regrowth/shifting cultivation using high-altitude aerial photographs (b) and Landsat MSS imagery (a) (Sierra Leone). Field checking confirmed the mapped boundaries using stereo-pairs of the aerial photographs to be correct. The errors in the boundaries mapped from Landsat were due mainly to confusing the spectral signatures of abandoned cultivation and young forest regrowth with high forest.

15.5 SMALL-FORMAT AERIAL PHOTOGRAPHY

In contrast to high-altitude aerial photography, small-format vertical aerial photography is usually at large to very large scales for the purpose of supplementing the information content of other imagery. As discussed earlier, the photography is taken with either 70 mm or more often 35 mm film, often in colour and usually from a light aircraft or occasionally from a helicopter. Small-format photography is attractive from the point of view of low cost; but as explained in Chapter 5, it has the disadvantage of the small area of ground coverage per photograph. A recent comprehensive review of Canadian systems of large-scale photography using helicopters and light aircraft is provided by Spencer and Hall (1988).

In consequence, small-format vertical aerial photography is not used to give complete coverage of an area, but to provide strips of photographs (i.e. strip photography or transect photography) or photographs of small target areas (i.e. pinpoint photography). The latter includes photo-sample plots as part of multistage inventory (Chapter 18), and experiments in choice of the film–filter combination and scale for specific management objectives and monitoring of forest damage. If stereoscopic coverage is needed, then as with conventional aerial photography, the minimum endlap is 55–60%. If stereoscopic viewing is not required, as for example in animal counts, then only a small endlap (e.g. 10%) is recommended for matching up the photographs in the same strip. The endlap should be kept to a minimum to conserve film and to reduce printing and analysis costs. The exposures should be controlled by an inter-valometer when advancing the film in the camera during flight. For dead/dying trees and animal counts per hectare, acceptable counts can be obtained using markers fixed on the aircraft wing (Figure 15.3), and supplemented with photographs as needed.

When small-format aerial photographs are taken at low altitude, accurate knowledge of the flying height may be needed for determining the photographic scale and the height of trees using the parallax formula. If the terrain is flat or undulating, reference to the aircraft altimeter may be sufficient, but on other occasions a radar altimeter is needed. Provided the flying is at low altitude, twin cameras mounted on a boom, fixed below a helicopter, may suffice. This provides a known photographic airbase, since the distance between the cameras is fixed, and hence the flying height can be determined. References include work by Lyons (1966) in western Canada and Spencer (1974, 1979) and Spencer and Howard (1971) in eastern Australia.

For each pair of photographs obtained from the two synchronized cameras, the ground airbase (B) is the horizontal distance between the two cameras and the photographic airbase can be measured on the photographs, as the distance between the principal point and the transferred principal point. The flying height (H) is calculated by substituting b/B in the formula (cf. section 5.2):

$$\frac{f}{H} = \frac{d}{D}$$

That is

$$\frac{f}{H} = \frac{b}{B}$$

or

$$H = \frac{fB}{b}$$

Large-scale, small-format photography taken from a light aircraft has proved to be reliable and in forestry as useful on many occasions as photographs taken with a 23 cm format camera. Much of the work using fixed-wing light aircraft was pioneered in Canada as part of two-stage forest inventory (Spencer and Hall, 1988). Flight control and location of the photographs can be obtained from satellite imagery, photo-mosaics,

(a)

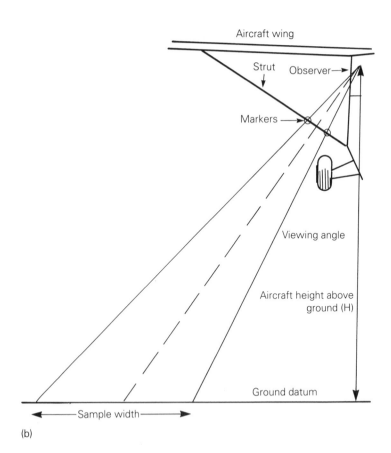

(b)

conventional photographs and maps. As reported by Zsilinsky and Palabekiroglu (1975), using a Wild A9 Autograph stereo-plotter, precise measurements of tree parameters were obtained from stereo-pairs of 35 mm photographs (fine grain, ASA 64, shutter speed one-thousandth second) taken with a 55 mm focal length lens at a nominal flying height of 228 m (750 ft). Small-format colour photographs have been used in tree species identification, updating plantation maps, monitoring annual planting and progress of logging operations, supporting forest inventory, supplementing satellite data, detecting and assessing weed growth, assessing disease and insect attacks and for many types of photographic research where the cost of standard-format photography would be inhibitive.

Tree species on macro-scale aerial photographs

Tree species are best identified on very large-scale aerial photographs; and results are based on studies mainly carried out in the 1960s and 1970s, particularly in Europe, Australia and North America. From the writer's experience, there is frequently a marked information content improvement in identifying the tree species at a scale of about 1 : 5000 or larger and for convenience this is termed macro-scale photography. In south-eastern Australia (Howard, 1967b), a study of several eucalypt species indicated that photographic tone is the most important characteristic at smaller scales and tree species identification did not improve as such between scales of 1 : 20 000 and 1 : 80 000. That is, when tone was the primary factor. The information accuracy tends to improve as the contact scale of the photographs is increased and probably reaches a maximum at a scale greater than about 1 : 2000; but other factors influence the choice of the best scale, including the film–filter–lens combination, the crown characteristics of each species, the species mix, the increasing cost of the photography with larger scales and constraints imposed by flying conditions. Most macro-scale photography is in small format and taken using a light aircraft.

Figure 15.3 (a) For dead/dying trees, wildlife and livestock animal counts by species, visual reconnaisance using a light aircraft is often adequate; markers may be fixed to the struts of the aircraft wing. For a fixed flying height, the number counted between the two markers can be expressed as stocking per unit of land area; if many animals are present in groups, it is useful to supplement the counts with 35 mm aerial photographs and to use a tape recorder for additional data; (b) diagram illustrating counting procedure using a light aircraft at 100 m to 200 m above the ground. The sample width is controlled by the flying height and the fixed markers on the wing strut. The sample length (transect) is controlled by the flight direction.

The trees observable on the macro-scale photographs should be examined individually and as members of a group both near the principal point (Chapter 10) and in profile towards the periphery of the photographs. Tree characteristics include crown shape, crown size, branching habit, variations within the crown of tone and colour (hue, value, chroma), tree height and phenology of the species. The image analyst should be familiar with the geographic distribution and site requirements of each species. As pointed out by Sayn-Wittgenstein (1960), identification begins by eliminating those species the presence of which in the area of the photograph is impossible or unlikely, due to climate, physiography or locality. Next, by a knowledge of the association formed by the common species and of their common requirements, the group or groups of species which are likely to occur in the area are established. The identification of species in a group depends finally on the tree-form, particularly branching habit and tree/crown form.

The use of tree-form to identify a species becomes less and less valuable as the scale of the photographs is reduced, until eventually this characteristic is replaced by photographic tone, texture and possibly shadow-pattern. Shadow differences within the crown can be helpful. Dark shadows cast on the ground indicate a compact crown and dense foliage, and a 'light' shadow, as with many eucalypt species, the opposite. Many northern hemisphere broad-leaved species provide rounded shadows, while many conifers, particularly when young, provide pointed shadows.

Considerable variations in tone may be expected for the same species across the photograph and particularly on the line about the axis joining the specular and reflex reflection zones (section 3.3). What is often more important than an exact colour or tonal definition for a species is its relative colour or tone in relation to other species (section 8.3). A grey-scale (Figure 8.6) may be used to correlate the tone of the species recorded on black-and-white photographs. If colour photography is used, then the tree crown image, as represented by its hue, chroma and value, may be related to standard Munsell colours (Figure 15.4, for several hardwood and conifer species in northern Minnesota) (Heller, Aldrich and Bailey, 1963). At scales between 1 : 1888 and 1 : 3960, they obtained highly significant differences based on colour and tree shapes.

As mentioned for evergreen forest in Chapter 14, the initial stratification of the forest into land systems can help in the identification of tree species on macro-scale photographs, since the stratification has the effect of reducing the number of already known species within a definable climatic zone. In general, the few published studies on using macro-scale aerial photographs in Australia indicate that the identification of eucalypt species is more difficult than for cool temperate species of the northern hemisphere. This, in the writer's experience, is due to the phenology of this evergreen genus and crown characteristics; but, on the other hand, researchers have achieved encouraging results (e.g. Howard, 1969; Myers, 1976, 1978). For tropical forests, there are

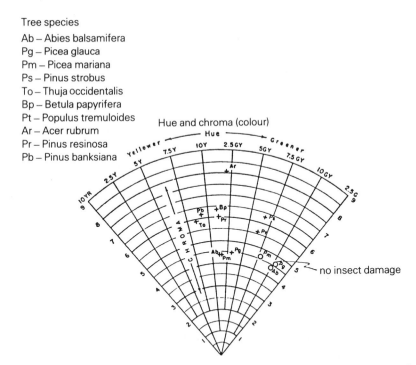

Tree species

Ab – Abies balsamifera
Pg – Picea glauca
Pm – Picea mariana
Ps – Pinus strobus
To – Thuja occidentalis
Bp – Betula papyrifera
Pt – Populus tremuloides
Ar – Acer rubrum
Pr – Pinus resinosa
Pb – Pinus banksiana

Figure 15.4   The Munsell colour notation in terms of hue, value and chroma used to help identify the images of tree species recorded on macro-scale aerial photographs.

also few studies, and again the results are encouraging. For example, Clements and Guellec (1974), using colour photographs at a scale of 1 : 5000 in Gabon, found that 73% of *Acoumea kleineana* (okoume) could be identified, although confusion with other species was easily made. Using natural and IR colour at scales of 1 : 2000 to 1 : 4500, Sayn-Wittgenstein, de Milde and Inglis (1978) were able to identify several tree species in Surinam; and in India, Tiwari (1975) was also able to identify several species.

Shrubs on macro-scale aerial photographs

In some circumstances the interest of the forest photointerpreter extends to shrubland. This may occur, for instance, when studying vegetation succession, in the management of marginal or degraded lands, and when survey and management extends to rangelands. Usually the visual image analysis techniques resemble those employed in the survey of open forest or woodland, but are modified for the small size of the shrubs. If interest lies only in delineating areas of shrubland as a distinctive physiognomic type, then standard photography is usually sufficient, as discussed in Chapter 5.

The small size of most shrubs compared with tree species necessitates macro-scale aerial photography, if each shrub is to be recorded separately on the photographs. Experience gained in several countries (e.g. Argentina, Australia, Kenya, USA) suggests that the preferred film–filter combination is colour IR, followed by normal colour as second choice. Colour enhances shrub species identification and assists in separating vegetation from surface litter and bare soil. Although it is not normally feasible to measure stereoscopically the heights of individual shrubs, taller shrub types/shrublands can often be placed in 1–2 m height classes.

Usually the photographs are taken in strips using small-format cameras with flight planning involving medium to small-scale standard 23 cm format photographs. If the shrubs are small in size then a scale of 1 : 1000 or even 1 : 500 may be needed to detect the shrubs and to achieve maximum species identification. Using a 152 mm lens, this necessitates flying heights between about 100 m and 600 m, and can result in the times of flying being restricted by air turbulence. As with the aerial photography of trees, the flight timing should provide photographs with the shadows of the shrubs present, but not too long; and the phenology of the shrubs should have been studied in the field to ensure that the aerial photographs are taken in the best season for photointerpretation.

Macro-scale photography of shrublands expedites field survey, provides useful information on surface soil characteristics and soil erosion, facilitates monitoring of wood biomass for fuel (e.g. Kilmeyer and Epp, 1983), facilitates the separation of shrubs, forbs and herbs, enables estimates of plant cover to be made and some shrub species to be identified. That macro-scale photographs can provide acceptable estimates of shrub cover is illustrated by the study of Carneggie, Wilcox and Hacker (1971) in Western Australia. For saltbush, they obtained cover estimates using dot grids of 7.8% and 12.6% as compared with ground estimates of 8% and 15%, and for mulga 22.8% and 10.8% and 24% and 11% respectively.

Success in species identification depends considerably on the crown characteristics of the species, the season of the photography and the influence of grazing. As with tree species, some shrub species can be accurately and consistently identified, while others are too variable in shape, colour and texture on both colour IR and normal colour photographs. Driscoll and Coleman (1974), for example, working in Colorado with four interpreters, were able to identify consistently on CIR photographs at scales between 1 : 800 and 1 : 1500 two shrub species at 100% (low sagebush, mountain mahogany), five shrub species above 80% and the remaining four shrub species at 25–78%. The best results were obtained by the skilled interpreter familiar with the area in the field.

15.6   FOREST DAMAGE ON PHOTOGRAPHS

Airborne sensing consists usually of visually sketch-mapping damaged forest at low altitude from a light high-winged aircraft or sometimes a helicopter; and strip and pinpoint macro-scale photography from a light aircraft using small-format cameras and/or medium to small-scale 23 cm format photography.

Murtha (1983) pointed out, using a t-test, that an aerial photographic scale of 1 : 4000 is sufficient for counts of dead trees on colour IR photographs, but a scale of 1 : 1000 was needed for the interpretation of subtle differences in stress. So far, the role of satellite sensing has been modest by providing a first synoptic view of large areas in which damage has or is occurring and by providing the frame within which aerial reconnaissance is developed. A useful bibliography on damage assessment using remote sensing data was provided by Henniger and Hildebrandt (1980).

Disease and insect damage

The question of choosing the most suitable film type requires some consideration. Colour IR film with a yellow filter is nowadays strongly favoured, followed by normal colour (Heller and Ulliman, 1983). Craig, as early as 1920, successfully used ortho-chromatic black-and-white film to map in northern Ontario boreal forest damaged by spruce budworm. In general, in the ensuing years until the 1950s and the availability of colour film (e.g. Wear and Bongberg, 1951), panchromatic black-and-white aerial photography was little used and confined to severe physiognomic damage including fire and windstorms. However, Spurr (1948) had reported that heavily defoliated and dying trees appeared conspicuously dark toned on black-and-white IR aerial photographs, but also noted that severe defoliation could be interpreted on black-and-white panchromatic photographs.

In the early 1950s, the advantage of colour aerial photography began to be recognized, as important in studying classes of damage caused by insects and disease. This was accompanied by several detailed studies in the USA of variations in spectral reflectance in the visible and near-IR between healthy and diseased foliage (e.g. Colwell, 1956). However, the early expectation has not been sustained of being able operationally to detect previsually the stress caused by insects and disease in forest stands. Nevertheless, visually observable stress of the upper parts of tree crowns can often be better seen and classed on colour IR aerial photographs than from the ground.

With the improvement in colour aerial films, many studies from the late 1950s onwards included comparisons of normal colour and colour IR photography, and conflicting opinions emerged on the usefulness of colour IR film (e.g. Benson and Simms, 1967). In general, colour IR film has the advantage of providing a record of spectral reflectance in both the visible and near infrared. Although on processed imagery, healthy foliage often appears red/reddish and dead or dying foliage appears blue/bluish-green, there is no exact hue, value and chroma indicator of disease. As pointed out by Murtha (1972), a study of literature shows that conflicting opinions were derived from different definitions of damage, the degree of damage and the photo-interpretative technique used. To these can be added the significance of the forest type(s), the atmospheric conditions when taking the photographs, the CIR film age and film processing.

In planning a survey of a particular type of damage, reference to literature may indicate the most suitable photographic scale and the best film–filter combination to be used for specific objectives and types of defined damage under recognizable regional conditions. Otherwise it is advisable to plan experimental photography with small-format cameras, and to be forearmed with knowledge of the spectral reflective properties of vegetation (Chapter 3), aerial photography and film processing and the characteristics of the damage. Where the physiognomy of the forest has changed, resulting from extensive damage, or the only information needed is that of the presence of dead trees or of conspicuous visual damage to individual trees, then panchromatic black-and-white photography may suffice at large to medium scales. If the analysis requires to be completed with several categories of damage clearly visible in the field, then colour, possibly normal (panchromatic) colour aerial photography, may be the most suitable; but, as pointed out by Spencer (1985), panchromatic black-and-white photography offers advantages in obtaining tree and stand measurements. The fact must not be overlooked that shadows are densest on IR photography and that panchromatic black-and-white photography is the easiest to acquire, and processing requires less skill.

Plate 9(a–d) provides a comparison of four film types at a scale of 1 : 5000, taken simultaneously with a 35 mm multiband camera by the author, of an 18-year-old *Pinus radiata* plantation suffering from dieback of the crowns due to severe drought. Plate 10 (Spencer and Howard, 1971) shows similar thinned radiata pine trees on normal colour film at a scale of 1 : 1300, which enables the individual tree crowns to be placed in one of the following five crown-condition classes:

A = Dead top, needles have fallen
B = Dead top, brown needles persist
C = Top dying, needles pale green to yellow, but not yet brown
D = Healthy green crown
E = Dead top with new green leader subsequently developed

Of the 43 trees, 38 (88%) were correctly classified on the black-and-white photographs. The main difficulty was in distinguishing between classes B and C and in detecting class E. On the colour photographs at approximately the same scale, 42 trees (98%) were correctly classified.

Aerial photographic scales used in North America range from about 1 : 800 to 1 : 160 000. Probably 1 : 2000 to about 1 : 10 000 are preferred as a compromise between cost and successful identification of insect damage. Insect infestations photographed in the USA have included spruce budworms in Maine, gipsy moth in Massachusetts, bark beetles in California and the south-eastern States, weevils in New York State and balsam woolly aphid on silver fir in Oregon. For the latter, 1 : 10 000 colour photographs were used to locate the boundaries of the outbreaks, and 1 : 2500 colour photographs for density counts of dead and dying trees. Little improvement was achieved by increasing the scale from 1 : 5000 to 1 : 2500 for damage to *Pinus coulteri* in California. The recording of damage on photographs may be possible within a few days of attack by defoliators and somewhat longer for sucking insects; but may not be discernible for one or two years in the case of bark beetles. In general, defoliated groups of trees tend to photograph in contrasty 'faded' hues on colour photographs.

Sometimes, oblique photographs are useful. For example, in south-eastern Australia, severe phasmid damage (*Didymuria* spp.) has been monitored using 35 mm oblique normal colour transparencies showing heavily defoliated eucalypt stands, which contrasted with the greenish hues of healthy trees. In the western USA, medium scale, CIR, panoramic photography (e.g. 1 : 26 000) taken at high altitude using a 600 mm focal length lens has been used in tree counts of dead and dying pines damaged by bark beetles and in planning salvage operations. Results compare favourably in accuracy with vertical aerial photography (Dillman and White, 1982) and even more favourably in cost (Klein, 1982, Table 15.3).

Pollution damage

Increasing consideration is being paid to using colour aerial photography in assessing air or soil pollution, which leads to foliage damage. This can result in the death of the trees along roadsides in urban areas and in wide-scale forest damage in the countryside. Kenneweg (1973), for example in Germany, determined by visual analysis the vitality of about 10 000 roadside trees using colour IR photography at a scale of 1 : 10 000; and others (e.g. Akca, 1971) have used densitometers for identifying damaged trees according to tonal and textural differences.

Air oxidant injury is frequently slight at first and can be easily overlooked on black-and-white film, but is much more readily detectable on infrared colour film. Some of these symptoms are subtle and affect the same species of trees selectively; that is, some trees are more resistant to air oxidant injury than others. An apparently healthy tree can be growing adjacent to a neighbour of the same species in its last stages of decline. Because of the absence of lush foliage on injured trees, bare branches may be visible on macro-scale colour

Table 15.3 Comparison of costs (US $) and estimates of numbers of trees and volume of lodgepole pine killed by the mountain pine beetle from two separate surveys using panoramic photography and vertical photography

Survey		Number of trees			Volume ($ft^3 \times 10^6$)					Cost
	Acres	Total	SE	SE (%)	Per tree	Total	SE	SE (%)		
Panoramic	520 640	1 891 510	194 804	10.3	0.0143	27 001	3682	13.6		37 001
Conventional	270 255	1 270 617	55 931	4.4	0.0191	24 237	2403	9.9		32 649

photographs and permit the identification of positively affected trees. Healthy conifers usually retain their needles for up to five years, appearing green/green-yellowish on normal colour prints, whereas smog-affected foliage may have only a current growth of short needles at low density, appearing yellow/yellow-reddish on the prints.

At the operational level in Germany, as developed by the University of Freiburg, CIR photographs are taken at large scale (e.g. 1 : 5000) in strips, and photo-plots are then established on the photographs. Sample trees are identified on the photographs by species and damage class using photointerpretation keys. All photographs are of high quality, being taken with 23 cm survey camera, preferably incorporating forward image compensation and using a 300 mm focal length lens.

One of the most extensive and severe occurrences of pollution damage (mainly by sulphur dioxide) of forest stands, is that in the Erzgebirge and Iserbirge mountains in the vicinity of the German and Czech borders (Hildebrandt and Kadro, 1984). This damage is readily seen and monitored using Landsat TM data (Plate 11). On the basis of the results, Kadro (1989) concluded that the satellite data provides two damage classes and a third class as healthy or slightly damaged.

Forest decline

The Commission of the European Communities drew attention in 1983 to forest decline in western Europe. As pointed out by Cielsa (1989), forest decline is a gradual process of tree and forest deterioration in condition and vigour that can eventually lead to the death of the trees. The syndrome results from an interaction of a number of factors, some of which may not be identifiable. Decline can be caused by obvious insect damage and disease or industrial pollution, but it may also be associated with climatic change, acid rain, etc. that regionally places the trees under increasing stress. Forest decline is not new. It was reported as being widespread in central Europe for silver fir (*Abies alba*) about 250 years ago (Kandler, 1985); and the widespread dieback in eastern Canada of birch (*Betula* spp.) early in this century has been attributed to regional warming trends (Hepting, 1963). The author is of the opinion that forest decline is resulting from climatic change/climatic variation acting as the predisposing factor to the effect of other adverse environmental factors (e.g. acid rain).

Forest decline is now a major silvicultural problem of the 1980s and 1990s in North America and Europe and appears to be affecting a wide range of species throughout the world. In Mexico, for example, about five years ago extensive natural forest (*Abies religiousa*), in the vicinity of Mexico City, developed stress symptoms and many trees have since died.

In Germany with the decline in the coniferous forests reported as exceeding

50% in some states, it is often first noticeable on the ground by a loss of tree vigour, possibly chlorosis, and a thinning of the tree branches in the upper canopy. This may be followed by loss of the older foliage, which usually remains on the coniferous trees for up to five years; and the development of a tufted appearance by the remaining foliage. Finally, most of the crown appears to consist of dead branches before the eventual death of the tree (Figure 15.5, Cielsa and Hildebrandt, 1986).

In the USA, the techniques follow those employed in insect damage surveys. The photographs are usually taken with colour IR film at large scale (e.g. 1 : 8000). Two-stage sampling is used, which combines regression analysis, in order to establish the correlation between dead and dying trees recorded on the photographs and those counted in the field. In the north-east, for example, 1 ha plots were used in studies of decline in the spruce–fir forests. In West Virginia, colour IR panoramic photography at a nadir scale of 1 : 32 000 was used to stratify the red spruce forest into damage-class polygons, of which a sample was examined in the field. The data on the mortality classes was then digitized and stored in a geographic information system for future reference.

15.7 SIDE-LOOKING AIRBORNE RADAR

Reference should be made to Chapter 6 (section 6.3) and to *Imaging Radar for Resources Surveys* (Trevett, 1986) concerning the technical principles of real aperture radar (RAR) and synthetic aperture radar (SAR). If choice of forest coverage is exercised in terms of one sensor, then preference would be given, in the writer's opinion, to aerial photography due to its superior image resolution, then to satellite sensing due to the low cost of the synoptic data per unit ground area, and finally to radar due to its near all-weather capability. The best radar image resolution now available operationally (i.e. 6 m) is several times coarser than even high-altitude aerial photography (i.e. 1–2 m); and several years ago the difference in image resolution was even more in favour of aerial photography (cf. section 6.3). In the planning in 1976 of the national remote sensing coverage of Nigeria for small-scale land use and vegetation survey, opinions differed on the choice of sensor. In retrospect, those, including the author, who advocated making maximum use of Landsat imagery in the dry north and confining radar to the wet cloud-prone south, were probably correct. For northern Nigeria, the usable information derived from the

---

Figure 15.5   Progressive stages of decline in Germany of Norway spruce, *Picea abies*: (a) tree in early decline stages (class 1); note thinning of branches in upper crown; (b) tree in more advanced stage of decline (class 2) characterized by branches which have a tufted appearance indicative of loss of older foliage; (c) final stages of decline (class 3) characterized by a preponderance of dead branches and presence of epicormic branches.

Landsat imagery appears in general to be comparable to that obtained from SLAR.

General

The question must be asked, therefore: how useful is SLAR in the study of forest resources? We must acknowledge the fact that there is considerable flexibility in SLAR techniques, due to being able to sense in one of several microwave bands, to sense sequentially in two or more of these, to use like- and cross-polarized channels, to vary the look angle of the sensor and to acquire imagery and multidate imagery on precisely planned dates regardless of most weather conditions. In analysis, use is made of a wide range of techniques extending from visual interpretation involving tone, texture and contexture and of digital image analysis involving pixel grey levels (tone), filtering to reduce image 'speckling', unsupervised/supervised classifications, polarization ratioing and overlaying SLAR data on digitized maps.

The value of SLAR is well recognized in geology and mineral exploration, where it can be informative on rock structures and surface lineaments; and in geomorphology/hydrology, where it can provide information on landforms, drainage patterns under forest cover, length of water courses, and, in more arid areas, the location and pattern of sub-surface prior streams.

As long ago as 1963, Fischer drew attention to the fact that very thin veneers of sand on the soil surface are far better detected on SLAR imagery than on aerial photographs. In soil survey, the usefulness of SLAR varies greatly due particularly to the influence of the forest canopy and 'false' soil-type boundaries being produced by changes in the surface roughness of the vegetation and the terrain, changes in their dielectric constants and the interaction of the two factors. For example, dry, light-coloured sand may appear dark toned on the SLAR imagery; while with the presence of soil moisture, the same soil series records light toned due to the change in its dielectric constant, and dark toned again with the presence of surface water.

In land-use application, studies have shown that, depending on the characteristics of the SLAR sensor system, some cover types can be identified; but this cannot be predicted ahead of the SLAR mission and examination of the imagery. For example, Brisco and Protz (1982) in Ontario (Canada) found using visual analysis with tone and texture as the main discriminants that woodland, roughland and maize (corn) could each be identified with an accuracy of about 90%, but pasture and other cereals were consistently confused (accuracy about 50%).

Forest application

The most important contribution of SLAR to forestry has been that of providing small-scale imagery for reconnaissance surveys. This is particularly the situation for tropical forests, when the area is large and unmapped or the maps are unreliable and it is difficult to map using other sensors due to the problem of cloud cover. In the inventory of tropical rainforest, which is often fraught with logistical problems and where enumeration is very slow, SLAR imagery can help in the positioning of pre-investment field studies and stratifying the forest ahead of field sampling. In addition, radar sensing, due to its almost all-weather capability, provides for precise timing of the remote sensing operation and yields a uniform image base of the entire area. For example, the whole of Nigeria was covered by SLAR RAR in X-band in two look-directions in 181 days, which included 14 night sorties. Previous conventional photography for photogrammetry had taken ten years; and there still remained small 'pockets' without photography in the cloud-prone south. High-altitude aerial photography was not suitable in the extreme south due to the very cloudy conditions throughout the year, the requirement that coverage was to be completed in a short period and the large size of the country (i.e. $933\,000\,\text{km}^2$).

Most of the early forest studies used real aperture radar (RAR) in K_a, K and X bands, and recent studies have used synthetic aperture radar in X, C and L bands (e.g. European SAR-580 project). In the earlier studies with direct continuous recording on to film, the negatives were at a scale between about $1:200\,000$ to $1:400\,000$. Details were interpreted using a simple magnifying lens or the negatives were enlarged two to four times (i.e. up to $1:50\,000$). Practice indicated that forest typing was usually best obtained using a two times enlargement.

Several cloud-prone forested zones have been covered in the tropics by SLAR at small scale. The first Darien Province in Panama was covered in 1968 (Viksne, Liston and Sapp, 1970). This was followed by 10 million hectares at a scale of $1:220\,000$ in a strip 100–150 km wide, from Panama across Colombia and into Ecuador (Sicco Smith, 1971, 1978). The massive Radam project (1973–79) using X-band covered the entire northern half of Brazil, including the vast Amazonian basin ($c.$ 5 million km^2). The imagery has been so useful in geological/mineral resource surveys that SLAR coverage has been extended southwards. Visual interpretation of the imagery has included studies of the geomorphology and vegetation and the production of land-use maps.

Visual analyses of SLAR imagery world-wide have shown that the forests can be stratified at small scale for inventory (Plate 12), that usually forest and non-forested areas can be separated, that water courses and some logging tracks and roads can be mapped, that some forest types based on the tone of the stand physiognomy and landforms are readily identified and delineated. Also, recently clear-felled areas and sometimes old burn patterns are recognizable, provided that they exceed 1–2 ha in area. Using HH polarized SAR imagery, coniferous and deciduous cover types can be separated (Hoffer et al., 1982). The tree line can usually be delineated. Stands as a forest type on hilltops are often identifiable and at low elevations tree communities growing along water courses can be mapped, including, in the tropics, mangroves and nipah palm.

In general for forest types, only a simple classification can be made. Banyard (1979) in Indonesia, for example, found that with the help of photographic keys the maximum was six forest types, namely hill forest, dry lowland forest, wet lowland forest, swamp forest, mangrove forest and secondary forest/ shifting cultivation. Also, topographic variations result in false type boundaries; and in rugged terrain shadow may unpredictably 'black out' key forest features, unless the imagery has been flown with two-look directions.

As the older SLAR is recorded with a resolution of 20–25 m or 40–60 m (Chapter 6), tree crowns seldom resolve individually, and what is observed on the imagery is texture provided by a group of coalescing crowns. An exception is the mature large-crowned *Araucaria Cunninghamii* along ridges on imagery of Papua New Guinea, which could be separately identified. Normally, however, an appreciation of the forest type is provided by the forest stand and not by the separate trees. Tree species are usually only identifiable if gregarious and forming large consociations or single-species plantations. In Nicaragua, for example, natural pure stands of *Pinus caribea* were successfully identified and separated into three density classes.

Recent trends in forestry

Probably the most encouraging development for forestry is the availability of finer resolution imagery (i.e. 6 m) and multiband data. Multiband/multi-specular SLAR sensing involves recording simultaneously multipolarized signals and signals recorded in two or more microwave spectral bands. Finer image resolution enables individual crowns and crown shadows to be resolved and thus improves textural differences to be observed between forest types. It also provides an impression for larger trees of crown size and hence tree size. However, the coarser resolution of older imagery may be preferred in geology, since fine detail is often distracting to the interpreter; and for agricultural crops, which provide relatively smooth surfaces, finer resolution may have no advantage and speckling may be much more conspicuous and distracting to the human eye.

Recent forest studies have related recorded SAR signals to definable forest stands. For example, Sader (1987, 1988) in his investigation of L-band SAR covering hardwood and pinewood sites in Mississippi, has reported that cross-polarized mean digital numbers of the signals recorded from pine stands were significantly correlated with green-weight biomass and stand structure; and that multiple linear regression, with five stand variables (biomass, tree height, diameter at breast height and basal area, number of trees), provides an integrated measure of canopy roughness. Highly significant correlations were obtained between the broad-leaved (hardwood) stands and the HV/VV ratio.

Recognizable canopy characteristics included branching pattern, crown area and crown volume; but changes in the soil moisture of fully stocked stands had no influence on the magnitude of the signals. **Digital numbers** (DN) refer to the transformed magnitude of the received signals encoded into a discrete number of steps.

In considering empirical approaches to recent developments, examples include temperate forest plantations and tropical natural forest. In the UK as part of the European SAR-580 project, imagery of C, L and X bands and polarized SAR was evaluated visually for several geographically different sites and compared with aerial photographs (Thallon and Horne, 1984; Horne and Rothnie, 1984). Digital analysis of the data provided final information with 'speckling' being a major constraint. Radar shadowing provided a constraint to visual interpretation, when attempting to examine on the imagery steep-sided valleys of the Exmoor National Park. In general, X band provided more information.

For a test site in the flat-to-undulating terrain of Thetford in eastern England nearly all road lines were detected on the aerial photographs, with omissions only where the tree canopy was closed. On the SAR imagery a number were not recorded at right angles to the flight line and in the radar shadow of taller forest stands. For the compartment/subcompartment boundaries shown on the stockmap, 90% were identified on the black-and-white aerial photographs, 94% on normal colour photographs, 85% on the X-band imagery and only 66% on C-band imagery. Omissions occurred where the height differences were less than 2–3 m (2–7 years in age classes), between mixed conifer crops and an adjoining stand of one of the species of the mixture and small intimately mixed stands. Both the aerial photographs and X-band SAR recorded boundary features not shown on the stockmaps including low stocking; but the longer wavelength C-band SAR (5.7 cm versus 3.2 cm in X-band) did not record these subtler details. L-band SAR was found to be most suited to the mapping of small clumps of trees and delineating forest boundaries in the perspective of general land use. For all three bands, tree counts in hedgerows shown on the imagery proved inferior to aerial photography and would preclude its future operational use in the census of non-woodland areas.

16

Airborne remote sensing application: tree and forest stand measurements from aerial photographs

From the introduction to photogrammetry in Chapter 9 and from the comments made in the last chapter, it will be appreciated that the main source of remotely sensed quantitative data of forests has been, is and will be at least in the near future the aerial photograph. SLAR, lacking the resolution of aerial photographs, is not suited to obtaining tree and stand parameters as used in forest practice; and linear and area measurements are not as accurate as when obtained from aerial photographs. Airborne multispectral scanning (MSS) remains experimental in forestry, with the exception of using the far IR in fire control (section 6.1).

This chapter therefore pursues the taking of tree and stand measurements from aerial photographs. The identification of tree species has already been examined. The attributes of the forest which can be observed and measured from the photographs depend primarily on the scale of the photographs; and, as the photographs are mainly used at management level, larger scales are preferred. This proviso on scale, in conjunction with the physiognomy of the forest, enables measurements to be made by forest types of individual trees and of the forest stand. The term **forest stand** is often loosely used as a synonym for forest crop; but more precisely it should be considered as descriptive of an aggregation of trees sufficiently uniform in composition to be distinguishable from adjacent crops. These variates can be species composition, age classes, size classes, etc. The stand is usually the smallest unit of forest management and thereby corresponds to the subcompartment or often to the **compartment**, which is a territorial unit of the forest for administrative purposes and description. The **subcompartment** is a subdivision of a temporary nature for special treatment and/or description.

Height and crown diameter can be measured with reasonable accuracy at scales larger than about 1 : 20 000 in many forest types, and in combination with crown closure (gross crown cover) are the most important photogrammetric parameters for stratifying the forest into merchantable volume classes. Under favourable conditions, tree counts may be combined with one or more of these parameters; but the determination of stem defect must always be made on the ground and consequently will remain a major constraint in some forests to making volume estimates using aerial photographs. Stem defect is particularly a problem to aerial volume assessment in forests in which termite damage or fire damage occur. These include the tropical woodlands of Africa and South America and the eucalypt forests of Australia.

16.1 TREE MEASUREMENTS

Tree height

Reference should be made to section 9.3. In comparing tree heights obtained by stereoscopic parallax difference measurements with those of standing trees in the field using a hypsometer, it must not be overlooked that instrument and human errors are introduced by the field measurements. Hence without carefully measuring a few selected trees, it cannot be assumed that the measurements of standing trees by field survey represent 'ground truth'. Particularly when the trees are cone shaped, as with many conifer species and young broad-leaved species, the tip of the crown will not resolve on the aerial photographs; and this will lead to underestimates of tree and stand heights using parallax difference measurements. This is corrected by adding a predetermined constant based on a reasonable number of measurements or preferably by regression analysis, which shows the relationship between photographic estimates of a range of tree heights and ground measurement of the same trees (Figure 16.1). Regression analysis is particularly important for adjusting tree heights obtained from small-scale aerial photographs. With deep shadow, the error results in the floating mark being placed below the ground plane seen in the stereoscopic model and results in an overestimate of tree height. In consequence the use of the simplified, less accurate parallax formula is often appropriate, i.e. (symbols as in section 9.3):

$$\text{Tree height}\,(h) = \frac{H(\Delta p)}{b \pm \Delta p} \simeq \frac{H(\Delta p)}{b}$$

If, however, the terrain is rugged with major elevational changes, and the photography is at large scale, the simplified formula should not be used. For example, with a decrease of ground elevation by about 350 m below the mean ground datum at the principal point and a flying height of 3000 m, the use of

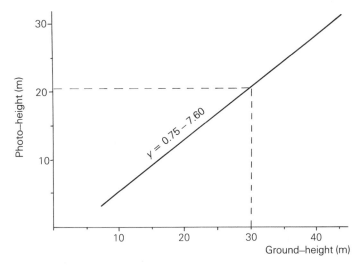

Figure 16.1 Relationship between photographic tree height from stereoscopic pairs of black-and-white panchromatic photographs at a nominal scale of 1 : 30 000 and tree heights measured on ground (evergreen forest, south-eastern Australia). Using a regression, meaningful photo-heights are obtainable (e.g. photo-height 21 m, regression adjusted height 30 m, field measured height 32 m).

the simplified formula results in underestimating the tree height by about 10%.

As also discussed in section 9.3, an alternative method of estimating tree heights is to measure on the single aerial photograph the shadow length of the individual trees at known sun-angles. The method is of very limited use, but has been used for many years in boreal forests and under favourable conditions gives satisfactory results. Obviously the trees have to be sufficiently widely spaced in the forest, at favourable sun-angles, for their individual shadows to be clearly resolved on the forest floor.

Crown diameter

Crown diameter is measured on stereoscopic pairs of aerial photographs with shadows towards the observer using either a transparent crown wedge or dot-type scales having circles of graded sizes for direct comparison with the stereoscopic view of the crown (Figure 16.2). It is often necessary to take two crown width readings at right angles when using a crown wedge. The two diverging lines of the wedge are placed tangent to and touching the two resolved edges of the crown in the stereo-model. Crown width, read off the scale in hundredths or thousandths of an inch, or the equivalent in millimetres,

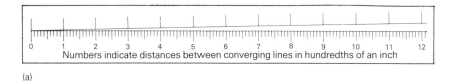

(a)

CROWN DIAMETER SCALE

CENTRAL STATES FOREST EXPERIMENT STATION

NUMBERS INDICATE DOT SIZE IN THOUSANDTHS OF AN INCH

(b)

Figure 16.2   Crown diameter measuring scales (US Forest Service): (a) crown wedge; (b) dot-type crown scale (reduced).

is converted to feet or metres using the nominal photographic scale, or in rugged terrain from the calculated local scale. Crown widths are more precisely estimated at smaller photographic scales using dot-type scales. Measuring becomes difficult at photographic scales smaller than about 1 : 20 000.

Crown diameter measurements obtained from the photographs will frequently not agree precisely with measurements made on the ground, as some branches may not resolve on the photographs and shadows can introduce errors. The perspective view of the crown is different as seen from the ground and as seen from above on a stereo-pair of photographs. Generally, aerial crown measurements of the same tree will be less than measurements taken on the ground (Spurr, 1960). Obviously some crowns may not be measured, due to being hidden and two or more crowns may coalesce to provide the impression of a single large crown. These problems are minimal in cool temperate forests and maximum in tropical rainforest. For example, in North America, 0.5–1.5 m crown classes are often possible at scales between about 1 : 8000 and 1 : 20 000 respectively.

The accuracy of measuring crown width can be improved using regression analysis. Figure 16.3 shows the linear relationship of photographic crown diameter, obtained from normal colour photographs at a scale of 1 : 11 000 and plotted against crown diameter measured on the ground for sawlog size messmate (*Eucalyptus obliqua*) (south-eastern Australia). The relationship approaches a straight line. The average percentage error of measurement of 20 trees was 12.6%.

In tropical rainforest, the crowns other than emergents are often so interlaced that it is impossible to measure crown diameter; but the crowns of dominant trees have been measured at a scale of 1 : 10 000 (e.g. Paelinck, 1958). Working with 1 : 40 000 photographs of Amazon forest, Heinsdijk

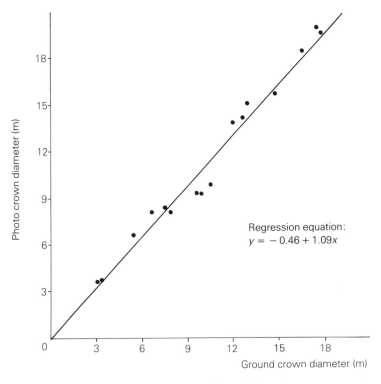

Figure 16.3 Regression relationship for *Eucalyptus obliqua* between crown diameter measured on aerial photographs and crown diameter measured on the ground (see text).

(1958) was able to recognize four crown diameter classes (10, 15, 20, 25 m). For three vegetation types in Guyana, Swellengrebel (1959) recorded crown widths on the photographs (1 : 10 000) and on the ground. For photographic measurement, best results were obtained with a crown diameter gauge, which provided class intervals of 1.6 m. This was better than could be obtained by measurement on the ground due to difficulty of locating the visible edge of the crowns.

Diameter at breast height

Crown diameter is known to be correlated with diameter at breast height (Figure 16.4). Unfortunately the precision of the estimates and the measurement of crown diameters on aerial photographs except at macro-scales have not been sufficiently accurate to satisfy the normal objectives of most intensive ground inventories. On the basis of ground measurement, Zieger as

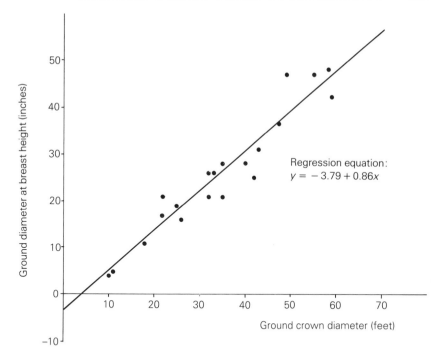

Figure 16.4 Crown diameter is known to be correlated with diameter at breast height. The graph shows the correlation for 20 measured trees (*Eucalyptus obliqua*).

long ago as 1928 demonstrated the correlation existing between crown diameter and diameter at breast height for Scots pine. Other workers, who have shown similar relationships for northern hemisphere conifers and hardwoods, include Ilvessalo (1950) for Scots pine in Finland, Eule (1959) for beech in Germany, Dilworth (1959) for Douglas fir in the western USA and Feree (1953) for hardwood species in the north-eastern USA. Davies (1954), working with photographs at a scale of 1 : 10 000 in New South Wales (Australia), was able to divide the red gum (*Eucalyptus camaldulensis*) into six diameter classes at breast height. In general, the regression of stem diameter on crown diameter tends to be linear for middle-diameter classes, but somewhat curvilinear for young and old trees.

Zsilinsky and Palabekiroglu (1975), working in boreal deciduous forest in Ontario, favoured the direct measurement of stem diameters from small-format macro-scale colour aerial photographs (35 mm). These were then combined with 100% tree counts by species and tree heights (using parallax measurements) to provide estimates of tree and stand volumes. In the sampling trial, the planning was based on conventional 23 cm format black-and-white aerial photographs at a scale of 1 : 15 840; but the macro-scale small-format photographs for the photogrammetric measurements were taken at a measured flying height of about 228 m (750 ft) above ground using a radar altimeter (scale approximately 1 : 4500). An 80% endlap between adjoining photographs provided five different perspectives of each tree. The images of the tree profiles were examined at ten times magnification. Breast heights of the stereoscopic models of the trees were determined by parallax measurement and the stem diameters measured directly on the photographs.

16.2 STAND MEASUREMENTS

Stand height

As with ground survey, only a small sample of the trees are measured on the photographs, which is sufficient to provide a reasonable estimate of the mean height of the forest stand. Under many circumstances, this overcomes the problem that the physiognomy of the forest does not permit the stereoscopic measurement of the heights of all trees recorded on the photographs. It permits the determination of stand height under conditions when some of the individual dominant and codominant trees cannot be measured due to shadows, crowns coalescing, dense canopy, etc. Suppressed trees and many subdominant trees will not be observable in the stereo-model, but usually these will contribute little to the timber volume. Thus the mean stand height determined using the parallax formula is that of the codominants or codominants and dominants and not the mean stand height of all trees, as could be measured on the ground. Care must be exercised that the measured heights of the selected dominant and codominant trees provide a representative sample of the forest stand. As with ground survey, the sample will not be representative if it is provided by trees along the edge of the stand (e.g. roadside).

Stand height is most readily measured photographically for boreal forest. In temperate closed forest, it is often necessary to search the stereo-model at low magnification under the stereoscope for gaps in the canopy and then at higher magnification (e.g. 3–6 times) to estimate the stand height from trees in the gaps using parallax measurements and a crude adjustment for the height of the understorey vegetation based on field experience.

In rainforest, it is normally impossible to measure stereoscopically the height of individual trees and often impracticable to measure stand height, since the ground is seldom seen and the crowns coalesce and are entwined with lianas. Nevertheless, the tops of the dominant trees are always exaggerated in the stereo-model (section 9.3), and this makes the separation of the crowns of dominant trees from the main crown layer feasible. In addition, the main canopy is usually much more even heighted, seldom exceeding 40–50 m; and is thus often not as important a factor in rainforest typing as in many temperate forests. Often a skilled interpreter using his/her visual skills and local knowledge can type out several volume classes.

Crown closure/crown cover

This is defined as the proportion per unit area of the ground covered by the vertical projection on to it of the overall tree crowns; and in combination with stand height, based on parallax measurements, it is frequently used to stratify the forest into timber volume classes. It is always the gross crown cover which is measured on aerial photographs.

Crown cover is determined using (a) stereograms covering the range of stand densities expected to be encountered on the photographs, (b) dot grids (Figure 8.6) with the spacing of the dots suitable for the scale of the photographs and with the colour or tone of the dots to contrast with the photographic background, and (c) crown closure/crown cover percentage cards (Figure 16.5) for comparison with the image of crown cover observed in the stereo-model and prepared with the texture of each cover class to correspond with the photographic scale and the physiognomy of the forest. Using crown closure cards an interpreter can quickly place a stand, as observed on a stereo-pair of photographs, in its appropriate crown closure class. Occasionally, as in western Australia (McNamara, 1959), canopy closure classes have been estimated for the upper storey and the total stand.

As pointed out by Spurr (1960), crown closure can be better and more quickly estimated from the view provided by the photographs than on the ground. He commented that for photographic scales between 1:7000 and 1:30000 in North America sampling on the photographs overestimates the crown closure by 5–10%. As observed by the writer for the same area of dry

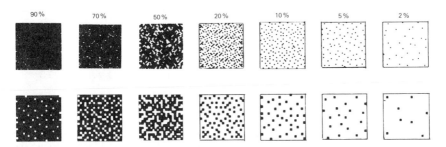

Figure 16.5 Crown closure/crown cover card used for estimating the crown cover observed on stereo-pairs of photographs. The texture of each percentage class should be similar to the textures encountered on the photographs – shown above are two texture classes. Dense is considered to be over 90%, moderate 50%, and open under 10%.

sclerophyllous forest in central Victoria (Australia), the sampling error and crown closure estimate increase with smaller photographic scales (1 : 2700 to 1 : 34 000 (Figure 16.6). In general, tall trees, polar-facing slopes, short focal lens, infrared photographs, early morning or late afternoon photography, areas towards the periphery of the photographs and small-scale photographs all favour the overestimation of crown closure by a photointerpreter.

As pointed out by Husch (1963), the weakness of crown closure as a descriptive characteristic of forest stand volume is that crown closure is not closely correlated with the number of trees or tree sizes. In the Netherlands, Stellingwerf (1963) determined crown closure for *Pinus sylvestris*, using 500

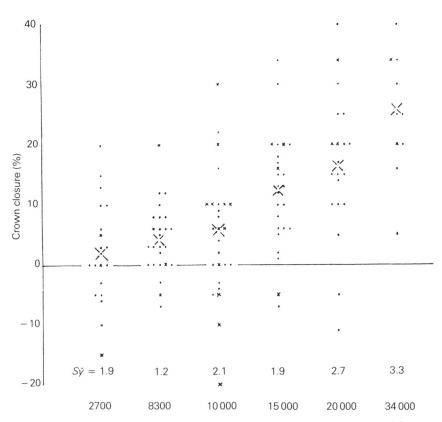

Figure 16.6 Crown closure/crown cover estimates are influenced by several factors including the film–filter combination, flying height and (lens) focal length. Using the common combination of panchromatic black-and-white film and a minus-blue filter, crown closure and the sampling error will be found usually to increase as the photographic scales become smaller. As shown above, the percentage differences between photographic and ground measurements change with scale.

points or about one point per 160 m^2, and concluded that although crown closure varied between 45% and 75%, it was not generally correlated with volume. It is important to appreciate also that the crown closure percentage of a single-storey stand often does not cover 100% of the ground area even when fully stocked. Usually a forest stand is at maximum stocking when the ground is not fully covered by the crowns. By age class, the crown closure of fully stocked stands of eastern white pine (*Pinus ponderosa*) in the USA on an average site has been reported as 100% at 40 years, 90% at 70 years and 65% at 100 years (Spurr, 1960).

Stocking

This is the number of trees per unit area (i.e. per ha, acre); and together with crown closure these are the only two direct measurements of stand density obtainable from aerial photographs. Angle crown counts by adapting Bitterlich's method (1948) of stem diameter counts remains experimental. Counting of trees from stereo-pairs of photographs depends on the skill of the interpreter, the scale of the photographs, the physiognomy of the forest, and the species and age of the stand. To some readers, it may be surprising to learn that the counting of trees on aerial photographs is often difficult, and it can result in substantial sampling errors, due to crowns coalescing, some crowns being hidden and small crowns not resolving on the photographs. The most accurate counts are made on large-scale photographs of boreal forests, open-grown woodlands and recently thinned plantations.

The use simply of number of trees as an estimate of stand density has two disadvantages. Firstly, as mentioned, crown counting from photographs may be difficult and inaccurate. For example, in making crown counts in Australia of radiata pine on 1 : 8000 photographs, it was found that a spacing of about 12 ft by 12 ft was needed for successful counting, and stand volume per acre was not closely correlated with number of trees per acre (Cromer and Brown, 1954). Secondly, as at a specified age natural stands at full crown closure frequently have a wide range in the number of trees per unit area, it is necessary to use this measure of stand density in association with some other parameter (e.g. height, basal area, crown diameter). Smith (1965), for example, commented that even with 100% crown closure in natural forest in western Canada, the number of stems in stands which were open-grown may be only 25% of the 'normal' number of stems and the timber volume may be only one-sixth of that in an average stand. The height/crown width ratio varied between 2.5 and 8.

When the number of stems is combined with another parameter, the two-dimensional measurement may reveal correlation with age and site quality. It is important to recognize that density or the crowding of individuals within an area influences timber volume, crown width, crown length, the crown width/diameter bh (breast height) ratio and the tree height/crown width ratio.

A crude but quick method of estimating the number of stems (N) per hectare from photographs is to use average crown width (C_w) in metres, crown closure (C_c) and the following formula as an alternative to attempting to count the number of stems:

$$N = \frac{10\,000 \times C_c\%}{0.785(C_w)^2}$$

(N.B. 0.785 is the decimal coefficient of a square and circle of the same diameter.)

16.3 TIMBER VOLUME ESTIMATES

The approaches to using aerial photographs in forest inventory may be summarized as follows: (a) in poorly mapped or unmapped areas, the individual photographs or ortho-photographs or constructed photographic mosaics, as discussed in Chapter 11, are used directly as a substitute for a map in the planning of fieldwork; (b) as an extension of (a) the forest boundaries and the principal forest strata are interpreted and delineated on the aerial photographs and/or on photographic mosaics as a guide to ground inventory and to avoid field crews traversing non-forest or unproductive forest; (c) as an extension of (b) the strata delineated on the aerial photographs are used as a sampling frame within which the ground samples are distributed according to some economic or volume criteria of each forest type and hence the number of field samples is reduced to achieve a predetermined level of accuracy; (d) stereograms of each expected volume class are prepared and used in conjunction with (c) and/or the forest inventory on the ground to estimate the timber volume of each stand/forest type; (e) as discussed below and in Chapter 18, plots (usually circular) are distributed on the aerial photographs and after stereoscopic analysis a percentage of these photo-plots are visited in the field to determine the timber volume per plot, per hectare and per stand/forest type; (f) photogrammetric parameters are used to calculate tree volumes; (g) as an extension of (f), tree aerial volume tables are prepared and used to estimate the timber volumes per sample photo-plot; (h) stand aerial volume tables are constructed and used to estimate the volume per photo-plot or stand.

It is questionable whether (a) should be considered as a photointerpretative technique since the photographs are being employed as a map and if a suitable map existed then the photographs would not be used. Examples, however, occur in publications on remote sensing applied to forestry and to mineral exploration in poorly mapped regions; and this approach extends to the use of satellite imagery.

Photographic stratification (b)

This approach has been widely used in forest surveys for many years, extending from the tropics to the boreal forests. In Australian evergreen forests, the stratification into sampling strata is made with the aid of photographs, but usually all volume data is collected on the ground. In Tasmania, for example, the eucalypt forest is classified into old growth and regrowth, with the former often having five height, four crown cover and four tree count classes. This photographic stratification of the forest in conjunction with a 1% or less ground cruise provides the timber volume estimate.

The approach extends to a timber buyer using stereo-pairs of photographs to determine which stands should be purchased and what is a suitable price. As an experienced timber buyer estimates visually by a cursory field visit the volume and value of a timber stand, so an expert photointerpreter appraises the timber volume of the stand recorded on the aerial photographs. Burns, for example, working in the USA in 1979, has shown that volume per acre can be estimated visually from stereo-pairs of aerial photographs within 15% of the field estimated values. The third approach, (c), has been widely used in forestry since the 1960s (e.g. Avery, 1964) and will be discussed in Chapter 18.

Stereograms (d)

The estimate of timber volumes from photographs using stereograms is frequently as good or better than volumes obtained by using aerial volume tables; but the precision cannot normally be expected to be as high as the precision obtained by ground measurements in conjunction with aerial photographs. Frequently each stereogram comprises a stereo-pair of about $2.5 \, \text{cm} \times 2.5 \, \text{cm}$ cut-outs mounted at a distance of about $5.5 \, \text{cm}$ centre to centre. Given practice, supporting statistical data and suitable stereograms, the image analyst in addition could probably estimate, as accurately as estimates of timber volume, the leaf dry weight of trees per unit ground area, total cellulose production, total biomass and the leaf area.

In the Netherlands, Stellingwerf (1963) applied the stereogram technique to timber volume estimates of Scots pine (*Pinus sylvestris*). Stereograms had volume differences of $50 \, \text{m}^3/\text{ha}$ and the area of each photo-plot was $500 \, \text{m}^2$ at an intensity of one per hectare. About $500 \, \text{ha/day}$ were sampled on the photographs when estimating volumes in $10 \, \text{m}^3$ units/ha. Some of the plots were double sampled on the ground, so that a regression equation between photographic volume and ground volume could be prepared. The mean photographic volume was $153 \pm 3 \, \text{m}^3/\text{ha}$ compared with a terrestrial volume of $149 \, \text{m}^3$. About fifteen ground plots representing the entire range of stand volumes were found to be sufficient for determining the equation. The regression equation was calculated as:

$$y = 86 + 0.61x$$

where y is the adjusted photo-volume and x is the unadjusted photo-volume.

Photo-plot sampling (e)

The forest is divided into strata, as in approach (b) and (c); but the strata are used for distributing a large number of plots on the aerial photographs and not for distributing field plots. A minor percentage of the photo-plots are then selected as field plots. As discussed in Chapter 18, the sampling design may be multiphase or multistage; and it may involve three-stage sampling using satellite imagery as the first stage.

A classic example of two-stage aerial photo-plot sampling in forest inventory was carried out in the vicinity of the Caspian Sea (Iran) by Rogers (1961). Basically this combines estimation of stand volumes from aerial photographs and regression sampling. As photographic variables, total tree height and crown closure were used. Regression constants and coefficients were computed from the data collected from 54 ground plots and the corresponding photo-plots, using the method of least squares. The ground plots were systematically selected at every forty-third photo-plot. The 2308 photo-plots provided the average stand height ($\bar{X}_1 = 18.7$) and the average crown closure ($\bar{X}_2 = 63.0$). These values were then used in the following regression formula to provide the average volume ($\bar{V}_1$) per hectare for the gross land area.

$$\begin{aligned}
\bar{V}_1 &= P_f(a + b_1\bar{X}_1 + b_2\bar{X}_2) \\
&= 0.8151(21.8297 + 3.6506\bar{X}_1 + 1.3565\bar{X}_2) \\
&= 143.09
\end{aligned}$$

where p_f is the proportion of forest land to gross land from aerial photographs, a is the regression constant, b_1 is the regression coefficient for height and b_2 is the regression coefficient for crown closure.

The average volume per hectare was multiplied by the forest area to give the total volume. The total volume was then distributed by forest types according to the volume classes of the ground plots.

Photogrammetric determination of tree volume (f)

An alternative approach to using stereograms is the direct use of photogrammetric variables to determine tree and stand volume; and it can be considered as an intermediate step to the construction and use of aerial tree volume tables. The same variables may be used, but no volume table is constructed.

The simplest, as developed by the writer in open sclerophyll forest in south-eastern Australia, is based on the classical 'mean tree' method of calculating stand volume. It has the same problem of selecting the representative mean tree in the stand, which is more difficult when making the selection from the stereoscopic model. On colour aerial photographs at a scale of 1 : 8000, the forest stand (messmate, *Eucalyptus obliqua*) was delineated, based on forest type, stand height and crown closure. Within each stand shown on the aerial photographs, crown diameters were measured, the mean crown diameter calculated and a tree with the approximate mean crown diameter identified. This tree was then located in the field; its diameter at breast height, upper stem diameter and log length were measured and its timber volume calculated. The volume per hectare was determined by multiplying the mean tree volume by the number of trees per hectare. For three stands, the volumes per hectare were 18%, 21% and 29% less than the volumes determined by ground inventory.

It seems likely that the volume estimates could be improved by introducing sample trees for each of several diameter classes. It has, for example, already been shown for beech stands in the Federal Republic of Germany (Hildebrandt and Kenneweg, 1969) that the distribution of stems by diameter classes at breast height (dbh) can be made based on the crown diameter distribution measured on aerial photographs (Figure 16.7).

If macro-scale photography is used, it is feasible to calculate the tree volume directly using stereoscopic measurements. Two Canadian publications will be cited. Sayn-Wittgenstein and Aldred (1967) pointed out that in typical mixed conifer/hardwood stands in eastern Canada that 85% of the tree species could be identified on photographs at a scale greater than 1 : 3000, that tree heights can be measured with a standard error of the estimate of about 1 m, well-formed tree crowns with an accuracy of about ± 0.6 m and tree number at

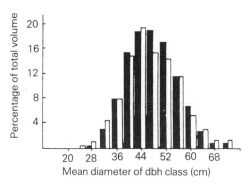

Figure 16.7 Comparison of tree volume distribution for beech in Germany, based on aerial photo-interpretation (black) and complete tree tally in the field (outlined in white).

about 96%. Since most of the trees not counted are of small size, the volume loss is very small (e.g. under 1.5%). Zsilinsky and Palabekiroglu (1975) demonstrated that accurate measurements can be made of stem diameter at breast height, upper stem diameter and stem length. When regression analysis is used as part of the image analysis, there is probably little justification for using more than three or four photogrammetric variables of the forest stand, since any potential minor improvement in accuracy is likely to be offset by measuring errors etc.

16.4 AERIAL VOLUME TABLES

Aerial volume tables have been widely used in the USA, but seldom favoured in Europe (cf. section 15.1). Aerial volume tables are analogous to conventional single tree volume tables and forest stand volume tables. Unfortunately they do not give the accuracy and precision obtainable by using carefully constructed conventional regional and local volume tables. As in the construction of conventional volume tables, there is a choice in selection of the variables to be used.

Aerial tree volume tables (g)

The approach outlined in the last section on the 'mean tree' method of estimating volume could be expanded into a tree aerial volume table. However, single tree aerial volume tables usually result from the modification of existing local tree volume tables by relating the independent variable crown diameter to diameter at breast height (dbh), and this to tree volume. Alternatively, crown width and tree height, as the variables, are related to volume. As outlined already in section 16.1, regression analysis is used to establish the relationship between crown diameters on the aerial photographs and the dbh of the same trees measured in the field.

If a local tree volume table is not available then an aerial tree volume table is prepared by correlating the dependent variable volume of selected trees measured on the ground with the independent variables of top height and bole diameter via apparent crown diameter of the same trees on the photographs. Details on the construction of a tree volume table were given by Feree (1953). As early as 1928, an aerial volume table was constructed by Zieger for Scots pine in Germany. This used tree height by parallax measurement, photographic crown diameter, dbh ratio of measured trees and (tree) form factor (Jacobs, 1932).

An aerial tree volume table is used to estimate the volumes of forest stands by the measurement of individual trees on selected plots located on the photographs. From the tree timber volume on these plots and the areas of the stands, the total volume of timber in the stand is calculated. Several writers (e.g. Spurr, 1960) have pointed out that the standard errors obtained when using tree volume tables are high; and in consequence aerial volume tables involving photographic measurement of forest stands and not trees are to be preferred. Usually bias is introduced in using either tree or stand aerial volume tables, when the user of the table is a person other than its constructor. To minimize this problem, the user makes an adjustment by introducing regression between plots of ground measured volumes and the aerial photographic volumes of the same plots for a sub-sample within the forest.

Aerial stand volume tables (h)

Since the Second World War over 40 publications have become available on aerial stand volume tables. These are mainly from the USA, but a few are available from other countries (e.g. Taniguchi in Japan, 1961; Nyyssonen in Finland, 1955; May in Australia, 1960; Stellingwerf in the Netherlands, 1973). In western Europe, the existence of intensive management data has contributed to the reduced interest in an application of aerial volume tables.

As pointed out by Meyer and Worley (1957), three types of error contribute to errors of aerial stand volume tables: (a) systematic and random errors of the photographic measurements, (b) error of the estimate in the aerial volume table and (c) sampling error of the selected plots. In applying the aerial volume table, the sample photo-plot estimates involve human interpretative errors as well as sampling errors. If the resultant error is doubled by using an aerial volume table, four times as many photo-plots may be needed to give the same accuracy as the ground survey. However, photo-plots may take only 1/10 the time to measure and cost 1/100 as much as a ground plot (Avery and Meyer, 1959). Thereby, this makes the use of photo-cruises for direct determination of volume attractive.

Conventional stand volume tables are usually based on the product of stand height and stand basal area, although a third variable may be incorporated in the tables. Aerial stand volume tables are based on the correlation between stand volume and variables of the stand, which can be measured on aerial photographs. In the ideal case of an even-aged, fully stocked stand on a uniform site, basal area remains relatively constant, and stand volume varies with stand height alone. Normally, stand aerial volume tables are compiled from measurements of sample plots on the ground, which cover the range of variation likely to be met with in the use of the table. Ground measurements should be made in a similar way to photo measurements. For example, height measurements of young conifers, with finely tapered crowns, should be made to the visible crown height resolved on the photographs. This may be considerably below the tip of the tree. Stand height is best expressed as dominant height, that is, the mean height of a given number of the tallest trees per plot. The photo-plots are generally circular for convenience, and vary in size according to stand density, photographic scale and the information required from them. Plots are seldom less than 1/100 ha or larger than 5 ha, the larger sizes being necessary for very open mature stands or stands suffering from severe damage. Since the photographic scale varies with topography, constant plot size can only be obtained by varying the photo-plot radius with major changes in elevation; but this is not usually necessary unless elevation varies considerably in terms of flying height (Chapter 9).

An example of the mathematical steps involved in preparing a theoretical one-variable aerial stand volume table was summarized concisely by Stellingwerf (1973) as follows. The linear equation may be written as:

$$Y = a + b(x_i - \bar{x})$$

where Y is the regression estimate of y, y is the individual value dependent on individual x_i value,

$$a = \bar{y} = \frac{\sum_1^n y_i}{n}$$

n is the number of plots, i.e. number of y and x values, and b is the regression coefficient. y_i is the volume on individual field plot in m^3 ($= v_{it}$), $\bar{y}$ is the mean volume ($= \bar{v}_t$), x_{it} is the crown cover percentage on corresponding photo-plot ($= c_{it}$), $\bar{x}$ is the mean crown cover percentage ($= \bar{c}_t$), and n_i is the number of plots.

The equation, which is used to tabulate aerial photo-volumes, may be written:

$$V = \bar{v}_t + b(c_{it} - \bar{c}_t) \tag{16.1}$$

The regression coefficient is given by:

$$b = \frac{\sum_1^{n_i}(c_{it} - \bar{c}_t)(v_{it} - \bar{v}_t)}{\sum_1^{n_i}(c_{it} - \bar{c}_t)^2}$$

The standard error of $\bar{v}_t$, or rather that of v_{it} when $c_{it} = \bar{c}_t$, is

$$S_{\bar{v}_t} \quad \text{and} \quad S_{\bar{v}_t}^2 = \frac{s_{vc}^2}{n_1}$$

The standard error of b, S_b is given by:

$$S_b^2 = \frac{s_{vc}^2}{\sum_1^{n_1}(c_{it} - \bar{c}_t)^2}$$

and the standard error of estimate, s_{vc}, by:

$$s_{vc}^2 = \frac{\sum_1^{n_1}(v_{it} - \bar{v}_t)^2 - b^2 \sum_1^{n_1}(c_{it} - \bar{c}_t)^2}{n_1 - 2} \tag{16.2}$$

Since c_{it} is not subject to sampling error, the variance of the regression estimate, V, corresponding to c_{it} is given by:

$$S_V^2 = S_{\bar{v}}^2 + (c_{it} - \bar{c}_t)^2\, S_b^2$$

or:

$$S_V^2 = \frac{s_{vc}^2}{n_1} + s_{vc}^2\, \frac{(c_{it} - \bar{c}_t)^2}{\Sigma_1^{n_1}(c_{it} - \bar{c}_t)^2} \tag{16.3}$$

If the table is applied in new work to n_2 photo-plots the following steps may be taken to check the bias:

1. Determine c_1 with the same class interval used in the aerial volume table on n_2 photo-plots of the same size as used for n_1 plots in the table;
2. Calculate mean crown cover:

$$c_{n_2} = \frac{\Sigma_1^{n_2} c_i}{n_2}$$

3. Substitute in the aerial volume equation the value $c_{it} = \bar{c}_{n_2}$, which gives the unadjusted mean volume per plot:

$$\bar{v}_{n_{2(unadjusted)}} = \bar{v}_t + b(\bar{c}_{n_2} - \bar{c}_t) \tag{16.4}$$

As $\bar{c}_{n_2}$ is subject to sampling error, the variance of $\bar{v}_{n_{2(unadjusted)}}$ is given by:

$$S_{\bar{v}_{n_{2(unadjusted)}}}^2 = S_{\bar{v}t}^2 + S_b^2(\bar{c}_{n_2} - \bar{c}_t)^2 + b^2\, S_{\bar{c}_{n_2}}^2$$

or:

$$S_{\bar{v}_{n_{2(unadjusted)}}}^2 = \frac{s_{vc}^2}{n_1} + s_{vc}^2\, \frac{(\bar{c}_{n_2} - \bar{c}_t)^2}{\Sigma_1^{n_1}(c_{it} - \bar{c}_t)^2} + b^2\, \frac{s_{c_i}^2}{n_2} \tag{16.5}$$

Of the three terms at the right-hand side of the equation, the first two refer to the volume table, while the third is caused by the n_2 photo-plots.
The variance s_{ci}^2 can be calculated as:

$$S_{c_1}^2 = \frac{\Sigma_1^{n_2} c_i^2 - \{[\Sigma_1^{n_2} c_i]^2/n_2\}}{n_2 - 1}$$

Most aerial stand volume tables have been prepared for a single species or a group of similar species for the purpose of statistical accuracy; but Avery and Meyer (1959), for example, provided a 'composite' volume table for conifers and hardwoods in southern Arkansas, since these species could not be separated on the photographs (scale 1:20 000). Part of the table giving volumes by stand height and crown cover (10% intervals) is shown in Table 16.1. In preparation of the table, nine stand variables were tested, being combinations of the three basic variables: height, crown cover and crown

Table 16.1 Example of an aerial stand volume table (see text)

Average total height (ft)	Crown closure									
	5%	15%	25%	35%	45%	55%	65%	75%	85%	95%
	(Gross ft³ per acre)									
40	175	215	250	290	325	365	400	440	475	515
45	240	295	345	400	450	505	555	610	660	715
50	330	395	460	525	590	655	720	790	855	920
55	395	490	580	675	765	860	950	1045	1140	1230
60	480	600	715	830	950	1065	1180	1300	1415	1530
65	585	725	860	1000	1135	1275	1410	1545	1685	1820
70	715	865	1020	1175	1330	1480	1635	1790	1945	2100
75	860	1025	1195	1360	1530	1695	1865	2030	2200	2365
80	1020	1200	1380	1555	1735	1910	2090	2270	2445	2625
85	1205	1390	1575	1760	1945	2130	2315	2500	2685	2870
90	1410	1600	1785	1975	2165	2350	2540	2730	2915	3105
95	1635	1820	2010	2200	2385	2575	2765	2950	3140	3330
100	1875	2060	2245	2430	2615	2800	2985	3170	3355	3540
105	2140	2315	2495	2675	2850	3030	3210	3385	3565	3745
110	2420	2590	2755	2925	3095	3260	3430	3600	3765	3935
115	2725	2880	3030	3185	3340	3495	3650	3805	3960	4115
120	3045	3180	3320	3455	3595	3730	3870	4005	4145	4280

diameter. As average total height was shown to be the most valuable single variable for predicting gross stand volume, the table was compiled with 5 ft height classes. Crown cover was found to be the next most important variable; and crown diameter was omitted from the final table, since it did not contribute to the volume estimate in the five-term regression.

16.5 CONSERVATION MEASUREMENTS FROM AERIAL PHOTOGRAPHS

General

In addition to obtaining tree and stand parameters from aerial photograph, with the primary purpose of determining the utilizable timber volumes, other useful measurements can be made. These are mainly concerned with the forested land, as requiring protection and biomass as distinct from timber volume.

Stands of unique ecological and floristic interest can be identified and delineated on the aerial photographs or photographic mosaics, their areas estimated using dot grids and access routes identified and traced from the stereoscopic model. Within the forest stand, trees of special interest can be identified on the photographs in terms of their nearest neighbours, their size seen stereoscopically, their health condition assessed, their crown diameters and crown area measured using dot grids and their heights estimated using a parallax wedge or parallax bar.

The boundaries of water catchments can be delineated and the areas measured using dot grids or a planimeter. Drainage lines can be traced and measured using an engineer's scale and slopes measured by parallax differences (section 9.3). Erosion and erosion potential can be assessed in terms of the estimated tree crown cover and type and incidence of erosion related to slope and soil type. For areas with little or no tree cover, the incidence and type of erosion is best seen on large-scale aerial photographs; and often colour IR photographs are to be preferred in order to distinguish between the vegetation and bare soil (Figure 16.8).

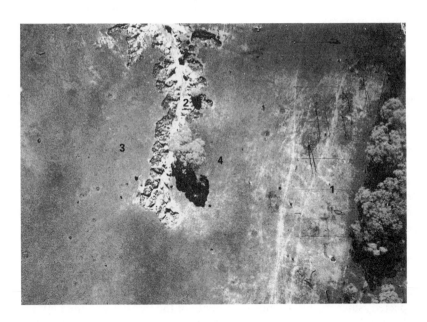

Figure 16.8 In addition to obtaining statistical data on the forest stand, aerial photography can be used to locate eroded areas and types of erosion. Note above the overall loss of forest cover, the area of sheet erosion (1), the extent and depth of gulley erosion (2) developing sink-holes indicating subsurface erosion (3) and tree crown and shadow (4) [scale 1 : 7500, reduced from 1 : 5000 colour (infrared)].

Biomass

On the basis of well-established mathematical relationships existing between tree diameters and biomass, it is possible to establish similar relationships between crown diameters or crown cover on the photographs and the area of tree foliage, branch dry weight, leaf dry weight, leaf green weight, total dry weight, etc. of a forest stand. The estimates can be expected to be at least as accurate as estimating bole volumes by aerial volume tables.

In establishing correlations, there is a choice of two techniques established through measurements on the ground. Material may be collected and weighed as is customary in crop plant and fruit tree studies (e.g. Moore, 1968) and as used by Ovington (1957) in the study on the ground of forest stands in the United Kingdom. Alternatively a mathematical relationship between the size of different branch members and foliage etc. may be determined (e.g. Attiwill, 1962) and expressed in the general form:

$$\log_{10} y = a \log_{10} x + \log b$$

where a and b are constants and x and y are weight or dimensions (e.g. upper diameter of stem).

A useful review of literature of studies carried out on individual trees on the ground was provided by Kittredge (1944), who also studied ten species and twenty-eight stands in California. He concluded that the relationship between leaf weight and diameter (at breast height) is applicable to trees of different sizes, densities, crown classes and ages at least up to the culmination of growth and beyond that age for tolerant species in all-aged stands. Satoo, Kunugi and Kumekawa (1956) and Cable (1958) extended Kittredge's studies to leaf area per tree and oven-dry weight using aspen poplar and ponderosa pine respectively by a least-squares solution; and they showed that surface area in square centimetres can be estimated from dry weight of needles. Both the branch dry weight and leaf area of even-aged stands of *Eucalyptus obliqua* have been calculated from branch girth, demonstrating that the girth of a branch depends on the weight of material supported by the branch and that the crown weight is a function of the square of the stem diameter at the base of the crown and at breast height.

With respect to using satellite data to estimate forest biomass, few studies have been confined to woody vegetation; but significant findings can be expected, since the reflectance/reflectivity of the forest, as pointed out earlier (Chapter 3), is influenced by both height and crown cover. For example, Herwitz, Peterson and Eastman (1990) concluded that the Landsat TM sensor (bands 4/3 ratio) may be a better guide to moderate changes in the LAI of closed canopy pine plantations at local scales than field measurements involving allometric equations. However, Sader *et al.* (1989) from a study of Landsat TM data of the Amazonian rainforest caution against using the vegetation index for predicting total biomass or carbon storage of wet tropical forest. Hellden and Olsson (1982) working in the savanna, shrubland and open woodland of the Sudan, concluded that woody biomass can be estimated and described as a function of the Landsat MSS 6/MSS 5 spectral band ratio and tree/bush height for a given physical environment (cf. land subprovince, Chapter 14). They established the following relationship for woodland/shrubland, but excluding grassland.

$$Y = -45.5 + 37X_1 + 2.78X_2$$

where

Y = wet weight of woody biomass

X_1 = MSS 6/MSS 5 pixel ratio

X_2 = tree height

Forest damage

With respect to forest damage, this has been considered in section 15.6. A brief comment will be made here on relevant measurements. Estimates of crown closure can be made using a density scale and the respective forest areas delineated and measured on the aerial photographs. In the study of individual damaged trees on large-scale aerial photographs, crown diameters can sometimes be measured with a crown diameter wedge, or more often the accuracy is improved by estimating crown area by using a dot grid or by comparisons with a crown area chart. Often, the edges of the crowns of damaged trees are not adequately resolved on aerial photographs even at large scale.

The parallax formula is applied in estimating tree and stand heights of damaged trees; but difficulty is often encountered on placing the floating mark correctly on dieback crowns and partly defoliated crowns. If the volume of dead trees and dead forest stands is required, a special aerial volume table can be constructed. For example, Weber (1965), using 70 mm colour photographs at a scale of 1 : 1584, measured crown closure and stand height of the dominant and codominant dead white spruce and balsam fir trees, and obtained a high correlation with the ground-measured cubic volume of the dead trees. The aerial stand volume table for dead trees was then constructed using multiple regression and the two variables of crown closure and stand height.

17

Satellite remote
sensing application

On the basis of the forest detail to be obtained from aerial photographs, as considered in the last two chapters, it is unlikely that satellite remote sensing (SRS) will be able to provide the same quality of usable information. The role of SRS will remain mainly complementary; and its attractiveness will be the comparative low cost per unit land area of data acquisition, and the lower cost per unit land area of image analysis. No doubt, this will favour the expanding operational use, whenever feasible, of SRS in forest resources studies, particularly in large area surveys and for monitoring frequently at national, continental and global levels changes in forest cover, forest density and land use. For example, as reported by Myers (1980), a land resources survey at small scale in northern Thailand, using Landsat MSS imagery, took 240 man-days to complete, while a comparable survey using aerial photographs would have required 4200 man-days. The cost ratio per km^2 was approximately 1 : 8.

Aerial photography is usually too expensive to justify the coverage of the same forest more often than once every 5–10 years. Also, it seems advisable when a large area is to be covered by aerial photography to acquire firstly earth resources satellite imagery in order to determine its role, since this may modify proposed aerial photographic specifications and help to reduce overall costs. The role of environmental satellite data has yet to be adequately identified in forestry; but, due to its daily repetitive coverage (section 7.3) and very low cost per unit of land area, it may prove attractive for several purposes, especially 'change' monitoring at continental and global levels.

Research studies over a period of almost two decades, since Landsat was launched as ERTS-1 in 1972 (section 7.1), have ensured that the present and near future operational role of satellite remote sensing by earth resources satellites is satisfactorily understood and used in forestry. At times the information expectancy of Landsat has been overstressed and, in retrospect, the operational role of Landsat has been modest in forestry. We must, however, not overlook the fact that the earlier Landsats were always considered by NASA as experimental satellites; and that in 1972 there was no experience world-wide in applying the information content of the data provided by a continuously orbiting earth resources satellite. Without this gained understanding and experience, the application of the spatial, spectral and radiance qualities of the second generation of earth resources satellites would not be so advanced and the digital analysis techniques not so well developed.

We should always remember that irrespective of the currently used electro-optical sensors on board a satellite (section 7.1), the pixel provides an average radiance value of an extensive ground area, as compared with the many grey-scale values within the same area on an aerial photograph. Varying with the aerial photographic scale, there will be different densities within the images of the tree crown, in the shadow of the crown and of the bare soil, litter layer or herbaceous vegetation between two contiguous trees. Even within the same ground area covered by a single pixel of the fine resolution of SPOT, there can be 10–20 tree crowns. Whether in some circumstances the averaging effect on the recorded spectral reflectance provided by the satellite pixel can be advantageous in improving image analysis is debatable; but satellite imagery has the advantage over aerial photographs of providing for an identified image class more uniform grey-scale values (density values).

As described in sections 7.1 and 7.3, pixel size varies from 100 m^2 for SPOT to about 16 km^2 for NOAA AVHRR (GAC). Adjoining pixels with the same grey-scale values may have quite different natural surfaces when observed on the ground or examined on stereo-pairs of aerial photographs. Thus if one pixel comprises 25% of a highly reflective surface (e.g. bare sandy surface) and 75% of forest with low spectral reflectivity, its averaged signal may equate with an adjoining pixel of intermediate grey-scale value, which is entirely covered with green grassland and scattered trees. In practice, this can result in misclassification when using computer-assisted techniques; and it follows that in selecting training sets for supervised classifications, the operator should have a sound knowledge of the reflective properties of natural surfaces. This will help in categorizing initially the likely spectral classes to be encountered in analysing the scene, and serve as a useful guide when using aerial photographs to select more correctly representative samples for supervised classification. Even if the sensor's resolution is fine, its usefulness for spectral classification may not be greatly improved, since the extracted information still relies basically on spectral differences between images, and not on the wide choice of diagnostic elements available to the aerial photointerpreter (section 8.3).

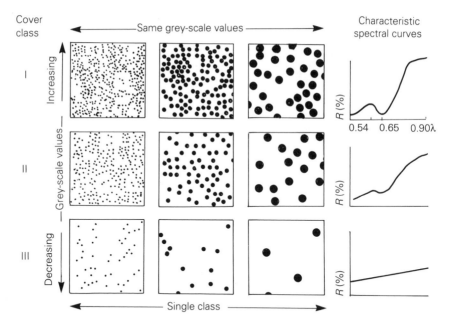

Figure 17.1   Diagram illustrating three classes of tree crown cover (I, II, III) within individual pixels. Within a single class of crown cover the spectral grey-scale values may be similar, although the crown sizes and timber volume per hectare vary greatly. The dots represent the crown sizes within the pixels and cross-wise the crown cover is constant. The spectral curves to the right are indicative of their characteristic shapes for declining crown cover above a uniform soil.

Figure 17.1 helps to illustrate these comments. Three decreasing crown cover classes (I, II, III) are shown hypothetically against a background of uniform bare dry soil. Crosswise in the diagram within a crown cover class, as might be recorded within a single pixel of the satellite's sensor, the vegetation is represented by different-sized crowns having the same leaf area index. For example, class I might be the green crowns of saplings, class II pole-sized trees and class III mature sawlog sized trees. In these circumstances, the class spectral curve would be expected to be similar and the radiance values the same, although the stand structures vary greatly and would be immediately recognized on aerial photographs. This helps to explain the problem often occurring with digital satellite image analysis in separating timber classes and classes related to regeneration following clear felling or with shifting cultivation in tropical forests. At various stages of development, the tropical regrowth forest rapidly attains a similar high leaf area index (LAI) to the surrounding mature forest.

17.1 VISUAL IMAGE ANALYSIS APPLIED TO FOREST STANDS

This presupposes no ready access to a digital image analysis system, and that the analysis is made using standard products supplied by a distribution centre (i.e. bulk products). Intermediate between analysis considered in this section and digital analysis considered in section 17.2 are 'tailored' hard copies provided by locally available digital image analysis sytems. The purchased standard products (section 7.4) will usually be colour composites in three spectral bands, as prints or transparencies; but single spectral band black-and-white photographs will also be available. Also, depending on the image source and the distribution centre, the imagery will have been to varying degrees spatially corrected and may have been digitally colour enhanced (section 7.2). A choice of scales is available as commented on in Chapter 7.

The advantage of visual analysis of satellite imagery is that it can be intimately associated with fieldwork and local knowledge. Although it may be supported by specialized equipment including an overhead projector, additive viewer or diazo printer (section 8.6), most studies will be qualitative in approach as compared with aerial photographic interpretation being both qualitative and quantitative. The smallest classifiable areas are likely to be groups of nine pixels or possibly 25 pixels, although under very favourable conditions provided by the colour contrast smaller groups may be identified. In Chile, for example, eucalypt plantations and radiata pine plantations of 1 ha or larger were separately mapped from Landsat MSS imagery. For forest in Michigan, over 50% of interpretation has been found to be associated with areas of 2 ha or less in size (Karteris, 1985). In Brazil, (Baltaxe, 1980) the minimum area for mapping eucalypt and pine plantations from Landsat MSS imagery was determined as 20 ha. In São Paulo state using black-and-white imagery, the area of eucalypts was overestimated by 2.5% and that of pines by 5.3%.

Due primarily to the relatively large size of the satellite pixels (section 7.1), there is not the same need to consider the interpretative techniques and results according to the zonal composition of the forests, as has been made in Chapter 15 when using aerial photointerpretation. If a zonal distinction is desirable in developing satellite imagery analysis, then it is probably best considered on the basis of the usefulness of large area surveys or in terms of forest management. The application of satellite imagery to forestry is least beneficial in European countries with a long tradition of detailed forest management and having large-scale stockmaps; and it is most useful in many developing countries with their needs for surveys at reconnaissance level and lacking up-to-date small-scale maps (Chapter 11).

Developing countries

FAO, for example, has been associated with the use of satellite imagery in over 100 countries. If reliable maps do not exist, then the satellite imagery readily provides an important low-cost substitute, namely as a sketchmap or small-scale planimetric map (section 11.2). It also serves as a base for defining the area to be covered by aerial photography. With the finer resolution data provided by Landsat TM and SPOT, more man-made and natural features are readily identified on the imagery and these can be used to orientate field crews and to plan the field survey (Chapter 18).

Access routes to the forest can be identified, forest separated from non-forest and the forest divided into two or more strata. As discussed in Chapter 14, the land cover on Landsat MSS imagery can be classified at level I and in some cases at level II; and on the finer resolution imagery at level II, or occasionally level III (section 14.2). Vegetation can usually be typed at the level of plant formations and possibly one or more of these divided into subformations, and particularly on the finer resolution imagery, land systems can be mapped. For example, in an FAO project in Bolivia, in the absence of aerial photographs and suitable maps, Landsat MSS black-and-white imagery (red, IR bands) at a scale of 1 : 250 000 was used in the inventory of tropical high forest, as a simple planimetric map, to position field samples and camp sites and to stratify the forest into broad types. The latter were classified on physiognomy and site as palm forest, riverain forest, swamp forest, forest on drained sites and savanna on seasonal swamps.

In general in the tropics, extensive and varied use has been made of Landsat MSS imagery. Application has ranged from specific uses, including refining at the FAO Remote Sensing Centre in Rome, climatic isolines for agro-ecological zone mapping in Kenya, to the world's largest forest cover monitoring programme in Brazil. The latter initiated by INPE in 1979 used about 200 Landsat MSS scenes to cover the Brazilian Amazonian basin and the results were compared initially with Landsat coverage repeated in 1980. This gave a vivid representation of the deforestation process for policy-makers and planners. Areas of deforestation and the main vegetation types were identified and, in association with fieldwork, were aggregated in one-degree grid cells at a scale of 1 : 5 000 000 on government standard map sheets. From the repeated survey, the total area of current deforestation, the average area being deforested annually, rate of expansion of human settlements, etc. were obtained. In addition, individual areas were thematically mapped on existing topographic map bases at scales of 1 : 250 000 and 1 : 500 000.

The growing importance of using satellite imagery to provide a ready means of updating maps and providing current thematic information is also well illustrated by recent work in Sri Lanka (Itten *et al.*, 1985). A serious backlog exists in map revision, which cannot be overcome entirely by aerial photography, since its use is constrained by cost and the availability of first-order stereo-plotters. Overall photogrammetric control is being provided by the existing 1 : 63 000 old topographic maps, which form the base for the new national series of land-use maps incorporating forest data. The new land-use maps at a scale of 1 : 100 000 involve extensive fieldwork and using aerial photointerpretation and the interpretation of digitally enhanced satellite imagery. The latter was initially Landsat MSS, but this is being replaced by Landsat TM, available from the satellite data receiving and processing stations in India and Thailand. The interpreted land-use cover types delineated from the satellite imagery include dense forest, open forest, forest plantations, mangroves and scrubland.

Finally, it must not be overlooked in countries with a low per capita income and where the forest area is large that visual image analysis (analogue image analysis) is often much more cost effective per unit area of land than digital image analysis; and that bottlenecks caused by the non-availability of expensive equipment and maintenance problems can be avoided.

17.2 COMPUTER-ASSISTED IMAGE ANALYSIS APPLIED TO FOREST STANDS

General

Most forest activities, involving the digital image analysis of computer-compatible tapes (CCTs) of primary processed data, have been carried out in North America and western Europe. Increasingly, however, digital image analysis is being used in Australia, in several Asian countries including Bangladesh, China, India, Japan, Philippines and Thailand and in Latin America particularly Argentina, Brazil, Mexico and Peru. The interactive digital analysis is used, for example, to provide informative image classification, to improve geometric accuracy, to enhance the imagery for field use, to update maps and to provide layered information as inputs to geographic information systems.

Techniques have been discussed in Chapters 10 and 12, and implementing a typical project involves, as pointed out by Kalensky (1986), the following steps:

1. selection and review of input data;
2. decision on classification level and type of desired classes;
3. radiometric and geometric corrections of digital imagery;
4. image enhancement and/or unsupervised classification (clustering);
5. ground truth data collection;
6. supervised classification;
7. post-classification filtering of classes;

8. accuracy analysis;
9. planimetry of classes;
10. cartographic completion (co-ordinate grid, annotation, legend);
11. hard-copy outputs (imagery, statistical tabulations and/or thematic maps).

As provided by the many published articles on computer-assisted image analysis in forestry, most techniques focus on supervised classificatory methods. These in many circumstances achieve an accuracy of 80–90%, when classifying cover classes at levels I and II (section 14.2). Of growing importance, however, is spectral band ratioing (section 3.4), particularly in change monitoring and in biomass studies at the regional level, and the increasing interest with the launching of second generation satellites of texture analysis. The latter is already used to improve the mapping of boundaries.

Supervised classification

Supervised methods were discussed in section 10.3 and have been well tested for forest application. Unsupervised classification, with recognized human control (i.e. guided clustering), is sometimes useful in the initial stages of a forest study. For forest areas about which there is little or no information, unsupervised classification can provide a print-out which serves as a preliminary sketchmap, and which helps with forest stratification and the selection of training sets for supervised classification.

Maximum-likelihood classifiers are popular in which each pixel vector is measured against the sample means of the training sets, and are assigned and clustered on the basis of minimum distances as compared with Gaussian maximum-likelihood classification requiring much more processing time. Using array processors and this type of maximum-likelihood rule, a four-band scene comprising about 150 000 pixels can now be classified (80 classes) in four minutes real-time as compared with about 100 minutes without the array processor (Townsend and Justice, 1981). In forest studies, the number of final classes is usually 8–16, which is partly the result of equipment design, partly due to the accuracy and precision with which forest classification can be made and partly due to the composition of the forests.

The pixels of a training set need to be representative spectrally and radiometrically of the forest type that the site represents. Subject to constraints resulting from the area of the forest type or compartment, reasonable compromise in what constitutes a reasonable number of pixels in the training set is to use 9–25 pixels; but locating homogeneous areas of forest of this size can be a problem, particularly when, as in Europe, the compartments are small. It is often useful to weight the number of pixels in the training set according to the importance of the strata, as will be discussed in Chapter 18, when weighting the number of sample plots per strata in the field according to timber volume, etc.

In studies of forests in mountainous terrain, shadowing is the cause of major radiometric inaccuracies in the image analysis. The effect can often be reduced modestly by adjusting the pixel values for aspect and slope interpolated from topographic maps. In change monitoring involving digital data of two different years or seasons, correction may be needed also for the sun-angle. Hoffer *et al.* (1979), for example, tested three methods of combining Landsat MSS data with topographic data, and concluded that the employment of a layer classifier algorithm gave the better results (i.e. 15% improvement at level III). The first layer comprised spectrally distinct cover classes, and the second layer of topographically derived terrain data was used to further classify the major cover classes (cf. section 12.4).

In poorly mapped regions, due to the lack of topographic data, a visual impression of the terrain in three dimensions can be obtained using a digital elevation algorithm; and when topographic maps are available the elevational terrain model can be controlled and viewed. Plate 13 (Schnurr, 1988) provides a comparison between Landsat MSS, Landsat TM and SPOT after using an elevation algorithm to improve the image analysis. The technique is particularly useful in mountainous terrain, for planning forest inventory, in stratifying forest areas and locating computer training sets and inventory field plots. In the tropics, for example, the introduction of digital terrain modelling into the analysis can avoid confusion between mangrove forest and hill forest caused by the influence of shadows on the spectral signature.

Textural analysis

This has been little used for forest resources in the past. It now requires much more attention, due to the improved resolution of satellite-collected data, and because supervised spectral classification has severe limitations, since it is based only on aggregations of pixels having similar grey levels between, say, 0 and 64 or 0 and 256, in several spectral bands. Texture, viewed as the spatial or statistical distribution of grey tones within an image, adds a new dimension to image analysis. Unfortunately, computational time, processing techniques and memory storage are all constraining factors to research in textural analysis in forestry.

The basis of classifying a pixel or an identified point of a satellite scene by textural analysis is that the differences in radiant values are compared between that of the pixel with its neighbours. Difference and not similarity is the criterion. As with supervised spectral classification, the scene is segmented into subsets of different types and then one of several methods of texture analysis is applied. These include 'parallel' and 'segmentation' techniques. Average grey-level differences, which are a measure of coarseness and contrast of the images, are calculated by screening the data in windows of $2n \times 2n$ windows (e.g. 9, 16, 25). The data is merged into texture features taking into consideration calculated values of pixel separation. A fine-texture feature results when there is little or no spatial pattern and the grey-tone variation between features is large. In the United Kingdom, Drury (1989) carried out a texture analysis of part of the Forest of Dean using simulated SPOT data. He obtained encouraging results, but was constrained by the complexity of the forest, the small size of the forest stands and limited data storage.

Landsat MSS

From the launch of Landsat 1 (ERTS 1) in 1972, computer-assisted image analysis has formed an integral part of research and application of satellite remote sensing in forestry. As mentioned earlier, unsupervised classification (Gaussian maximum likelihood) was used in the USA immediately following the launch. This involved forest cover typing (Landgrebe et al., 1972). In ensuing research of several forest types algorithms for 'maximum likelihood' and 'minimum distance to the means' were used. Heller et al. (1975) in point-by-point computer comparisons obtained respectively 90% and 80% accuracies for forest cover at level I (section 14.2); and the forest area error was within 6%, with the exception of coniferous forests. Similar accuracies could not be obtained at level II using either visual or digital analysis.

Over the ensuing years, however, reference to publications indicates improvement in results at level II. For example, Hoffer et al. (1982), working in Colorado, obtained accuracies of 91–94% at level II for conifer and hardwood forest types and on one site achieved an accuracy of 76% at level III (i.e. ponderosa pine, Douglas fir/spruce, oak, aspen). The area error on estimating cover types at levels I and II varied between 6% and 10%. In their conclusions they pointed out that the cost of computer-assisted analysis at level I for small areas (e.g. 23 000 ha) was twice the cost of visual interpretation, but for large areas (e.g. 1 million ha) the computer-based classification was about four times cheaper.

With respect to mapping scale, Mead and Meyer (1977), working with boreal forest types in Minnesota, concluded that 1 : 125 000 was the smallest acceptable scale when working with Landsat MSS data. They also pointed out that at this scale each pixel can be observed on the video (CRT) screen, and that mis-registration is often about 1½ pixels (i.e. about 120 m). Although upland conifers, lowland conifers and mixed broad-leaved forest types could be adequately classified, at the species level accuracy did not exceed 70%, which is not acceptable to the forest manager. Kalensky and Scherk (1975), working in similar forest types in Canada, obtained accuracies between 67% and 81% for coniferous forest, deciduous forest and non-forest and Kalensky (1974) found that multidate imaging improved the results by only 2%.

Developing countries

In developing countries, digital image analysis, when available, has been shown to provide more information than can be extracted from 'bulk product' imagery and to be more useful than equivalent analysis in the industrialized countries. For example, in Columbia a comparison of the bulk product satellite scene (Plate 12(b)) and computer-processed data using supervised classification (Plate 12(c)) revealed the very much improved information content of the machine-processed data related to the forest. Forty training plots, each of ten pixels, were distributed among the 12 forest types, identified in the field using aerial photographs (Draeger *et al.*, 1974). Applying supervised classification, this provided imagery showing 11 classes of vegetation.

Probably no better example of the usefulness of digital analysis of Landsat MSS data, in helping to update forest information in developing countries, is that provided by studies in China. For example, in Jilin Province in northeastern China following principal component analysis (Xu *et al.*, 1985), 76 spectral training samples were carefully chosen taking into consideration minimum spectral distances, land-cover type and timber volume. In the supervised analysis, the simpler box classifier gave results comparable with maximum likelihood classification and was up to six times quicker. Image classification included stand type (four classes), age group (three classes), canopy density (three classes) and timber volume distribution. For 208 000 ha, the accuracy of classifying forest and non-forest land was determined as 88.2%.

Landsat TM and SPOT

Improved results in digital analysis of land use and forest cover types can be expected with Landsat TM and SPOT, although the machine time needed for processing the data of the same ground area as covered by Landsat MSS is greatly increased. The latter, however, is not always relevant to the traning sets of supervised classification, since the estimated statistical accuracy is related more to the number of pixels in each sample set than to the area of each pixel.

The operational advantages accruing to forest resources from the analysis of the data of the second generation satellites appears to hinge both on the finer resolution (i.e. SPOT) and on the improved spectral qualities (i.e. Landsat TM). Nevertheless, this opinion remains to be fully explored and there are circumstances when the results indicate the reverse, due probably to variations in the forest physiognomy and stand size. Schnurr (1988), for example, in applying a hierarchical clustering algorithm in combination with F-tests as measures of separability of the training sets followed by image classification using discriminant analysis, concluded that spectral resolution has a much stronger influence than geometric resolution for separating vegetation types. He obtained more significant F-values for Landsat TM than for Landsat MSS and SPOT. However, as pointed out by Townsend and Justice (1981), classification accuracy can sometimes be improved by the coarser resolution of sensor data, since internal variations (i.e. scene noise) decline within each single class. Recorded variations caused by the reflectivity differences between tree crowns and their shadows are increased as the sensor resolution is made finer, but are averaged when the sensor resolution is coarse. It is therefore useful sometimes to degrade the recorded signals by averaging groups of pixel values (e.g. 3×3 window) within each segment which is being digitally analysed; but the major risk is that more mixed pixels are introduced into the classification.

That spectral classification in forestry is improved by incorporating the mid-infrared band between about 1.55 and 1.75 μm is increasingly recognized. Hoffer $et\ al.$ (1975) demonstrated, when analysing Skylab data (13 channels), that the most effective classification of forest types was obtained using a combination of the visible, near-IR and mid-IR bands. More recently Benson and Gloria (1985), working with TM data of forest-cover types in California, reached a somewhat similar conclusion, namely that the colour products should normally contain a spectral band of the visible, near IR (band 4) and the mid-IR (band 5) for maximizing the interpretability of vegetation types, and the TM colour composites should contain band 4 in all cases. However, they also pointed out that for some vegetation types MSS colour composites are superior for interpretation when band 4 is excluded. The incorporation of the mid-IR in spectral analysis is seen not so much as improving species identification, but as being sensitive to density changes in forest stands.

As discussed earlier (section 7.1), the much finer resolution provided by the second generation satellites compared with Landsat MSS favours forest mapping at a scale of 1 : 100 000 and map revision at a scale of 1 : 50 000 or larger; and this is stimulating its use for the revision of forest maps at a scale of 1 : 25 000 and sketch mapping at even larger scales. In Mexico (Monterey), for example, when geometrically rectifying the images to a Universal Transverse Mercator (UTM) registration with 21 ground points per scene, the residual mean square errors (RMS_{xy}) were determined for Landsat MSS, Landsat TM and SPOT to be respectively between ± 60 m, ± 24.7 m and ± 18.2 m (Schnurr, 1988). In Ireland (MacSiurtain $et\ al.$, 1989), Landsat TM data has been demonstrated experimentally as suited to forest stock mapping at a scale of 1 : 25 000, but with the species information mainly provided from other sources; and for stockmap revision at scales up to 1 : 10 500 of windthrow and clear felling using a GIS overlay of forest compartment and subcompartment boundaries.

There is no doubt that the combined impact of the improved spatial, spectral and radiometric characteristics of Landsat TM is providing information more readily acceptable to the forest user. There may be little improvement in the mapping of forest species, but often there is a better discrimination of forest types and density classes at level III (section 14.2). For example, Forstreuter (1988) working in the rainforests of Guinea found that density classes could be typed and used to stratify the forest, which was not possible using Landsat MSS imagery. This trend is also supported by the results of earlier TM simulation studies prior to the launching of the satellite.

In much of central and western Europe, the problem of improving forest cover classification using satellite data is frequently exacerbated by the small size of the forest stands and the influence of management on those stands. Even under conditions where subcompartments are classified as uniform on the basis of basal area or volume, species composition, canopy physiognomy and canopy density may vary and result in differences in spectral reflectivity, which provides the single parameter of satellite remote sensing. In contrast in aerial photointerpretation, at larger scales, although the eye can discriminate only several grey levels, other remotely sensed characteristics of the stand show up differences occurring within the stands, which otherwise would be considered uniform.

In considering Landsat TM and SPOT applied to intensively managed forest plantations, the classification value is seen to be towards obtaining information equivalent to high-altitude aerial photographs; but, as we know, the cost is much less and it thus permits costwise more frequent coverage subject to constraints of cloud cover. Three examples will be quoted of second generation satellite data applied to plantations. In Kenya (Howard and Lantieri, 1987), SPOT imagery was used to identify and to delineate the boundaries of the land systems. Within the land system near Njoro, the plantation boundaries were mapped at a scale of 1 : 100 000. Within the forest compartments, the planted species, height classes (age classes) and crown cover could often be inferred by variations in hue (colour), tone and texture. Consistently, four plantation classes were identified, which appear to be associated with stand structure and stand density. These were: *Pinus patula* (30 m in height and smooth), *Juniperus procera* (10–15 m, coarse textured), mixed *Juniperus/Cupressus* (10–15 m, smooth) and young *Juniperus/Cupressus* (5–10 m).

In a study in Switzerland Bodmer (1988), using TM data of an average stand size of about 0.5 ha, covered with mixed and pure conifers (spruce, pine, silver fir) and deciduous broad-leaved species (beech, ash, oak), found reconciling the classified data with the existing stand map to be difficult and often impossible. This was primarily due to the many mixed pixels. Spectral radiance was found to change with the 'age' of the stands (i.e. saplings, poles, timber size, mature) (cf. Figure 17.2). It was concluded that a minimum stand area of about 3 × 3 pixels (c. 1 ha) is needed to avoid the problem of shadows along the stand edges; and subdivision of the categories was difficult due to

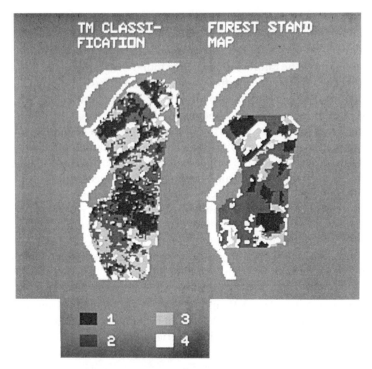

Figure 17.2   Comparison of (a) Landsat TM classification with (b) part of the forest stockmap having an average stand size of about 0.5 ha. 1: old conifer forest; 2: mixed old forest; 3: young forest, deciduous, conifers; 4: clearings, regeneration and water surface.

different thinning grades, stand densities and species composition. The Landsat TM data was initially integrated into a geographic information system with a raster resolution of 25 m × 25 m, but the forest/non-forest mask was delineated manually (visually). The subscene (200 × 80 pixels) was classified using minimum distance and a maximum likelihood classifier with TM bands 1–5. Colour IR aerial photographs at a scale of 1 : 9000, a TM colour composite (bands 3, 4, 5), a stockmap and field checking were all used in the selection of training sets.

In the United Kingdom, the National Remote Sensing Centre assessed in 1986 the woodland cover of the Isle of Wight using Landsat TM data. This has shown that a satisfactory woodland cover map can be produced at a scale of 1 : 50 000. Supervised classification techniques were used to identify broad-leaved and conifer woodland classes with training areas based on Forestry Commission stockmaps (scale 1 : 10 560) and aerial photointerpretation (Horne, 1984). The map was prepared to overlay exactly the 1 : 50 000

Ordnance Survey Map and is based on image classification using Landsat TM bands 3 and 5. Band 5 was used to separate conifer woodland (451 ha) and broad-leaved woodland (3131 ha) of areas greater than 0.25 ha. The total woodland area (3582 ha) was close to the FC Census area of 3095 ha (1982). Woodland blocks were also placed in ten size classes between 0.25 ha and 640 ha, as useful to environmental studies.

17.3 SYNTHETIC APERTURE RADAR (SAR)

As yet, there is only scant experimental information on satellite sensing using SAR. The results, however, for forestry are interesting. The resolution of the imagery obtained from outer space can be favourably compared with the imagery recorded by airborne real aperture radar (section 6.3). This implies operational uses related to deforestation and hydrology similar to that achieved by this type of airborne radar, but with the added economic advantage of low cost of coverage per unit of land area achieved by satellite sensing. However, as experienced with the processing of Seasat data, the demands on computer time are considerable and are far greater than for Landsat MSS data.

With respect to tropical forest cover, imagery of SIR-A and SIR-B was used in the states of Rondonia and Amazonas in Brazil to study deforestation. Figure 17.3 provides a comparison with SLAR. For areas cleared and being used for agriculture, the area estimates using SIR-A imagery were found to be within 10% of the areas as shown on the maps prepared with the aid of a video camera. In contrast, abandoned agricultural areas returning to forest were not reliably recorded on the SAR imagery (Stone and Woodwell, 1985). This is understandable in view of the fact that regrowth is rapid and varied in tropical forests; and presumably, at least in part, the regrowth had attained a green biomass and possibly a surface roughness comparable with the virgin forest. The study draws attention to the complexity of the problems that will be encountered when monitoring forest cover by future satellites using radar.

Exposed water surfaces and water surfaces under forest cover have also been examined on SIR imagery. Previous comments on imaging with SLAR (section 15.6) appear to be applicable. Further, comparison of Landsat MSS and Seasat imagery acquired over forested wet lands suggests that swamp forest is better delineated on SAR imagery and additional information is obtainable on the water regimes (Joyce and Sader, 1986). In eastern Australia, for SIR-B, examination of imagery of red gum forest (*Eucalyptus camaldulensis*) on the Murray flood plain (Richards, Woodgate and Skidmore, 1987) suggests that the major components of the backscatter of the signals towards the sensor are canopy volume (biomass), scattering from the canopy towards the ground and specularly reflected back in the manner of dihedral corner reflection from buildings (section 3.3), a similar scattering from the crown and diffuse reflection from the ground.

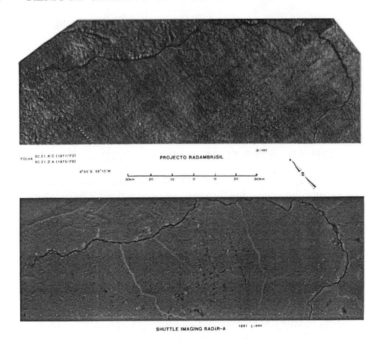

Figure 17.3   Comparison of X-band side-looking airborne radar (SLAR) and L-band synthetic aperture radar (SAR) taken from outer space (below). The synthetic aperture radar (SAR) recorded by the space shuttle (SIR-A) is more informative, show tropical rainforest (Brazil) with areas cleared along the river for agricultural crops and access routes to the river.

17.4   ENVIRONMENTAL SATELLITES

The technical specifications of environmental satellites have been discussed in section 7.3 and band ratioing/vegetation index in section 3.4. NOAA AVHRR imagery in the format of global area coverage (GAC) has been archived since 1979 and is therefore available for historical studies. Also since 1982, NOAA AVHRR data in the visible and near-IR bands have been ratioed at NASA (Maryland) to generate weekly or 10-day products providing world-wide and Africa coverage respectively in the vegetation index format. Two advantages of these products is that, due to the frequency of coverage, the problem of cloud cover is largely overcome by the computer-processed composites and that band ratioing using the normalized vegetation index reduces the effect of varying sun-angles across the scene and the effect of atmospheric haze/aerosols. A major disadvantage, often overlooked, is that detailed ground sampling to correlate the data is problematic, since the radiance value of each GAC pixel results from a small systematic sample on board the satellite of the

1.1 km × 1.1 km LAC pixels (Chapter 8). The GAC pixel is interpolated and not a direct recording of the radiance within its ground area, as is the case for LAC pixels and the pixels of earth resources satellites.

As the environmental satellites provide imagery of very low spatial resolution, they have received little attention so far in forestry, but will be of increasing interest in the future for zonal and global monitoring, due to their high temporal resolution and the very low cost of imagery per unit ground. The imagery provides best a synoptic view at the continental or sub-continental levels; but may also serve as a base at the national level on which to correlate finer resolution imagery. Schade (1985), for example, observed when working in Angola that a major constraint for a national or regional cover assessment using Landsat imagery was the lack of cloud-free scenes; and he concluded that, to overcome this problem, a multistage approach is needed using environmental satellite imagery with its daily repetitive coverage, as the first stage for stratifying the vegetation of a developing country. This is then followed by the selection of available cloud-free Landsat imagery in each stratum as the second stage. Alternatively, the initial stratification could be based on major land units observable on the NOAA imagery and followed by land systems delineated on Landsat TM or SPOT imagery (section 14.3).

Since the mid-1980s, there have also been several NOAA AVHRR studies covering forest vegetation classification at continental level and vegetation cover/green biomass change at the global, continental, sub-continental and country levels. These had been preceded by meteorological and climatic studies. For example, Barrett, as early as 1975, prepared contoured annual rainfall maps of northern Sumatra, which was partly forest covered, using archived NOAA imagery, historical range gauge recordings and regression analysis. Continental digital image composites using NOAA AVHRR GAC data were prepared at the level of plant formations/plant panformations for Latin America (Justice *et al.*, 1985) and Africa (Tucker, Townsend and Goff, 1985). The image composite of Africa (Plate 14) was prepared using feature space classification of first and second principal components for eight periods of 21 days of the year 1982–83. The second principal component represented seasonal cover types. The feature space used to label the land cover classes is shown in Figure 17.4. Some dissimilarities in boundaries were to be observed between the finalized image composite and published maps on African vegetation. Seasonal colour variations occur on the images for tropical rain-forest, tropical seasonal evergreen and semi-deciduous high forest, tropical savanna woodland and tropical grassland. The seasonal spectral reflectance variations of evergreen broad-leaved high forest, as distinct from the well-known seasonal variation of deciduous forest, have been commented on elsewhere as based on spectral reflectance studies by the author (Figure 3.15).

On the ratioed LAC imagery, many linear features provided by large rivers and major highways, most towns and the larger deforested areas are

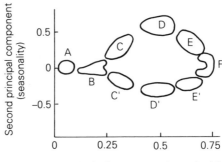

Figure 17.4   Feature space used to label land cover classes for Plate 14. The resulting labelling delineates classes of water (A), desert (B), semi-arid seasonal grasslands (C and C′), dry savannas (D and D′), wet savannas (E and E′), and tropical forest (F).

identifiable. In Brazil, for example, large rainforest areas in the Amazonian state of Rondonia, which are being cleared for agriculture and rough pasture, can be identified and mapped (Tucker *et al.*, 1985; Mallingreau and Tucker, 1987). In India (Agrawal and Sharma, 1983) forest and non-forest areas were generally mapped in flat and undulating areas, when the tree cover is in full foliage. However, hill shading is a major problem in the mountainous north (Howard and Reichert, 1986) and the normalized vegetation index values may be adversely affected by the atmospheric conditions at the approach of the monsoon (Subbaramaya, Bhanukumar and Babu, 1982). This problem has also been observed by the writer when planning aerial photography using NOAA imagery, and can be associated with the presence of thin high altostratus clouds. In West Africa, an initial LAC study of the forest cover provided by the high closed forest of the state reserves to the west of Kumasi (Ghana) has been compared with the results obtained using Landsat TM (Paivinen and Witt, 1988). Using unsupervised maximum-likelihood classification the cover was favourably determined as 29% and 28% for TM and AVHRR respectively, but ground-based data was not available.

18

Statistical considerations in forest fieldwork using aerial photographs and satellite data

This chapter brings together remote sensing and the necessary fieldwork. The chapter draws attention to selected statistical methods, which are commonly used in forest surveys employing aerial photographs and satellite data. The methods take into consideration sampling on the ground and what can be obtained from the images, as useful to forest survey. The very purpose of applying statistical techniques, which combine field sample plots and photographic plots, is to improve accuracy and precision, save time and reduce considerably the overall cost of the forest survey. Careful attention to the sampling design will often also result in other advantages. For example, a sound statistical approach may avoid the expense of having to prepare thematic maps.

Many well-proven statistical techniques used in forest inventory are highly cost effective; and often those developed with the aid of aerial photographs are equally applicable to satellite-acquired data. Careful and sometimes lengthy consideration needs to be given to stratification and sample design as covering sampling within a specified tree population. All too often, insufficient time is devoted to designing the stratification best suited to the forest survey and thus improving overall efficiency. Attention needs to be given to the sample distribution and the determination of the sample size, the sample selection procedure and the estimation of the population means, variances and confidence limits.

The application of statistical techniques requires a working knowledge of standard deviation, standard error, sampling error, confidence limits of the sample mean, correlation coefficients, t- and F-tests, covariance, regression analysis, solving simultaneous equations, principal component analysis and, if the data is qualitative, possibly the chi-square test. For an understanding of how to use these techniques, a suitable textbook on applied statistics should be consulted. These include Husch (1963), Loetsch and Haller (1964), Cochrane (1977), Snedecor and Cochran (1978) and Gregory (1978). In the writing of this chapter, it is assumed that the reader is already conversant with common statistical methods used in forest inventory.

18.1 FOREST STRATIFICATION

Once the availability of the remotely sensed data has been established, the first step will be to acquire sufficient hard copies to determine, in association with field checks, the nature of the stratification (Figure 18.1) and the type of sampling to be used. Careful perusal of the entire area to be investigated, as recorded on the hard copies, is essential to formulating a step-wise approach to the sampling design. If no up-to-date aerial photographs are available, then a decision is needed on whether to delay the survey until new photography is available or to use satellite imagery supported by old photography.

The purpose of stratifying a ground area into two or more units is to reduce, when sampling, the variation occurring within each stratum, and to increase the variation between strata by separating initially the area into more homogeneous classes. Alternatively, stratification can be viewed as giving a

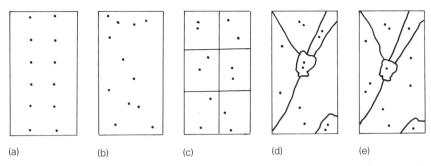

(a)   (b)   (c)   (d)   (e)

Figure 18.1 Forest stratification: each dot represents the location of a field sample placed on the aerial photograph, photo-mosaic or satellite imagery: (a) systematically arranged line or strip samples but no stratification; (b) randomly located samples but no stratification; (c) stratified paired random samples (i.e. box stratification); (d) random stratification of forested ground area based on delineated forest types (i.e. type stratification); (e) the same as (d) but the allocation of the plots depends on the area of each delineated forest type (i.e. sampling probability proportional to area). As discussed in the text, (e) may be further improved by weighting the distribution of the samples according to expected timber volumes per stratum or the market value of the timber per stratum.

statistical reliability similar to unstratified sampling but with fewer samples. Thus, if the population being sampled is large and can be subdivided into strata within each of which the statistical variance of the attribute(s) is relatively small, a considerable saving can be expected in effort, time and cost. If a minimum of two samples are placed in each stratum, the within and between variances of the strata can be statistically tested for significance.

The principle of physically stratifying an area into primary sampling units and possibly secondary sampling units remains the same, whether the area being surveyed is a single forest, or the entire forest estate of a country, or the forest reserves in a geographical region. The first phase/first stage data can be in any form (e.g. spectral radiance values), provided there is correlation between these and the next level of data. Advantage is taken of such correlations, for example, in regression analysis and two-phase sampling. The boundaries of the strata are usually either geometric in outline, which may be termed **box stratification** and is favoured in digital analysis; or the strata may follow boundaries formed by natural features or forest types on the imagery, which may be termed **type stratification** and is favoured in aerial photointerpretation (Figure 18.1).

Stellingwerf and Lwin (1985) compared the two types of stratification for estimating the timber volume of spruce stands using 1 : 30 000 ortho-photographs enlarged to 1 : 10 000. For the same standard error, both methods consumed the same number of man-days. With the box design (stratified cluster sampling) more man-days were required in the field, due to requiring 72 against 55 field plots; but with forest type stratification (two-phase stratification), more time was needed in the office working with the photographs etc.

In type stratification, the forest type boundaries are carefully delineated on the aerial photographs/satellite imagery in terms of forest cover or cover classes (e.g. stand density) or the physiognomy of the forest (i.e. plant formations, plant subformations) or their phytogeomorphic characteristics (e.g. land systems). The stratification must be designed to suit the purpose of the forest inventory or monitoring programme and the quality and scale of the available imagery. For Landsat TM and SPOT, strata can be envisaged to cover forest and non-forest, evergreen, deciduous and mixed forest and two to three density classes. With large-scale aerial photographs, the stratification for timber volume classes would involve some tree species identification and the stand parameters discussed in Chapter 16 including several density/volume classes.

Area sampling frame

Of increasing interest to foresters is the area sampling frame used throughout the USA for obtaining information on land use and agricultural crop statistics.

Basically, efficiency in cost and statistics is obtained through the use of stratification and two-stage sampling involving remote sensing (Vogel, 1986). Two major differences, between stratification/sampling methods used in forestry and the area sampling frame used in agriculture, are that much more time is devoted to developing an efficient stratification and that the segments (field samples) are much larger and different in shape from the commonly used line transects and cluster samples of forest inventory. In agricultural areas of the USA, 1 km^2 would be a typical size of a segment, and this approach lends itself to carefully basing the positioning of samples on remotely sensed imagery.

The construction of the area sampling frame, carried out in several steps, requires the field checking of no more than about 0.5% of the land area and provides in the USA a sampling error for agricultural crops of between 2% and 4%. The first step is to identify the categories of land use that will improve sampling efficiency and to delineate on the maps or imagery the areas of similar land use. Highly cultivated areas (e.g. above 75%) are separated from the less cultivated areas. Typical land-use strata are intensive cultivation, extensive cultivation, rangeland or pasture, forest land, agro-urban and urban. The cultivated lands may be divided into several categories depending on the area under cultivation.

Within each delineated land-use stratum, the area is further divided into several primary sampling units (PSUs). The size of each is sufficient to contain 5–15 final sampling units or segments. Care is taken to select natural or man-made boundaries on the maps or imagery that are easily located in the field. Each PSU is as distinctive as possible on the basis of the cover types it contains. PSUs are then randomly selected within the stratum with probability proportional to size. Each selected PSU is subdivided into a reasonable number of final sampling units. One segment within each PSU is selected at random and visited in the field. Each segment consists of a block of one or more fields (tracts) delineated on an aerial photograph. The random sample for statistical analysis comprises the total of the selected segments and the data collected on those samples. As a general rule each segment should be small enough to be assessed in one day.

Improving efficiency

Assuming the same number of samples are used to test the relative efficiency of stratifying an area, a useful indicator of efficiency is provided by the variance ratio or F-test, which is obtained by dividing the variance of the between strata by the variance of the within strata. The variance (S^2) is obtained by squaring the standard deviation (S) of the parameter being examined. If the several strata represent a single population, the variances will be similar. Other techniques using variance have been suggested. For example, Moessner (1963) suggested an approximate method of rating a number of stratification schemes by providing an estimate of the number of samples needed to give similar reliability using different methods of stratification. The pooled within-stratum variance of each scheme $(P_i S_i^2)$, obtained by weighting the variance of each stratum within the scheme by its area (P_i expressed as a percentage of total area), is divided by the variance of the unstratified scheme (S^2). This is then subtracted from one, i.e.

$$1.00 - \frac{\Sigma P_i S_i^2}{S^2}$$

The remainder is used as a rating of the efficiency of the method of stratification, and as an estimate of the reduction in the number of plots needed to give the same reliability as an unstratified scheme.

As summarized by Avery (1964), several aerial photographic methods are available for estimating the variances and required number of field plots. The most obvious method involves laying out a series of random samples from photo-strata in the field and using these to calculate the variance of each stratum prior to a full survey. This, however, is costly and often impracticable due to the terrain conditions. A modification of the method involves the allocation of field plots directly to photo-strata according to the sampling precision of similar surveys elsewhere; but there are statistical objections to this approach, since the plots may belong to two different populations. The second method involves the full computation of the variances from photo-plots. This is useful when aerial volume tables are available.

A third method calls for the delineation of strata on the photographs, and then by stereoscopic examination an estimation is made of the range of volumes in each stratum. The range in each stratum is divided by four to indicate what the standard deviation is likely to be; and this is squared to provide a crude estimate of the variance. The method assumes the samples are normally distributed; and, as pointed out by Avery, the reason for dividing by four is readily appreciated, if one recalls that a dispersion of 95% of the samples encompasses ± 2 standard deviations.

The crude estimations of the standard deviation and variance so obtained are then used to calculate how many plots should be allocated to each stratum. For sampling in the field to give approximately the same standard errors of the mean, the calculated variances are summed (ΣS_i^2) and the variance of each stratum is expressed as a percentage of the total:

$$\frac{S_i^2 \times 100}{\Sigma S_i^2}$$

For a predetermined number of plots (n), the number (n_i) to be allocated to the ith stratum to provide a similar standard error of the mean will be:

$$\frac{S_i^2 \times 100 \times n}{\Sigma S_i^2}$$

The standard error of the mean will then be:

$$\sqrt{\left(\frac{\text{variance}}{n_i}\right)}$$

In practice this method may result in too few plots being allocated to a stratum estimated as having the greatest range. Also, it may be desired to place varying confidence limits on the mean of each stratum according to other criteria. For example, each stratum based on tree species and stand volume may have a different market value; and, as a result, it is desirable to vary the standard error of the mean with value of the forest stand as ascertained by stereoscopic examination of the photographs. Under such circumstances, the crude estimate of the standard deviation (S_i) is used to calculate the number of plots to be allocated to each stratum for a specified sampling error. Sampling error is the standard error $(S_{\bar{x}})$ expressed as a percentage of the mean $(\bar{X})$, i.e.

$$\frac{S_{\bar{x}}}{x} \times 100$$

The estimated mean of the stratum multiplied by the sampling error will give a crude estimate of the standard error of the mean (i.e. $S_{\bar{x}} = \bar{X}$ (sampling error)); and hence an estimate can be made of the number of plots (n) to be allocated to each stratum for a predetermined sampling error. For the ith stratum

$$n_i = \frac{(S_i)^2}{(S_{\bar{x}_i})^2}$$

i.e. at the 68% probability level, or:

$$n = \frac{(S)^2}{(S_{\bar{x}})^2}$$

for an unstratified population, or

$$n = \frac{(S)^2}{(S_{\bar{x}})^2} \frac{(N-n)}{N}$$

for an unstratified small finite population where N is the maximum number of samples in the population. If the probability level is increased to 95%, 3.84 times the number of plots will be required, as t^2 for a large number of samples is $1.96^2 (= 3.84)$.

The method of weighting the allocation of samples by area or sometimes by the expected volume of the product contained in the stratum has been commonly used in forest inventories. Obviously strata weighted by area and covering large areas will be allocated more sample plots than strata covering smaller areas; and as a cursory examination of a t-table will show, the t values used in fixing the confidence limits of the standard error of the mean vary according to the number of samples.

In computer-assisted satellite image analysis, the unit of stratification will be the pixel; and using unsupervised classification, the pixels are stratified into a specified number of strata on the basis of their spectral radiance values (section 10.5). Several attempts may be made at obtaining the most suitable stratification by varying the number of strata in the unsupervised classification and this may be accompanied by histogram stretching, 'smoothing' etc. (section 10.3). Supervised classification of satellite data will necessitate visits to the field or at least checking by reference to aerial photographs.

If a geographic information system is available, attributes other than radiance values may be selected from the computer file for developing the strata. In developing computer-based land cover classification with a minimum accuracy of 85%, Fitzpatrick-Lins (1981) in Florida obtained 26 categories as against only six using visual interpretation. The optimum number of samples allocated to each stratum will depend again on the estimated within variances of the strata, as explained previously, as well as being based on the cost of collecting field samples. For the first phase/stage, samples allocated to the strata on satellite imagery, point samples are usually favoured.

18.2 SAMPLING

The sample design involves not only the careful stratification of the forest into as many homogeneous units as is feasible, but also decisions on the sampling procedure within the strata which have been carefully delineated on the aerial photographs, mosaics, satellite imagery or maps. Sample design involves applied statistics covering sampling within an identified population, sample distribution, sample size, sample selection procedure and calculation and examination of the population means, variances etc. It involves choosing between systematic and random sampling (Figure 18.1), the use of cluster sampling to reduce fieldwork and combining field plots and photo-plot data by sub-sampling.

Often overlooked, in sample design incorporating remotely sensed data, is that systematic sampling may give a better estimate of the mean than an equal number of simple random samples or stratified random samples. In random sampling the placement of the samples is located entirely by chance, as for example drawing the random plot numbers and map co-ordinates from a hat or using a published table of random numbers.

Also often overlooked is that efficiency, using standard error as the criterion, will not be greatly increased by using more than forty to sixty field plots per stratum unless the number of plots is greatly increased. Possibly the error will be halved by quadrupling the number of sample plots. The hypothesis may be formulated that within any one stratum the number of field plots should usually not exceed forty to sixty. The word 'usually' is used since if the population is suspected of being heterogeneous the number of plots may be increased beyond these limits in expectation that upon being statistically analysed the data will be broken down into two or more strata. For example, if a stratum is suspected of being heterogeneous and likely to provide three homogeneous strata after the forest-survey, then up to 120 plots (i.e. 3×40) might be allocated initially.

In forest surveys using aerial photographs and irrespective of the size of the area, 0.25–2.5% of the total area will be covered by sample plots in the field, as compared with 5–20% when aerial photographs are not available. Normally, using photographs, the ground area likely to be covered by the sample plots will be under 1%.

Systematic sampling

This is a type of sampling in which the first sampling unit is chosen randomly, but all following units are at a constant sampling interval. An example is a dot grid with its systematically arranged points, which is placed randomly on a photograph. This has also been termed systematic sampling with a random start, so as to distinguish it from systematic sampling with a subjective start. In the latter case since the starting-point is chosen, for example, on arrival in the field, personal bias is likely to be introduced, and on this account the subjective approach should be avoided. In random sampling, each sample has an equal chance of selection. If the interpreter selects what he considers to be representative, then the sample is no longer at random and the method would again be termed subjective sampling.

Since probability theory is based on random selection, the theory and techniques of random sampling no longer apply to the systematic sample. The calculation of standard error of the mean, variance ratio, fiducial limits and other statistical parameters is no longer valid and therefore remains unknown. In the practice of visual image analysis, the photographs for interpretation are often selected randomly, particularly if there are many photographs. The sample points on the selected photographs are then located using a dot grid. Each set of points on the grid will then constitute a cluster and the interpreted data can then be further analysed as described in the next section.

Cluster sampling

If a group of plots provides a fixed pattern in relation to each other, and the group as a unit provides part of another pattern formed by other units, then each group of plots is referred to as a **cluster**. As mentioned, the use of a dot grid with a random start provides a cluster sample in which all the dots on the grid form the cluster. The degrees of freedom are provided by the random starts. For example, in Thailand, Loetsch (1957), using photographs at $1 : 48\,000$, employed clusters of seven systematically arranged squares (camp units) as a compromise between maximum dispersion of plots and forest conditions. One random photo-point only was needed, being that of the centre-square. Forty-eight ½ ha sampling areas were located around the periphery of each square to form the cluster.

Provided each cluster of plots is randomized in relation to other clusters, statistical parameters can be calculated even though the plots within a cluster are systematically arranged. The number of degrees of freedom, however, will be one less than the number of randomized clusters (i.e. $m - 1$) and not one less than the number of plots. Assuming there are m clusters and n plots in each cluster, then the variance (S^2) is:

$$S^2 = \Sigma \frac{(\bar{X}_j - \bar{X})^2}{m - 1}$$

where X_j is the mean of the jth cluster and $\bar{X}$ is the mean value for all plots, i.e.

$$\bar{X} = \frac{\Sigma X}{mn}$$

The standard error of the mean for all plots in all clusters will be:

$$S_{\bar{x}} = \sqrt{\left(\frac{S^2}{m} \frac{M - m}{M} \right)}$$

where M equals the maximum number of clusters in the population and corresponds to N samples in a finite population. If this is compared with the standard error, assuming all plots are randomized, i.e.

$$S_{\bar{x}} = \sqrt{\left(\frac{S^2}{n} \frac{N - n}{N} \right)}$$

it will be observed that many degrees of freedom are lost; and therefore, statistically, the method is not so efficient as using randomized single plots, but the cost of fieldwork is reduced.

Repetitive sampling

For many years, repetitive sampling has been used in field surveys. In repetitive surveys, an important design feature is whether or not the same sample plots are selected (i.e. matched). If the estimates are from the current time period of the same population, optimization results from choosing unmatched samples; but in change detection, this is normally achieved with matched samples. In Austria, de Gier and Stellingwerf (1988) obtained similar results when estimating the periodic timber volume of thinned Norway spruce (1974–84) in a comparison of the volumes measured in the field using random plots and measured using a single set of permanent first-phase photo-plots and dependent second-phase photo-plots. This involved two-phase sampling with regression estimators and black-and-white aerial photographs at a scale of 1 : 10 000. Crown cover percentage (crown closure percentage) per sample plot was found to be the effective independent photographic variable.

Repetitive sampling for remote sensing was summarized by Rosenfield, Fitzpatrick-Lins and Ling (1982) as follows in the context of estimating change in land use. Reference should be made to Cochran (1977) for the theoretical background. Let the designations x and y represent the categories of land use and land cover classification at the earlier time period (population X) and the later time period (population Y). Then the sample estimate of the population ratio R is given as

$$R = \bar{y}/\bar{x}$$

If x_i is the value of y_i at some previous time, the ratio method uses the sample to estimate the relative change Y/X that has occurred since that time. Thus, the population ratio represents the change in land use and land cover. Further, according to Cochran (1977) the ratio estimate is consistent and of negligible bias in sampling sizes exceeding 30, and if the coefficients of variation of $\bar{x}$ and $\bar{y}$ are both less than 10%.

When point sampling for the proportion of land use and land cover categories in the multinomial distribution, the sample proportion p represents the sample mean $\bar{y}$. The sample estimate of the population ratio R_j for the jth category is then

$$R_j = (p_y/p_x)_j,$$

where p_y, p_x are computed for the X and Y populations, respectively.

An estimate of the variance of the population ratio R_j is computed by error propagation as

$$v(R_j) = [1/(p_x^2)_j]\,[v(p_y)_j + R_j^2\,v(p_x)_j]$$

where $v(p_y)_j$ and $v(p_x)_j$ are computed for the Y and X populations, respectively. Confidence limits are computed in the normal form.

An unbiased estimate of the change in the population total Y_j is computed in the manner

$$\Delta\hat{Y}_j = \hat{Y}_j - \hat{X}_j,$$

where $\hat{X}_j$ is computed similarly to $\hat{Y}_j$ given above, but for the X population.

An unbiased estimate of the variance of the change in population total Y_j

is computed by error propagation as

$$v(\Delta \hat{Y}_j) = v(\hat{Y}_j) + v(\hat{X}_j)$$

where $v(\hat{X}_j)$ is computed for the X population. Confidence limits are again computed in the normal form.

An unbiased estimate of the change in the population total Y_j is computed in the manner

$$\Delta \hat{Y}_j = \hat{Y}_j - \hat{X}_j$$

where Y_j is known, and X_j is computed similarly to Y_j given above, but for the X population.

An unbiased estimate of the variance of the change in the population total ΔY_j is computed by error propagation as

$$v(\Delta Y_j) = v(X_j)$$

where $v(X_j)$ is computed for the X population.

If desired, the estimate of the population ratio R_j can be computed on the basis of proportions, as

$$R_j = (p_y/p_x)_j$$

where $(p_y)_j$ is the known population proportion and $(p_x)_j$ is the sampled estimate of the population proportion for population X.

An estimate of the variance of the population ratio R_j can be computed by error propagation as

$$v(R_j) = [1/(p_x^2)_j][R_j^2 v(p_x)_j]$$

where $v(p_x)_j$ is computed for the X population.

18.3   SUB-SAMPLING

Since sampling on imagery/photographs, as pointed out earlier, is far quicker than ground sampling, it is advantageous to sample in the field only a fraction of the plots sampled on the imagery; and then to collect field data, which cannot be otherwise obtained. This is termed sub-sampling. Two methods of sub-sampling are used. If the size of the individual photo-plots is retained, but only a percentage of the plots is sampled in the field, the method has been termed **multiphase sampling** or double sampling or photo-plot sampling (section 17.3) or two-phase sampling. Three-phase sampling involves satellite imagery, aerial photographs and ground samples, as shown in Figure 18.2.

If the sub-samples in the field form but a part of each photo-plot, then the term **multistage sampling** is used (Figure 18.2). In the field the process may be repeated by sampling the herb layer, as part of the plot used for tree measurements in order to obtain an estimate of available animal feed etc. The

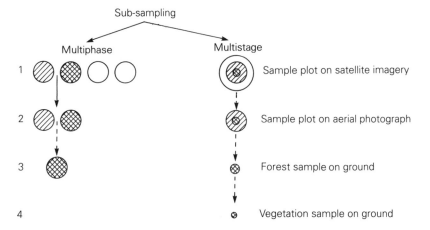

Figure 18.2   Multiphase and multistage sub-sampling (see text).

plots may be assigned first to satellite imagery and then to aerial photographs. This again provides an example of multistage sampling. Sometimes multiphase and multistage methods are combined. The effectiveness of double sampling large areas was demonstrated in the 1960s in Maine, where 39 817 one acre photo-plots were examined and of these 2146 1/5 acre plots were visited in the field (Young, Call and Tryon, 1963). Only 476 acres were actually measured on the ground for 17 100 000 acres of forest.

The availability of satellite data has stimulated interest in both multistage and multiphase sampling techniques. A disadvantage of three-phase sampling is seen that its use has required full aerial photographic coverage of the area under investigation; and that the introduction of satellite imagery, as the third phase, has contributed little to the efficiency of the survey. As pointed out in Part II, complete repetitive coverage with aerial photography at frequent intervals is not cost effective. Existing aerial photographic coverage can often be expected to be too old to use in three-phase sampling; and if new photography is commissioned, it is likely that two-phase (photo-plot) sampling would be preferred.

Two useful tools of double sampling are the separate calculations of the correlation coefficient and of the regression between data from the photographic sample plots and from the same plots in the field. The determination of the correlation coefficient between photo-plots and field plots is helpful in indicating whether a double-sample survey is likely to be as acceptable as a field survey without photographs. The correlation coefficient provides an estimate of the (linear) association between the photo-plot data and the corresponding field plots. If the association is poor, then the coefficient tends towards zero; and conversely if the correlation approaches 1, the correlation is high.

Regression analysis applied to photointerpretation depends on (a) the recording of measurements from a relatively large number of photo-plots (e.g. of tree height, stand volume, crown diameter, number of dead and dying trees) and (b) for a portion of the photo-plots, the corresponding field measurements of the same plots. The measurements of the smaller number of plots in (b) are used in calculating a regression constant and regression coefficient. The commonest form of the regression equation, assuming a linear relationship, is as follows:

$$y = a + b(\bar{x}_1 - \bar{x}_2)$$

where y is the adjusted mean photo-measurement of all photo-plots, a is the regression constant, the mean of all field plot measurements, b is the regression coefficient, calculated for field on photo-measurements, $\bar{x}_1$ is the unadjusted mean of measurements of all photo-plots, and $\bar{x}_2$ is the mean of the measurements of the photo-plots also sampled in the field.

Multiphase sampling

There have been many successful forest surveys involving two-phase sampling using medium to large-scale aerial photographs and ground samples (cf. section 16.3). Also, in the future, comparable results may be achieved in large-area volume inventories, but using smaller scale aerial photographs. An indication of this possibility is provided by timber volume surveys using two-phase sampling with small-scale aerial photographs. For example in Oregon (Maclean, 1981), comparisons have been made between two-phase sampling using black-and-white panchromatic aerial photographs at a scale of 1 : 63 000 and ground samples. For each photo-plot its timber volume was estimated using a two-way aerial volume table (i.e. stand height, crown closure). For every 16 photo-plots out of a total of 23 400 photo-plots, one was measured in the field. Plots were distributed within the delineated strata with probability proportional to area. Using the variance ratio as the test of efficiency, double sampling with nine strata produced estimates with twice the precision of ground sampling.

A further example of successful multiphase sampling is provided by the Alaska multiresource vegetation inventory (LaBau and Winterberger, 1988). The sample design uses three levels of remote sensing information: Landsat MSS digital classification, small-scale colour IR aerial photographs (1 : 60 000) and macro-scale colour IR photographs (1 : 3000 to 1 : 7000). The fourth phase is provided by the field plots. The sample design is illustrated in Figure 18.3 showing plots at 5 km spacing for Landsat, 10 km for the small scale IR aerial

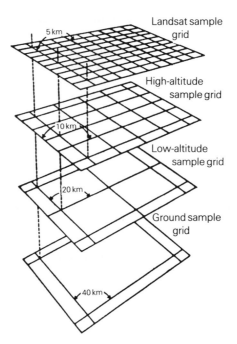

Figure 18.3 Multiphase sampling design (see text).

photographs, 20 km for macro-scale IR aerial photographs and 40 km for field plots (ground truth).

The percentage of land cover estimated by the classes at each phase level is given in Table 18.1. Needle-leaf vegetation (i.e. conifers) is shown to be conspicuously underestimated by Landsat as compared with ground measurements, while broad-leaved forest, mixed forest and herbaceous vegetation are overestimated. An examination of the table will also suggest similarity of results between Landsat and the small-scale photographs and between the macro-scale photographs and the field plots. Probably, as suggested by the authors, the differences are related to image resolution and in the recording of spectral reflectance.

Multistage sampling

The first forest study of multistage sampling involving satellite data was completed in 1969 using Apollo-9 colour IR photographs (Langley, Aldrich and Heller, 1969), and provides a suitable example on which to base future studies. The multistage survey encompassed about 4 million hectares of pine forest, broad-leaved forest and mixed forest in the Mississippi valley and

Table 18.1 Percentage of land cover by classes, as determined for each of the four phases in the Tanana River basin multiresource inventory (Alaska)

	Phases of inventory			
Land-cover class	Landsat MSS	Small-scale photography	Large-scale photography	Ground plots
	Percentage of area			
Needle-leaf	23	26	37	38
Broad-leaf	33	39	34	28
Mixed forest	16	14	7	11
Herbaceous	10	4	6	6
Barren ground	16	11	13	15
Water	2	2	3	2
Missing data*	0	4	1	0
Total	100	100	100	100
Sample size (plots)	5535	1343	331	88

* Missing data due to clouds covering the plot area.

resulted in a sampling error of only 13%. This could have been reduced further if improved aerial volume tables had been available. Without the benefit of the land-cover stratification provided by the satellite imagery, the sampling error would have been 30.7%. The use of the satellite imagery reduced the sampling error to 22.5%, and probability sampling in selecting the primary samples reduced the error to 13%. For such a large area, it would not have been practical to stratify using aerial photographs.

The estimated timber volume (v) in each stratum was calculated using the following derived formula:

$$v = \frac{1}{m} \sum^{m} \frac{1}{p_1 n_1} \sum^{n_1} \frac{1}{p_j} \frac{A_j}{a_c} \frac{1}{p_p t_p} \sum^{t_p} \frac{v_k}{p_k}$$

in which v_k is the measured volume of the kth sample tree on a selected ground plot, p_k is the probability of selecting the pth plot from the cluster, p_p is the probability of selecting the pth plot from the cluster of plots delineated on the $1:2000$ scale $70\,mm$ photos in a strip, p_j is the probability of selecting the jth sample strip in a sample 4×4-mile square area, p_1 is the probability of selecting the ith sample square, a_c is the area covered by the cluster of $1:2000$ scale $70\,mm$ photographs within a strip, A_j is the total area of the jth sample strip, t_p is the number of sample trees measured on the pth plot, n_1 is the number of sample strips in the ith 4×4-mile square, and m is the number of 4×4-mile squares included in the primary sample.

The calculation of the timber volume involved arbitrary probabilities of selection at every stage. Thus the estimates are unbiased no matter what degree of correlation is between the predictions and the variables. The sampling probabilities are formulated from additional information available at each stage by virtue of the increasingly fine resolution of the remote sensing imagery; and the last stage measurements are obtained on the ground and entered in the above formula.

As the first stage in the image analysis, the Apollo photograph was gridded into 4×4-mile squares, which provides the population from which the primary units are randomly selected with probability proportional to the prediction. Each square was classified according to its forest type, and the timber volume was estimated on the assumption that over an area of this size timber volume and forest cover are highly correlated. Each selected primary sampling unit was covered by aerial photographs at a scale of 1 : 60 000 with the purpose of laying out strip transects for 70 mm photography at a scale of 1 : 12 000 and 1 : 2000. Presumably the stage using small-scale photographs could be omitted nowadays due to increased experience and the improved image quality provided by Landsat TM and SPOT. The 1 : 12 000 photographs were assembled as a mosaic and used to position the 1 : 2000 photographs for location in the field of the photo-plots. The 1 : 2000 scale 70 mm format photographs were examined stereoscopically in triplets; and the centre photograph was partitioned into four squares to form plots of about 0.6 acre (0.24 ha). These were of a convenient size to serve as field plots; but only one of these plots was randomly selected per flight strip to be inventoried in the field. For each flight strip, the timber volume was estimated for all photographs using stand height, crown closure and crown diameter and using the available aerial volume table.

19

Towards the global monitoring of forest change using satellite data

As mentioned in Chapter 1, the world-wide monitoring of forest change on a country basis has been compiled periodically by the Food and Agriculture Organization of the United Nations (FAO). This has been based on questionnaires sent out to each member state; but as indicated by the last census in 1980–81, it is feasible to supplement the collected information with data collected by satellites. In some instances this will improve the accuracy and in other circumstances it will provide information not otherwise obtainable from the least developed countries. That monitoring, involving satellite-acquired data, will be needed urgently in the future is readily appreciated in view of the fact that 150 million hectares of the tropical forest and 76 million hectares of tropical woodland, existing in 1980, are likely to have been deforested by the year 2000 (FAO, 1981); and that forest dieback, associated with pollution and probably climatic changes, is already serious in some of the most important economic forests of the world (e.g. Germany, section 16.6).

In its narrowest context, monitoring from satellites includes the detection and evaluation over a short period of time of sudden forest damage; and in its broader context, it may be considered as concerned with continuous processes of change, being comparable with 'continuous forest inventory'. As a developing skill, satellite monitoring can be viewed as separate from forest survey supported by remote sensing, although the employed remote sensing techniques will depend considerably on experience gained in forest inventory. Forest inventory is essentially a stock-taking exercise to determine the composition, area, volume and condition of the growing stock; while monitoring is concerned essentially with a comparison of results acquired at different times.

19.1   TECHNICAL CONSIDERATIONS OF MONITORING BY SATELLITE

Unlike within-country remote sensing, which attempts to develop and to adapt space technology to achieve well-specified national objectives, attempts at global monitoring are likely to be dominated by both national and regional interests. For this reason, it is important to view global monitoring as a marrying of technical capabilities with political constraints. This chapter will therefore proceed by reviewing firstly the state of the art and science of satellite remote sensing for monitoring, and then comment on international considerations. Existing scientific skills are already considerable; but, as recognized by the 1989 IUFRO conference in Venice on global natural resources monitoring and assessment, preparatory activities will be complicated, intensive technical experimentation will be needed and realistic objectives carefully defined.

Monitoring by earth resources satellites

So far, change monitoring has been confined mainly to comparing two temporal sets of Landsat data, to comparing old aerial photographs and satellite imagery often taken several years apart, and to comparing satellite imagery with existing old maps. A disadvantage to using maps is that the sought thematic information may be inadequately or differently portrayed. The availability of Landsat imagery covering much of the globe, several times from 1972 onwards, adds to the attraction of using primarily satellite records; and the availability of earth resources satellite data seems secure up to the first years of the next millennium (Table 19.1).

The temporal satellite data is used to monitor changes which are (a) sudden (e.g. windblow, fire damage) or (b) gradual over a short period (e.g. bark beetle damage) or (c) gradual over a long period of time (e.g. forest decline). Sometimes after the initial image analysis, the satellite data may be found not to have sufficiently fine resolution and/or new satellite imagery cannot be acquired at the times needed, due to cloud cover or heavy haze. Often, at the within-country level, the satellite imagery, after an initial appraisal, has to be supplemented by aerial reconnaissance or replaced by aerial photography, despite the much higher cost.

For ease in the visual analysis of imagery acquired on two different dates of the same ground area, the two scenes should be at the same scale or able to be brought to the same nominal scale using a zoom stereoscope or overhead projector (Chapter 11). Black-and-white prints at two different dates and

Table 19.1 Proposed launch programmes using earth resources satellites (1990–2000 AD)*

Country	Agency	Spacecraft	Scheduled launch date
–	ESA†	ERS-1	1991
USA	NASA	Landsat 6	1991
France	CNES	Spot 3	early 1992
Japan	NASDA	JERS-1	mid-1992
USA	NASA	Landsat 7	mid-1994
Canada	CNSA	Radarsat	mid-1994
–	ESA	ERS-2	late 1994 (tentative)
Japan	NASDA	ADEOS	early 1995
France	CNES	SPOT 4	early 1995
USA	NASA	NPOP-1	late 1996
–	ESA	EPOP-1	mid-1997
USA	NASA	NPOP-2	mid-1998
Japan	NASDA	JPOP-1	mid-1999
France	CNES	SPOT 5	mid-1999
–	ESA	EPOP-2	mid-2000

* Often launch dates are delayed because of technical problems.
† ESA: European Space Agency.

at identical scale can be viewed (non-stereoscopically) under a mirror stereoscope. The viewing of multidate transparencies is achieved using an additive viewer or by matching diazo transparencies on a light table (Chapter 8). By allocating a single colour (hue) to each multidate black-and-white transparency, imagery can be compared which has been taken on several different dates. For the digital analysis of the satellite data, software programming needs to cover the spatial matching of pixels of the two or more different data sets. In addition, in order to improve results, it may be necessary, although expensive, to correct the radiometric values of the pixels of the two sets of data in order to minimize the effect of atmospheric haze and differences in sun-angle and hill shading according to the time of the year. In many circumstances, change monitoring will involve both visual and digital image analysis techniques.

If detail is needed, it may be worth while to examine the feasibility of commissioning new aerial photography to provide the commencing baseline (benchmark) to which the satellite monitoring can be related in later years. The aerial photography may be in the form of complete coverage or covering selected areas or strip photography. Observations of temporal change obtained by comparing imagery of two different dates relate, for example, to inundation in which dry land at one date is compared with it at a later date when flooding has occurred, the combination of seasonal data to separate deciduous and evergreen forest, the change from forest to non-forest and changes in stand structure caused by damage. New plantations can usually be identified by their geometric shape and colour contrast. However, as previously indicated in Chapter 17, the differences between tropical natural forest and young forest regrowth may be too subtle or variable to map accurately using satellite data; and therefore before commencing a temporal study, this problem requires to be locally examined.

Forest damage

So far, temporal studies related to insect damage and disease have been on a country or within-country basis or on shared country boundaries. These indicate that satellite data needs to be restricted to occasions, when there are major differences in the spectral signatures between the healthy and unhealthy vegetation covering extensive areas. It is important not only to be able to detect and map damaged areas, but also to recognize the host forest type, which on satellite imagery is often difficult.

More than twenty data transformation techniques, including multitemporal band ratioing, have been tested in studying insect damage (cf. Nelson, 1983). These indicate that, since damage is associated with individual tree species and differences in stand density, the classification of the forests using satellite data needs to be better than level II (USGS cover classification); and that the major constraint to the wider use of satellite data in disease monitoring and insect infestation monitoring is the failure to be able to identify the host tree species at an acceptable level of accuracy. Also, as pointed out by Heller (1978), most vegetation under stress begins to appear chlorotic in the yellow–orange band of the visible spectrum ($0.58-0.62\,\mu$m); and this spectral resolution so far has not formed a sensing band of the earth resources satellites (cf. Table 7.1).

Forest cover

The assessment of change on a country or within-country basis from forest to non-forest, using Landsat data, has contributed to the world-wide appreciation of the problem of the diminishing forest cover. Studies have been carried out using simple visual analysis. Several examples will be quoted to illustrate the usefulness of the technique. A study of the Gambia using Landsat MSS imagery taken in 1973 and 1978 indicated a loss of forest cover exceeding 20% (Howard, 1979). A study of central Sumatra, in which the cover class delineated on Landsat imagery was compared with existing old maps, indicated that the agricultural lands, resulting from the loss of high forest, had increased by 45% or 300 000 hectares (Schwaar, 1975). Plate 15 shows the loss of forest cover in central Java as provided by comparing Landsat MSS imagery and the existing topographic map (Soerkerno, 1976).

In eastern Thailand, a comparison was made by the Forest Service between the forest cover shown on the Landsat MSS imagery of 1972–73 and 1975–76. Black-and-white imagery in the visible (green) and near-IR spectra at a scale of 1 : 250 000 was used in conjunction with 1 ha field plots. This showed an annual rate of deforestation by administrative provinces of 1.2–12.5%. In Brazil, in an initial study of deforestation in the Amazon Basin, Landsat data of 1973 and 1976 were compared visually and digitally (Pacheco and Novo, 1978). In Mato Grosso state five plant formations/subformations were mapped (i.e. semi-deciduous forest, gallery forest, savanna (cerrado), grassland, flood plain). Results indicated that about 115 000 ha of forest had been exploited in the three year period, which was a 74% increase in the area of recent deforestation; and an analysed subsample of the imagery suggested that only 65% of the cleared land was being utilized.

Digital image analysis applied to forest cover monitoring has the advantage of being fast and able to handle large quantities of data, and is increasingly being used. For example, three or more temporal satellite data sets have been compared for the purpose of monitoring the change in forest cover (e.g. Brazil, Philippines, Thailand). The results suggest that Landsat MSS has been adequate for this purpose, when the periods are short (e.g. 1–3 years) and provided imagery can be obtained annually by regional data-receiving stations. In addition, when employing digital analysis, the registration of temporal sets of pixels has been found not to be a major problem, and an acceptable accuracy is achieved.

Also, the use of up-to-date aerial photographs to help in establishing a baseline, against which satellite data can be compared and used in monitoring, is gaining wider acceptance despite its relatively high cost. In Sri Lanka, for example, comparisons have been made for the period of 1976–79 using a 'benchmark map' at 1 : 100 000 prepared with aerial photographs and 1979 Landsat data; and for the period 1979–83 using only Landsat data (Schmid, 1985). For the digital analysis, all non-forest areas at the beginning of the monitoring period were masked out, which improved the interactive analysis.

In the future, closer association can be expected between satellite monitoring and geographic information systems, since it is not only relatively easy to store and recall earlier satellite data sets for comparison with future acquired data and to mask out those areas that are not relevant, but also the spatial changes can be associated quickly with other categories of filed data (e.g. climatic factors, soils, transportation routes). Nevertheless, the sophistication of the GIS will depend primarily on the cost of 'capturing data' and in many circumstances this may prove prohibitive.

Monitoring by environmental satellites

In circumstances under which cloud-free imagery of a geographic region or a country is readily obtained, it may still be necessary in regional and future global studies to exclude the data acquired by earth resources satellites, due to the cost of purchasing repetitive coverage. To this must be added the increased cost of analysing the very large quantities of digital data. This includes matching adjoining spatial pixel sets, and the physical effort of handling and interpreting all the imagery.

It will be recalled from Chapter 5 that the cost of aerial photography can be expected to be at least quadrupled when the scale is halved. Increases in cost with improved resolution also applies to satellite data. Although with satellite sensing the technology is different, one Landsat MSS scene covers less than 1/250 the ground area covered by a single NOAA AVHRR (GAC) scene and at much higher cost per unit area. One compromise, taking into account the superior resolution of the earth resources satellite data, is to supplement the environmental satellite data with selected scenes of the former. In addition, this helps overcome problems associated with the wider look angle of the environmental satellite (55° for NOAA AVHRR), which results in imagery with excessive earth curvature, differences in sun-angle and shadow lengths across the scene and wide variations in atmospheric factors.

Probably the first project integrating both Landsat and NOAA imagery, and taking into consideration the cost factor, was initiated by the writer in 1974 for desert locust monitoring (Howard, 1975; Hielkema and Howard, 1976). It was concluded, for monitoring ground conditions favourable to the development of the desert locust in north Africa, that cost precluded the complete and continuous coverage by Landsat; and that it should be used only in part as the second stage, followed up by field checking of high-risk areas as the third stage (Howard, Barrett and Hielkema, 1979). The high-risk areas were mapped in ½° grid cells from the NOAA imagery and based on estimated precipitation, which was assessed broadly twice daily with weekly aggregations using a regression model developed by Barrett.

Various meteorological models, now developed world-wide mainly for agricultural use at regional and global levels, could be applied to forest resources, if the need arises. Their importance may lie in detecting gross changes over extensive forest areas. This extends from fire risk to climatic changes.

The models are not confined only to precipitation estimates, which include determining the number of rain days a month; but also involve near ground temperature estimates based on experience in oceanography in estimating (thermal) sea surface temperatures; net solar radiation; degree days as being the accumulation in degrees of temperature above a certain base temperature and as used in agriculture in assessing plant growth stages; solar radiation or cloud factors with the aid of Smithsonian Meteorological Tables in models to determine net solar radiation; the surface albedos of the dominant land cover types, and potential evapo-transpiration and soil moisture.

Usually repetitive environmental satellite coverage is acquired two or more times daily. A resolution of 2.5 km is about minimal in order to identify the precipitation-bearing clouds (e.g. cumulo-nimbus) and to estimate and follow spatially the cloud temperatures. Soil moisture estimates require background information obtainable from small-scale soil maps (e.g. FAO/UNESCO Soil Map of the World). Real-time estimates of the potential evaporation can be calculated crudely using the precipitation and temperature estimates in well-tried ground-based models such as Thornewaite's, but the methods may not be sufficiently sensitive for monitoring climatic changes affecting forest vegetation on a global and continental scale. There is therefore impetus to develop new models, and to associate these with the silvicultural needs of forest practice.

An example of combining meteorological satellite data with the distribution of vegetation is provided by the hydrological model based on precipitation and maximum temperatures, which was developed by Trenchard and Artley (1981); and applied to plant formations/subformations in southern Texas by Norwine and Gregor (1983). The latter mapped the zone using NOAA AVHRR data in the format of the normalized vegetation index. When the normalized vegetation index was regressed against the hydrologic factor, as a test of the reliability, the resultant correlation was 0.86. The hydrologic factor (HF) was defined as:

$$HF_i = HF_{i-1} + P_i - (EP(TX_1 \times TN_1) \times HF_{i-1}/CAP.30)$$

where HF_i is the hydrologic factor on day i in inches, P_i is the precipitation on day i in inches, EP is the pan-evaporation function (estimated), TX_i is the maximum temperature on day i (°F), TN_i is the minimum temperature on day i (°F), CAP is the HF capacity in inches, and $0 \ll HF \ll CAP$.

As yet there is little information available on monitoring at continental levels forest cover change using environment satellite data. Technical methodology is available, as indicated in section 17.4, and encouraging results have been obtained related to food production, phenology and drought (e.g. Hock, 1984; Howard, 1984b; Howard et al., 1984; Howard and Odenyo, 1986; Justice et al., 1985; Hielkema et al., 1986). The analysis techniques include multi-channel density slicing, unsupervised classification, supervised classification and data compression using principal component analysis and vegetation indices. Multistage sampling is likely to require fine resolution earth resources satellite data, as a check when monitoring at global and continental levels, and this type of imagery combined with macro-scale aerial photography when monitoring at the national level.

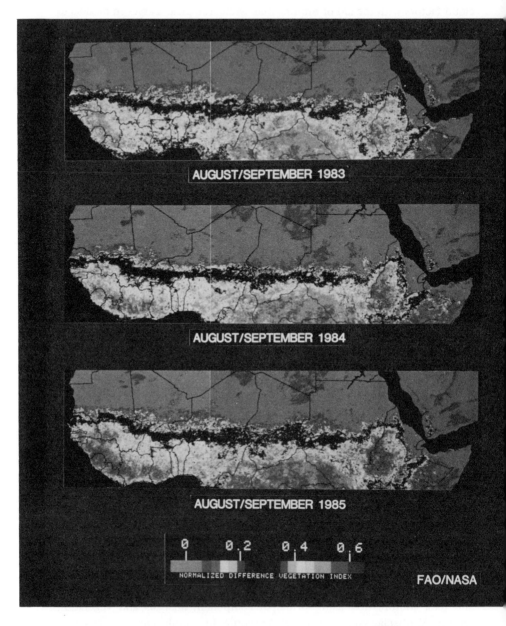

Figure 19.1 NOAA AVHRR GAC data in vegetation index format showing the north/south change of green plant cover/biomass of savanna/sahelian zones of Africa during the 1983/85 drought. By September 1985 the region was recovering. Country boundaries are shown (from colour print).

Examples of monitoring zonal vegetal changes at the earth's surface using NOAA AVHRR GAC data include the seasonal 'green wave' spreading from south to north in North America in the spring and early summer, as the green biomass of the agricultural crops, grasslands and forest cover increases. In West Africa, GAC data has been used in vegetation index format to monitor the catastrophic drought of 1983–85. As the severity of the drought spread from north to south, the green biomass/vegetal cover rapidly decreased in the north to south direction (Figure 19.1).

19.2   INTERNATIONAL CONSIDERATIONS

Any move towards the global monitoring by satellites of the earth's resources and forest decline must be in accordance with acceptable international strategies, which have yet to be defined. Within this ambit, the technically feasible systems, as indicated in the last section, require to be fitted. At the national and sub-national levels, programmes need to be developed; but, as indicated in Chapter 13, the lead time for developing a remote sensing application system is considerable. A lead time of at least 3–5 years should be envisaged for establishing with foreign aid a self-supporting image analysis centre in a developing country (Howard, 1983, 1984b); and the lead time for a regional or continental programme using advanced technology is likely to be much longer. For example, the operational lead time for the regional programme of desert locust surveillance using satellite data, conceived by the writer in 1974, was about 15 years.

Based on the experience gained from international programmes over many years, it may be preferable when contemplating global change monitoring to plan its development initially on country co-operation at sub-continental/ regional or continental levels. There are already well-established international organizations with country groupings in this form (e.g. regional offices of the United Nations family). At the sub-continental level, co-operation between technical institutes is often easier; and existing regional skills within countries may be similar, more easily adapted to local conditions and more easily organized.

Within the ambit of international co-operation, whether it be global or regional, are the problems associated with the acquisition, distribution of satellite-collected data and image analysis. Whether or not a country or a group of countries has a right to acquire information by satellite on land use and forest conditions in another country without its prior agreement is debatable and needs to be taken into consideration when planning global monitoring. That this limit should be on the basis of the boundaries of outer space is not easy to accept. An attempt at the United Nations to define outer space in physical terms, as commencing at 100 km above the earth's surface, was found to be unacceptable as the criterion. Military satellites with a reduced life expectancy already operate for short periods at lower altitudes, while most satellites used for the peaceful collection of data operate at much higher altitudes (sections 7.1 and 7.3; Chapter 15). Also, the finest image resolution, now available for civilian use from satellite systems (i.e. 5–6 m), provides usable information related to forest resources, approaching that of micro-scale aerial photography (Chapter 9). In contrast, consensus at the United Nations accepts the unrestricted distribution of imagery with an image resolution coarser than 50 m and places no restriction on the distribution of environmental satellite data.

Further, from meetings at the United Nations and its agencies, at the UN Conference on the Peaceful Uses of Outer Space in Vienna in 1982 and at the IUFRO conference on global monitoring in Venice in 1989, two distinct international views emerge on regional and global remote sensing by satellite. That is, most industrialized countries prefer the unrestricted collection and distribution of satellite data, which has technological, economic and administrative advantages and embraces a 'one world concept' of action. On the other hand, most developing countries and those which had centrally planned economies favour each country controlling the distribution of finer resolution data collected over that country. These differences in opinion may necessitate that future steps towards regional and global monitoring by satellite be planned on a country-by-country basis. However, it must be borne in mind that airborne remote sensing, which often provides the base line for satellite monitoring, is the exclusive prerogative of individual nations; and that a country has the final control over the effectiveness of the image analysis, since the collection of associated field data and field checking can only be achieved through the full support of that country.

19.3 THE NEXT DECADE

Major changes in remote sensing applications can be expected in the next decade, although these are not likely to be as dramatic as the developments that occurred in the period of the mid-1960s to the mid-1970s. With the long history of aerial photointerpretation, its present day pre-eminent role in local forest resources is unlikely to change; but the impact of analysis techniques used in geographic information systems on API may be significant. Whether comprehensive geographic information systems will be widely used remains to be decided; but simplified systems, based initially on existing maps, are likely to be popular in applying remote sensing to forest resources management. Already, for example, the successfully developed land information base, covering 52 million hectares of forested land in British Columbia, is being 'down loaded' to regional and district forest offices, and the system is being used in forward planning (Heygi, 1985).

Also, over 30 papers have been published in recent years on the employment of combined variables in calculating tree volumes and stand density using aerial photographs; and these increasingly indicate the greatest emphasis will be placed on tree and stand height as the first variable, followed by crown cover and crown area. Current research also suggests that, where crown diameters are easily measured, a further stand density photographic variable can be obtained by adapting the Bitterlich method of angle gauge counts (e.g. Gering and May, 1990).

At the national level of using remote sensing in forestry, its role to support forest management and strategic planning can be expected to expand. This may be stimulated by the interfacing of remote sensing and land information systems/geographic information systems and the resulting two-way flow of data. The trend towards the availability of satellite data with finer pixels can be expected to favour its use to replace micro-scale aerial photography and the development of new and faster methods for merging analogue and digital data.

Already SPOT provides imagery with a resolution of 10 m and Salyut/Soyuz outer space colour photography has a ground resolution of about 5 m. In the author's opinion, it is unlikely that satellite sensing for civilian purposes will have a spatial resolution much finer than 5 m, due to political constraints and for technical reasons mentioned previously. Unfortunately, this will not be conducive, except for special situations, to satellite sensing replacing aerial photography in the intensive management of forest plantations and the up-dating of small compartment records. The spatial, spectral and temporal resolution of the data provided by existing earth resources satellites, however, seems well suited to national surveys, particularly in developing countries; and includes the preparation of thematic maps at a scale of 1 : 50 000 and map revision at 1 : 25 000. Also proven satellite technology is available for systematic stereoscopic coverage for topographic mapping and elevation models with contour intervals of 20 m (Colvocoresses, 1990).

Provided the spectral resolution of the satellite data is improved, new opportunities may occur for its use in assessing damage; but the constraint in this respect is usually not being able to identify the host tree species from the satellite data. The situation of adequately identifying tree species is not likely to improve even with a spatial resolution of 5 m; but as already shown, sensing in the mid-IR can improve the identification and mapping of stand density classes and this in itself may lead to improvements. Also it must not be overlooked that, even with airborne multispectral scanning, the pixel size needed must be very fine to provide operationally acceptable results. For example, in a study in western Canada of the reddish-brown crowns of dead pines attacked by bark beetles (Kneppeck and Ahern, 1989), a pixel resolution of 1.4 m was required to give results considerable better than obtained by colour aerial photographs (a ratio of 1.36 : 1). The average crown diameter was 3.5 ± 0.5 m. With a pixel resolution of 3.4 m, the detection ratio in favour of aerial photography was 0.71 : 1; and with a pixel size of 6 m the recently killed trees were not generally recorded on the scanner imagery.

Irrespective of whether future studies in change monitoring are at the local level or more ambitiously at the continental or global levels, computer-assisted data merging techniques will need to be improved and adapted to recent innovations. This covers the temporal data of the same series of satellites, the merging of data of different sensors and the merging of data taken at different altitudes. Video camera recordings using low level flying are likely to be much more widely used to complement satellite remote sensing. The use of GPS navigation will facilitate the positioning of flight strips of macro-scale miniature photography, as has already been tested in Western Australia (Biggs *et al.*, 1989). The merging of aerial photographic images with the corresponding satellite data and high quality hard copy output of the processed data is probably at the present time restrained by the lack of suitable low cost equipment; and, as pointed out by Jaakola and Hagner (1988), a substantial improvement can be achieved in the spatial resolution of satellite imagery by merging the satellite multispectral data with digitized black-and-white aerial photographs.

In addressing the problem of monitoring changes in forest cover at the continental level, cost will necessitate the use of coarse resolution satellite data for overall coverage and confine fine resolution satellite data (e.g. Landsat MSS, TM) to selected sites; but, as indicated by recent studies in the United States (e.g. Nelson, 1989), the merging of the data is not simple and the accuracy of the estimates is influenced significantly by variations in the density of the forest cover. Also, with the advent of cloud-free, fine resolution, microwave satellite remote sensing in the next few years (e.g. ESA ERS-1, Canadian Radarsat), new possibilities will emerge for merging quite different types of data acquired by different types of satellite sensor. As already recognized from studies of SLAR data (e.g. Joyce and Sader, 1986), significant linear relationships can exist between polarized SAR data and forest green biomass and stand structure, which are not present using electro-optical sensing and photographic sensing.

Thus, the next decade is likely to be a period of consolidation in applying existing remote sensing collection techniques to forest resources, and will be often best achieved using multistage/multiphase sampling. On the other hand, data analysis and the preservation of the analysed data may be radically changed in some circumstances. Impetus for these changes is already occurring through advances in computer technology and the accumulation of ever larger quantities of remotely sensed temporal and spatial data related to a world of increasing population pressures.

Appendix A

Units of measurement

Unit	to convert to	divide by (÷)/multiply by (×)
hectare	acres	÷ 0.404
hectare	square metres (m^2)	× 10^4
kilometres (km)	miles	÷ 1.609
kilometre	metres	× 1000
metre (m)	centimetres (cm)	× 100
metre	millimetres (mm)	× 1000
metre	feet (ft)	× 3.25
metre	microns (μ)	× 10^6
centimetre	inches (in)	÷ 2.54
centimetre	millimetre	× 10
centimetre	metre	÷ 100
square kilometre (km^2)	square miles	÷ 2.59
kilometres	nautical miles	÷ 1.85
metre	micrometres (μm)	× 10^6
metre	nanometres (nm)	× 10^9
metre	Ångstroms (Å)	× 10^{10}
micrometre (wavelength, λ)	Hertz (frequency)	× $2.998 \times 10^{14} \div \lambda$
degrees Celsius/ centigrade (°C)	Fahrenheit (°F)	× 9 ÷ 5 + 32
degrees Celsius	Kelvin (K°)	add 273
radians	milradians	× 1000
radians	degrees (angular)	÷ 0.0175

Appendix B

Tri-lingual terminology – selected terms in English, French and Spanish

English	French	Spanish
	A	
aberration	aberration	aberración
absolute orientation	orientation absolue	orientación absoluta
absolute parallax	différence de parallaxes absolues	paralaje absoluta
absorption	absorption spectrale	absorción
accuracy	exactitude	exactitud
achromatic	achromatique	acromático
across-track	trace en travers	pista transversal, trazado transversál
active sensor	capteur actif	sensor activo
additive colour viewer	synthetiseur de couleurs	visor de color aditivo
additive viewing	combinaison de couleur	visión aditiva
aerial camera	chambre de prise de vues aériennes	cámara aérea
aerial camera mount	suspension de chambre de prise de vue aérienne	montaje (suspensión) de la cámara aérea
aerial exposure index	indice d'exposition aérienne	indice de exposición (para fotografía) aérea
aerial film	pellicule pour photographies aériennes	película para fotografías aéreas
aerial photogrammetry	photogrammetrie aérienne	fotogrametría aérea
aerial (air) photograph	photographie aerienne	fotografía aérea
aerial photography	prise de vue aeriénne	fotografía aérea
aerial photointerpretation (API)	photo-interpretation aérienne	fotointerpretación aérea
aerial triangulation	aerotriangulation	triangulación aérea
aerospace	aerospatial	aeroespacial
age-class	classe d'age	clase según edad

English	French	Spanish
air base (airbase)	base de prise de vues (aeriennes)	base aérea, base de cámara
air co-ordinates	coordonnées aériennes, coordonnées	coordenadas aéreas, coordenadas
	photographiques	fotografías
albedo	albédo	albedo
altimeter	altimètre	altímetro
altitude (flight)	altitude de vol	altitud de vuelo
angle of convergence	convergence binoculaire	convergencia óptica
angle of coverage (angular field)	angle de champ	ángulo de campo (cobertura)
aperture	ouverture	abertura de cámara
apogee	apogée	apogeo
appearance ratio	coefficient d'hyperstéréoscopie	apariencia relativa
automatic picture transmission (APT)	transmission automatique des images	transmisión automática de imágenes
attenuation	atténuation, affaiblissement	atenuación, amortiguación
attitude	attitude de vol	posición de vuelo
automatic classification	classification automatique	clasificación automática

B

English	French	Spanish
backscatter	retrodiffusion	retrodispersión
bandwidth	largeur de bande	ancho de banda, anchura de banda
bar scale	échelle graphique	escala grafica (de barra)
basal area	surface terriere	
base–height ratio	rapport base–éloignement	escala de base, relacíon de base
		a altura
base line	ligne de base	línea base

English	Français	Español
base map	carte de base	mapa base, carta fundamental, mapa básico
beam (radar)	faisceau (de radar)	haz (de radar)
beam width	ouverture de faisceau	abertura del haz
black-and-white film	émulsion (pellicule) noir et blanc	película en blanco y negro
black body	corps noir	cuerpo negro
block adjustment	compensation des blocs	compensación de bloques
bole	fut, tige	tronco
boundary	limite	límite, termino
branch	branche	rama
bridging	triangulation aérienne	triangulación
brightness scale	échelle de luminance	escala de luminosidad
broad-leaved	arbre feuillu	latifoliadas

C

English	Français	Español
cadastral survey	levé cadastral	levantamiento catastral
calibrated focal length	distance principale d'étalonnage, distance focale calibrée	constante focal de la cámara, longitud focal calibrada
camera	chambre photographique (chambre de prise de vues)	cámara fotográfica
camera, multiple lens	chambre à objectifs multiples	cámara multiple
camera, single lens (frame)	chambre à objectif unique	cámara unilente
camera lucida	chambre claire	cámara clara
camera, multiband (multispectral)	chambre multibande	cámara multibanda (multispectral)
C-factor (altitude–contour ratio)	rapport hauteur courbe de niveau	relación altitud/curvas de nivel
channel (spectral)	canal spectral	canal
charge-coupled detector (CCD)	détecteur a transfert de charge (DTC)	dispositivo de acoplamiento de cargas

English	French	Spanish
classification	classification	clasificación
clean (clear) fell	couper a blanc	claro, desmonte
clearing	défrichement	claro
cloud cover	couverture nuageuse	cobertura nubosa
cluster	ensemble, groupe	aglomerado
clustering	groupage, groupement	agrupación
colour balance	équilibre des couleurs	balance de color
colour composite	composition colorée	coloreado artificial
colour display	affichage en couleur	representación visual
colour enhancement	accentuation des couleurs	refuerzo de colores
colour film	film (émulsion, pellicule) couleur	película en color
colour imagery	imagerie en couleur	imágenes en color
colour infrared (CIR) film	film infrarouge couleur	película infrarroja de color
colour photography	photographie en couleur	fotografía en color
colour registration	superposition de couleurs	sobreposición de colores
colour rendition	rendu des couleurs	rendimiento de colores
colour vision	vision chromatique	vision cromática
comparator	comparateur	comparador
compartment	parcelle	departamento
complementary colours	couleurs complémentaires	colores complementarios
computer-compatible tape (CCT)	bande magnétique compatible avec l'ordinateur (BCO)	cinta compatible con/por ordenodor (computador)
computerized map	carte infographique	mapa establecido por computador, mapa infográfico

English	French	Spanish
conifer	conifère	conífero
contact print	épreuve (tirage) par contact	copia de contacto
contour interval	intervalle de courbes	curva de intervalo
contour line	courbe de niveau, isohypse	curva de nivel
contrast	contraste	contraste
control point	point de canevas, point de controle	punto de control (apayo)
co-ordinate grid	réseau de coordonnées	retícula de coordenadas
coverage	couverture	cubrimiento, cobertura
crab, drift	dérive	deriva, crab, desviación
crown	cime	cumbre
crown class	classe de cimes	classe ségun la cona
cut-off filter	filtre de coupure, filtre à elimination, filtre coupe-bande	filtro de separación, filtro elimina-banda

D

English	French	Spanish
data	données	datos
data acquisition, data capture, data collection	saisie de données	adquisición de datos
data analysis	analyse de données	analisis de datos
data bank	banque de données	banco de datos
data base	base de données	base de datos
data coding	codification des données	codificación de datos
data collection platform (DCP)	plate-forme de collecte de données	platforma de obtencíon de datos
data correction	correction de données	corrección de datos
data display	affichage (visualisation) de données	representación visual de datos
data flow	circulation de données	flujo de datos
data handling	manipulation de données	manipulación de datos
data management	gestion de données	gestión de datos

English	French	Spanish
data processing	traitement de données	procesamiento de datos
data reduction	réduction (condensation) de données	conversión de datos
data retrieval	recherche de données	recuperación de datos
data smoothing	lissage de données	suavización de datos
data storage	stockage de données	almacenamiento de datos
datum level	surface du canevas de nivellement	nivel del datum
densitometer	densitomètre	densitómetro
densitometry	densitométrie	densitometría
density	densité	densidad
density slicing	isodensitométrie, equidensitométrie	isodensitometría
depression angle	angle de depression	angulo de depresión (de profundidad)
detector	détecteur	detector
diapositive ('slide')	diapositif	diapositiva
diffuse radiation	rayonnement diffus	radiación difusa
diffuse reflection	reflection diffuse	reflexión difusa
digital image	image numerique	imagen digital
digitized image	image numérisée	imagen digital de computadura
digital recorder	enregistreur	indicador numérico
digital-to-analog(ue) converter	convertisseur numérique-analogique	convertidor numérico-analógico
dip angle	angle de dépression à l'horizon	depresión del horizonte
displacement	filé	desplazamiento
Doppler shift	effet (dérive) Doppler	efecto (cambio) Doppler
drift	dérive	deriva

English	French	Spanish
filtering	filtrage, filtration	filtración
flight line	ligne de vol	linea de vuelo
flight plan (map)	plan de vol	mapa de vuelo
flight path	trajectoire de vol	trayectoria de vuelo
flight strip	bande de photographies aériennes consécutives	tira fotografica de vuelo, faja de vuelo
floating mark	index de pointé stéréoscopique, marque repére	marca móvil
flying speed	vitesse de vol	velocidad de vuelo
focal length	distance focale	distancia focal
focal plane	plan focal	plano focal
focus	foyer	foco
forest	forêt	selva, bosque
forest stand		rodal forestal
format	format	formato
frequency	fréquence	frecuencia

G

English	French	Spanish
gamma	gamma, facteur de contraste	gamma
geodemeter	géodimètre	geodémetro
geometric accuracy	exactitude géométrique	precisión geométrica
geostationary satellite	satellite géostationnaire	satélite geostacionario
geosynchronous satellite	satellite synchrone	satelite sincrónico
	satellite géosynchrone	satelite geosincrónico
grain	grain	grano

English	French	Spanish
granularity	granularité, granulation	granulosidad
grey body	corps gris	cuerpo gris
grey scale	gamme de gris	escala gris
grid (map)	réseau, quadrillage	red de mapa
grid cell	maille	cuadrícula
grid co-ordinates	grille de coordonnées	red de coordenedas
ground control	canevas de restitution au sol (canevas au sol)	control terrestre, control primario
ground control point	point d'appui	punto terrestre de referencia
ground data	données de terrain	datos terrestres
ground range	distance au sol	alcance en tierra
ground resolution	réseau au sol	resolución del terreno
ground station	station au sol, poste à terre	estación en el suelo
ground temperature	témperature au sol	temperatura del terreno
ground track	trace, trajectoire au sol	pista en tierra
ground 'truth'	réalité de terrain	apoyo terrestre

H

English	French	Spanish
H and D curve (sensitometric curve)	courbe caractéristique, courbe sensitométrique	curva caracteristica
habitat	habitat	medio
hardware	matériel	conjunto de equipos, material, maquinaria
hardwood	bois dur	madera brava, madera dura
haze	brume (brume sèche)	brume, velo
haze filter	filtre anti-brume	filtro de niebla
heat budget	bilan thermique	
height class	classe de hauteur	distribución de energia solar

English	French	Spanish
high oblique photography	photographie oblique	fotografiá oblicua alta
high pass filter	filtre passe-haut	filtro de circuito de frecuencia alta
high resolution infrared radiometer (HRIR)	radiomètre à pouvoir de résolution élévé dans l'infrarouge	radiómetro infrarrojo de alta resolución
horizontal control	référence planimétrique, canevas planimétrique	control horizontal
hot spot (reflex, reflection point, no-shadow point)	point chaud, plage hyperlumineuse	mancha caliente
hue	tonalité chromatique	matiz

I

English	French	Spanish
image	image	imagen
image analysis	analyse d'image	análisis de la imagen
image conversion	transformation d'image	transformación de la imagen
image co-ordinates	coordonnées-image	coordenadas de la imagen
image correction	correction d'image	modificación de la imagen
image degradation	dégradation d'image	degradación de la imagen
image density	densité de l'image, noircissement de l'image	densidad de la imagen
image distortion	distorsion de l'image	distorsión de la imagen
image enhancement	accentuation d'image	realce de imagen
image plane	plan de l'image	plano de la imagen
image processing	traitement de l'image	tratamiento de la imagen
image quality	qualité de l'image	calidad de la imagen

English	French	Spanish
image ratioing	rapport de luminance d'images numériques, mise en rapport d'images	racionamiento de la imagen
image registration	superposition d'images	superposición de la imagen
image resolution	limite de resolution sur l'image	resolución de la imagen
image restoration	restauration d'image	restauración de la imagen
imagery	imagerie	imágenes
increment	acroissement en volume	aumento, incremento
index map	tableau d'assemblage	map indicador
infrared (IR)	infrarouge	infrarrojo
infrared photography	photographie infrarouge	fotografía infrarroja
infrared imagery	imagerie en infrarouge	imágenes infrarrojas
instantaneous field of view (IFOV)	champ de visée instantané	campo de vista instantáneo
intensity	intensité	intensidad
interactive display	système interactif, système conversationnel	sistema recíproco
interface	interface, surface de séparation	superficie de separación
interpretation key	clef d'interpretation	clave de interpretación
intervalometer	intervallomètre, chrono-déclencheur	intervalómetro
irradiance	éclairement énergétique	densidad de flujo radiante
isocentre	isocentre, point focal	isocentro
isoline	isoligne, ligne d'echelle constante	isolínea, línea de escala constante

L

English	French	Spanish
land	terre	tierra
LASER (light amplification by simulated emission of radiation)	laser	laser
latent image	image latente	imagen latente
layover	déversement radar	desplazamiento vertical

English	French	Spanish
leaf	feuille	hoja
leafless	sans feuilles	sin hojas
lens	objectif photographique	lentes
log	bûche	
look angle	angle d'ouverture de faisceau radar	ángulo de captación
low oblique aerial photograph	photographie aérienne oblique basse	fotografía aérea oblicua baja
low pass filter	filtre basse-bas, filtre coupe-haut	filtro de circuito de frecuencia baja
luminance	luminance visuelle	luminancia

M

English	French	Spanish
magnetic tape	ruban (bande) magnétique	cinta magnética
mapping	cartographie, lever	cartografía
map projection	projection cartographique	proyeccion cartográfica
mask	masque, cache	mascara
master image	original	imagen de archivo
mature	mur	maduro
mean age	age moyen	edad media
mean scale	échelle moyenne	escala media
medium scale photography	photographie à moyenne échelle	fotografía de media escala
memory	mémoire	memoria
meteorological satellite	satellite météorologique	satelite meteorológico
mid infrared (MIR)	infrarouge moyen	
minus-blue filter (yellow filter)	filtre moins bleu, filtre jaune	filtro amarillo
model (stereoscopic) deformation	déformation du modèle, déformation l'image plastique	deformaciónes del modelo

English	French	Spanish
modified infrared	infrarouge modifié	infrarrojo modificado
monitor	moniteur	monitor
monitoring	monitorage, contrôle	control
mosaic	mosaïque, photo-mosaïque	mosaico
multiband system	système multicana	sistema multibanda
multispectral analysis	analyse multibande	análisis multiespectral
multispectral imagery	imagerie multibande	imagen multiespectral
multispectral scanner	scanneur multibande	barredor multiespectral
multistage sampling	échantillonage multiple, échantillonage varié	muestreo de etapas múltiples, muestreo plurifásico
multitemporal photography	prise de vue multidate	fotografía multitemporal

N

English	French	Spanish
nadir	nadir	nadir
near infrared (NIR)	proche infrarouge	casi infrarrojo
negative	négatif, phototype négatif	negativo
noise	bruit	ruido
non-imaging remote sensing	télédetection sans image	percepción sin imagenes
normal angle lens	objectif à angle normal	lente de ángulo normal

O

English	French	Spanish
oblique aerial photography	photographie aérienne oblique	fotografía aérea inclinada
off-line	autonome	autónomo
on-line	en ligne	en directo, en conexión
orbit	orbite	órbita
orientation	orientation	orientación
ortho-photography	orthophotographie	ortofotografía
ortho-photomap	orthophotocarte	ortofotomapa

English	French	Spanish
overlap	recouvrement, chevauchement	recubrimiento
overlay	superposition, procédé par transparence	revestimiento
	P	
panchromatic film	pellicule panchromatique, émulsion panchromatique	película pancromática
parallax bar	stéréomicromètre, barre de parallaxe	estereómetro
parallax difference	différence de parallaxe	diferencia de paralaje
parallax wedge	coin de parallaxe	cuña de paralaje
parallel processing	traitement en parallèle	procesamiento paralelo
parametric classification	classification paramétrique	clasificación paramétrica
passive sensor	capteur passif	sensor pasivo
pass (wing) point	point de transfert	punto de transferencia
pattern	forme, trame	modelo
pattern recognition	reconnaissance de formes	reconocimiento de modelo
payload	charge utile	cargo util
perigee	périgée	perigeo
perspective grid	grille pérspective	retículos perspectivos
photogrammetric control	canevas photogrammétrique	control fotogramétrico, control menor
photogrammetry	photogrammétrie	fotogrametría
photo-index	tableau d'assemblage	mapa indicador
photointerpretation key	clef de photo-interprétation	clave de interpretación
pitch	tangage, convergence	grado de inclinación

English	French	Spanish
pixel (picture element)	pixel	elemento mas pequeño de una imagen (cuadro)
plan positioning indicator (radar)	indicateur de gisement panoramique	indicador de plano de posición
planimetric map	carte planimétrique	mapa planimétrico
plantation	plantation	plantación
platform	plate-forme	plataforma
plotter (plotting machine)	appareil de restitution photogrammétrique	aparato de restitución
plumb point (nadir point)	point nadiral	punto nadiral
polarization	polarisation	polarización
polarizing filter	filtre polariseur	filtro polarizador
positive	positif, diapositif, phototype positif	positivo
precision	précision	precisión
pre-processing	pretraitement	procesamiento preliminar
primary colour	couleur fondamentale (primaire)	color primario
principal point	point principal, point central	punto principal, punto central
print	épreuve	impresión
processor	processeur	procesador
pseudoscopic viewing	observation pseudoscopique	observación seudoscópica
pulpwood	pâte de bois	
push broom sensor (CCD)	capteur en peigne	sensor secuencial con peine detector

Q

English	French	Spanish
quick look	épreuve minute	imagen instantánea

R

English	French	Spanish
radar	radar	radar
radar beam width	largeur de faisceau radar	anchura del haz de radar

English	French	Spanish
radial (line) triangulation	triangulation radiale	triangulación radial
radiance	radiance énergétique, luminance énergétique	radiancia, flujo energético específico
radiant energy	énergie rayonnement	energía de radiación
radiometer	radiomètre	radiómetro
ratio print	rapport de reproduction	relación de impreso
RBV (return beam vidicon)	vidicon à retour de faisceau (VRF)	vidicón con haz explorador restituido al cátodo
read only memory	mémoire morte	memoria fija
read-out	sortie de lecture	lectura
real aperture	antenna réelle	abertura real
real time	temps réel	tiempo real
receiving station	station recéptrice	estación de radiorrecepción
rectification	redressement	rectificación
redundancy	redondance	redundancia
reflectance	réflectance, facteur de réflexion	reflectancia
reflectance ratio	rapport de réflectances	relación de reflectancia
reflection	réflexion	reflexión
reflectivity	réflectivité	reflectividad
refraction	réfraction	refracción
relative grey-scale value	valeurs relatives de la gamme des gris	valor característico en la escala fotografica relativo a las tonalidades del gris
relative orientation	orientation relative	orientación relativa

English	Français	Español
relief displacement	déplacement de l'image d'une dénivelée, déplacement de relief	errores planimétricos ocasionados por desniveles, desplazamiento
remotely sensed data	données obtenues par télédétection	datos telepercibidos
remote sensing (RS)	télédétection	percepción remota, telepercepción
remote sensor	télécapteur	teleperceptor remoto
representative fraction	échelle numérique	escala numérica, fracción representativa
resection	relèvement	resección
resolution	résolution, limite de résolution	resolución
resolving power	pouvoir de résolution, pouvoir séparateur	poder de resolución
ride (track	laie, piste	vereda
ridge	crête	cima, sierra
roll	roulis	balanceo

S

English	Français	Español
sampling interval	pas d'echantillonnage	intervalo de muestreo
SAR (synthetic aperture radar)	radar à synthèse d'ouverture	abertura sintética
satellite	satellite artificiel, satellite d'observation	satélite
satellite imagery	imagerie par satellite	imágenes desde un satelite
scale	échelle	escala
scan, scanning	scannage	exploración, rastreo
scan line	ligne de scannage, ligne de balayage	línea de exploración
scanner	scanneur	rastreador, barredor
scene	scène	escena
screen	écran	pantalla
scrub	broussailles	matorral, breñal
seedling	jeune plante	planta de semilla

English	French	Spanish
semi-controlled mosaic	mosaïque photographique	mosaico semi controlado
sensor	capteur, détecteur	sensor
shade bearer	essence d'ombre	
shadow	ombre	sombra
shrub	arbuste, arbrisseau	arbusto
shutter	obturateur	obturador
sidelap	recouvrement latéral, recouvrement bande à bande	recubrimiento lateral
side-looking radar (SLR)	radar à visée latérale	radar de mira lateral, radar de vista lateral
side-looking airborne radar (SLAR)	radar aéroporté à antenne latérale (RAAL)	
signature analysis	analyse de signature spectrale	análisis de rasgos espectrales
single-lens camera	chambre à objectif unique	cámara unilente
size class	catégorie de grosseur	clasificación por tomaños
slant range	distance-temps	distancia real
slope	pente	falda, cuesta
small-scale (micro-scale) aerial photography	prise de vue aérienne à petite échelle	fotografía aérea en pequeña escala
software	logiciel	logical, software
soil	sol	tierra, suelo
solar angle	angle solaire, angle du soleil	ángulo solar
space co-ordinates	coordonnées spatiales	coordenadas espaciales
space craft	spationef	vehículo espacial
space shuttle	navette spatiale	transbordador espaciel
space station	station spatiale	estación espacial

spatial resolution	résolution spatiale	resolución espaciel
species	espèce, essence	especie, raza
spectral band	bande spectrale	banda espectral
spectral resolution	résolution spectrale	resolución espectral
spectral signature	signature spectrale, caractéristique d'une cible	firma espectral
spectral window	fenêtre spectrale	ventana espectral
spectrometer	spectromètre	espectrómetro
spectrum	spectre	espectro
specular reflection	réflexion spéculaire	reflexión especular
step wedge	coin sensitométrique gamme de gris	escala gráfica de gris
stereogram(me)	stéréogramme, stéréophotogramme	estereograma
stereoscopic model, stereoscopic image	modèle en relief, image plastique	imagen estereoscópica
stereoscopic pair	couple de clichés stéréoscopiques	par estereoscópica
stereoscopic parallax	parallaxe stéréoscopique	paralaje estereoscópica
stereo-plotter	appareil de stéréorestitution	aparato estereoscópico de restitución
stereoscopic vision	vision stéréoscopique, vision en relief	visión estereoscopica
stereoscopy	stéréoscopie	estereoscopia
storey	étage	piso
strip (photo-)	bande de clichés, bande d'images, bande photographique	faja, tira
strip photography	prise de vue en bande	fotografía en bandas
subsoil	sous-sol	subsuelo
subtractive colour system	procédé soustractif d'obtention des couleurs	sistems color sustractivo
sun synchronous satellite	satellite héliosynchrone	satélite síncrono con el sol

English	French	Spanish
supervised classification	classification dirigée, classification contrôlée	clasificación controlada, clasificación dirigida
surface roughness	rugosité de surface	irregularidades del terreno
surface temperature	température de surface	temperatura superficial
survey, surveying	arpentage, levé, lever, prospection	apeo, levantamiento
sustained yield	rendement soutenu	rendimiento sostenido
swath	couloir exploré	zona explorada
sweep	balayage, trace	barrido
swing	angle de déversement	diámetro máximo admisible
synthetic aperture *see* SAR	radar à ouverture synthétique	radar de abertura sintética

T

English	French	Spanish
tape recorder	enregistreur à ruban magnétique	magnetófono
target	cible	blanco
templet (template)	gabarit de triangulation, plaque de triangulation	plantilla, gálibo
terminal	terminal, unite d'entrée-sortie	puesto terminal, aparato terminal
terrain analysis	analysise de terrain	análisis del terreno
test site	site témoin	lugar de ensayo
texture	texture	textura
thematic mapping	cartographie thématique	cartografía temática
thermal imagery	imagerie thermique	imágenes con rayos infrarrojos, imágenes térmicas
thermal infrared (far **IR**)	infrarouge thermique (**IRT**)	infrarrojo térmico
thermal scanning	scannage en infrarouge	exploración infrarrojo

English	Français	Español
thermal sensing	détection thermique	percepción térmica
thinning	éclaircie	
tilt	inclinaison	inclinación
tilt, lateral (x-tilt)	angle de convergence	inclinación transversal
tilt, longitudinal (y-tilt)	angle de site	inclinación longitudinal
tilt displacement	déplacement dû à l'inclinaison	errores planimétricos, desplazamiento
timber	bois d'oeuvre	madera
tone	ton	ton
topographic map	carte topographique	mapa topográfico
topographic mapping	lever topographique	levantamiento topográfico
track	route, trace, parcours	trayectoria, curso
tracking	localisation	rastreo
transferred principal point (conjugate principal point)	point-image conjugée, point-image homologue	transferencia del punto central
transmission	transmission	transmisión
transmittance	facteur de transmission, transmittance	transmitancia
tree	arbre	árbol
true colour	couleur réelle	color real

UV

English	Français	Español
ultraviolet (UV)	ultraviolet	ultravioleta
uncontrolled mosaic	assemblage photographique, mosaïque simple	mosaico sin control
updating	mise à jour	actualización
vertical aerial photography	prise de vue aerienne à axe verticale, prise de vue aerienne verticale	aerofotografía vertical

English	French	Spanish
vertical axis (z-axis)	axe de lacet	eje de guiñada
video recorder	enregistreur vidéo	vídeo registrador
vidicon	vidicon	vidicón
visibility	visibilité	visibilidad
visible band	bande visible	banda visible
visual acuity	acuite visuelle, pouvoir separateur de l'oeil	agudeza visual
visual analysis, visual interpretation	interprétation visuelle	interpretación visual
volume	volume	volumen
WXYZ		
wide angle lens	objectif grand-angulaire	lente con gran angular
woodland	terrain boise	monte, bosque, selva
x-axis	axe des X	eje-X, eje de abscisas
x-parallax, horizontal parallax	parallaxe horizontale, parallaxe longitudinale, parallaxe X	paralaje horizontal, paralaje horizontal de abscisas
yaw	lacet, déversement	banzado, guiñada
y-axis	axe des Y	eje-Y, eje de ordenas
y-parallax	parallaxe transversale, parallaxe Y, parallaxe verticale	paralaje Y
z-axis	axe-des Z	eje-Z
zenith	zenith	cenit
zoom lens	objectif à distance focale variable	lente zoom

Note Further information on remote sensing terminology will be found in several publications, including those of FAO and the American Society of Photogrammetry (the latter publishes terms in German, Italian, Portuguese and Russian, edited by G. A. Rabchevsky).

References

Agrawal, N. K. and Sharma, M. K. (1983) Application of remote sensing techniques for forest resources assessment with particular reference to NOAA-7 satellite of USA. Forest Survey of India, Dehra Dun.

Akca, A. (1971) Identification of land use classes and forest types by means of microdensitometer and discriminant analysis. Application of remote sensing in forestry. Report, University of Freiburg, FRG, 147–64.

Aldrich, R. C., Bailey, W. F. and Heller, R. C. (1959) Large scale 70 mm colour aerial photography techniques and equipment and their application to forest sampling. *Photogramm. Engng and Remote Sensing*, 25, 747–54.

Aldred, A. H. and Blake, G. R. (1967) Photo mosaics for forest stand mapping, Forestry Branch Information Report, FMR-X-3, Ottawa.

American Society of Photogrammetry (1960) *Manual of Photographic Interpretation* (ed. R. N. Colwell), ASP, Washington DC.

American Society of Photogrammetry (1983) *Manual of Remote Sensing* (ed. R. N. Colwell), 2 vols., ASP, Washington DC.

Anderson, J. R., Hardy, E. E., Reach, J. T. and Witmer, R. E. (1976) A land use and land cover classification system for use with remote sensor data, USGS Professional Paper 964, Washington DC.

Attiwill, P. M. (1962) Estimating branch dry weight and leaf area from measurements of branch girth of eucalypts. *For. Sci.*, 8, 132–41.

Avery, T. E. (1964) To stratify or not to stratify. *J. For.*, 62, 106–8.

Avery, T. E. (1977) *Interpretation of Aerial Photographs*, 3rd edn, Burgess Publishing Company, Minneapolis.

Avery, T. E. and Meyer, M. P. (1959) Volume tables for aerial timber estimating in northern Minnesota. USFS For. Exp. Sta. paper 96.

Backstrom, H. E. and Welander, E. (1953) An investigation in diffuse reflection capacity of leaves and needles of different species. *Norrlands SkogsvFörb. Tidskr.*, **1**, 141–49.

Baltaxe, R. (1980) *The Application of Landsat Data to Tropical Forest Surveys*, FAO, Rome, Italy.

Banyard, S. G. (1979) Radar interpretation based on photo-truth keys. *ITC Journal*, **2**, 267–76.

Barrett, E. C. (1975) Rainfall in northern Sumatra: Analysis of conventional and satellite data for the planning and implementation of the Krueng Jreue/Krueng Baro irrigation schemes. Report, Geography Department, University of Bristol.

Barrett, E. C. and Curtis, L. F. (1976) *Introduction to Environmental Remote Sensing*, Chapman and Hall, London.

Becking, R. W. (1959) Forestry applications of aerial colour photography. *Photogramm. Engng and Remote Sensing*, **25**, 559–65.

Benson, M. C. and Simms, W. G. (1967) False colour film fails in practice, *J. For.*, **65**, 904.

Benson, A. S. and Gloria, S. D. (1985) Interpretation of Landsat-4 Thematic Mapper and Multispectral Scanner data for forest surveys. *Photogramm. Engng and Remote Sensing*, **51**, 1281–89.

Bernstein, R. and Ferneyhough, D. G. (1975) Digital image processing. *Photogramm. Engng and Remote Sensing*, **41**, 1465–76.

Biggs, P. H., Pearce, C. J. and Westcott, T. J. (1989) GPS navigation for large-scale photography. *Photogramm. Engng and Remote Sensing*, **55**, 1737–41.

Bitterlich, W. (1948) Die Winkelzahlprobe. *Allgem. Forest-u*, **59**(1), 4–5.

Blandford, H. R. (1924) The aero-survey and mapping of the Irrawaddy delta. *Indian Forester*, **50**, 605–16.

Blasco, F. (1984) The international map of vegetation. *Handbook of Vegetation Science*, **24**, Elsevier, Amsterdam.

Blecher, P. and Birchall, C. J. (1977) A reconnaissance appraisal of land systems of Sierra Leone for agricultural purposes. *FAO/UNDP-MANR Report No. 1*, Freetown, Sierra Leone.

Bock, A. G. (1968) A breakthrough in high altitude photographs from jet aircraft. *Photogramm. Engng and Remote Sensing*, **34**(7).

Bodmer, H. C. (1988) Forest stand mapping by means of satellite imageries (TM and SPOT) in Swiss Mitteland. *Proc. IUFRO Division 4 Meeting, Helsinki, Finland*.

Boon, D. A. (1956) Recent development in photo-interpretation of tropical forests. *Photogrammetria*, **12**, 382–86.

Bourne, R. (1928) Aerial survey in relation to the economic development of new countries with special reference to an investigation carried out in Northern Rhodesia, *Oxf. For. Mem. No. 9*, Clarendon Press, Oxford.

Bourne, R. (1931) Regional survey and its relation to stocktaking in agriculture and forest resources of the British Empire. *Oxf. For. Mem. No. 13*, Clarendon Press, Oxford.

Brachet, G. (1986) Improving your image. *Space*, **1**, 30–33.

Brandenberger, A. J. and Ghosh, S. K. (1985) The world's topographic and cadastral mapping operation. *Photogramm. Engng and Remote Sensing*, **51**, 437–44.

Brandenberger, A. J. and Ghosh, S. K. (1987) United Nations 1987 Status of the World, topographic and cadastral mapping. *Economic and Social Council Report*, New York.

Braun-Blanquet, J. (1947) *Carte des groupements végétaux de la France. Région Nord-Ouest de Montpellier*, Station Internationale de Geobotanique Méditerranéene et Alpine, Montpellier.

Braun-Blanquet, J. (1964) *Pflanzensoziologie: Grudzüge der Vegetationskunde*, 3rd edn, Springer-Verlag, Vienna and New York.

Brazier, A. J. (1985) Work experience of transit satellite positioning in developing countries. *Proceedings International Seminar on Surveying by Satellite for the Developing World*, Trinity College, Dublin, 81–93.

Brennan, B. (1969) Bidirectional reflectance measurements from an aircraft over natural earth surfaces. Goddard Space Flight Center, Maryland, X-622-69-216.

Brink, A. B., Mabbutt, J. A., Webster, R. and Beckett, P. H. (1966) Report of the working group on land classification and data storage, MEXE, Christchurch, UK.

Brisco, B. and Protz, R. (1982) Manual and automatic crop identification with airborne radar imagery. *Photogramm. Engng and Remote Sensing*, **48**, 101–9.

Cable, D. R. (1958) Estimating surface area of ponderosa pine foliage in central Arizona. *For. Sci.*, **4**, 45–9.

Carneggie, D. M., Wilcox, D. G. and Hacker, R. B. (1971) The use of large scale aerial photographs in the evaluation of Western Australia rangelands. Technical Bulletin 10, Department of Agriculture, Perth.

Carneggie, D. M., de Gloria, S. D. and Colwell, R. N. (1974) Usefulness of ERTS-1 and supporting aircraft data for monitoring plant development and range conditions on California's annual grassland. *BLM Final Report 53500-CT-266(N)*.

Christian, C. S. and Stewart, G. A. (1947) General report on land classification (North Australia Regional Survey), CSIRO, Canberra.

Cielsa, W. M. (1989) Aerial photos for assessment of forest decline – a multinational overview. *J. For.*, **87**, 37–41.

Ciesla, W. M. and Hildebrandt, G. (1986) Forest decline inventory methods in West Germany: opportunities for application in North American forests. Report 86-3, UDDA Forest Service, Fort Collins.

Clark, W. (1949) *Photography by Infrared*, Wiley, New York.

Clements, J. and Guellec, J. (1974) Utilisation des photographies aeriennes au 1 : 5000 en couleur pour la détection de l'Okoumé dans la forêt dense du Gabon, Revue Bois et Forêts des Tropiques.

Cochran, W. G. (1977) *Sampling Techniques*, 3rd edn, Wiley, New York.

Colvocoresses, A. P. (1990) An operational earth mapping and monitoring satellite system: a proposal for Landsat 7. *Photogramm. Engng and Remote Sensing*, **56**, 569–71.

Colwell, R. N. (1952) Photographic interpretation for civilian purposes. In *Manual of Photogrammetry*, American Society of Photogrammetry, Washington DC, 536–45, 553–55, 560–67.

Colwell, R. N. (1956) Determining the prevalence of certain cereal crop diseases by means of aerial photography. *Hilgardia*, **26**, 223–86.

Colwell, R. N. (1961) Some practical applications of multiband spectral reconnaissance. *Am. Scient.*, **49**, 9–36.

Colwell, J. E. (1974) Grass canopy bidirectional spectral reflectance. *Proc. 9th International Symposium on Remote Sensing of Environment, Ann Arbor, Michigan*, 1061–86.

Colwell, R. N. (1979) Remote sensing of natural surfaces – retrospect and prospect. *Proc. Remote Sensing of Natural Resources*, University of Idaho, Moscow, 48–68.

Commission of the European Communities (1983) Diagnosis and classification of new types of damage affecting forests (ed. F. Bauer), DG VI, EEC, Brussels.

Craig, R. D. (1920) An aerial survey of the forests in northern Ontario. *Canadian Forestry Mag.*, **16**, 516–18.

Cromer, D. A. and Brown, A. G. (1954) Plantation inventories with aerial photographs and angle count sampling. *Bull. For. Timb. Bur. Aust.*, **34**.

Curran, R. J. (1980) Satellite measurements of earth radiation budget for climate applications: space shuttle dawn of an era. *Advances in Aeronautical Sciences*, **41**.

Dansereau, P. (1951) Description and recording of vegetation upon a structural basis. *Ecology*, **32**, 172–239.

Dauzet, J., Methy, M. and Salager, J. L. (1984) A method for simulating radiative transfers within canopies, subsequent absorptance and directional reflectance. *Ecol. Plant.*, **5**, 403–13.

Davies, N. (1954) The assessment of Murray River red gum forests by aerial survey methods, NSW Forestry Commission, Sydney.

de Gier, A. and Stellingwerf, D.A. (1988) Two stage versus simple-stratified sampling using Landsat and orthophotos. *Proc. IUFRO Division 4 Meeting* (Forest Resources Inventory and Monitoring/Remote Sensing), Helsinki, Finland.

de Rosayro, R. A. (1958) Tropical ecological studies in Ceylon. *Proc. UNESCO Kandy Symposium*, 33–9.

de Steigeur, J. E. and Giles, R. H. (1981) Introduction to computerized land-information systems. *J. For.*, 734–7.

Devine, H. A. and Field, R. C. (1986) The gist of GIS. *J. For.*, 17–22.

De Wit, C. T. (1965) Photosynthesis of leaf canopies. Report 663, Centre of Agricultural Publications, University of Wageningen.

Diesen, B. C. and Reinke, D. L. (1978) Soviet Meteor satellite imagery. *Bull. Amer. Meteorol. Soc.*, **19**, 804–7.

Dillman, R. D. and White, W. B. (1982) Estimating mountain pine beetle-killed ponderosa pine over the front range of Colorado with high altitude panoramic photography. *Photogramm. Engng and Remote Sensing*, **48**, 741–47.

Dilworth, J. R. (1959) Aerial photo mensuration tables. Res. Note Oregon Agricultural Experiment Station 46, Oregon.

Draeger, W. C. *et al.* (1974) Forest resources analysis in the Macarena region, Columbia using ERTS-1 (UC, Berkeley). Report AGD/RSC, FAO, Rome, Italy.

Driscoll, R. S. and Coleman, M. D. (1974) Color for shrubs. *Photogramm. Engng and Remote Sensing*, 451–59.

Drury, J. D. (1989) The use of SPOT simulation imagery for forest inventory and management. PhD thesis, University of Aston, Birmingham.

ECE (1986) *The ECE Timber Committee Yearbook – XL(1)*, United Nations Economic Commission for Europe, ECE/FAO Agriculture and Timber Division, Geneva.

Elliel, L. (1939) Analysis of error present in planimetric prints. *Photogramm. Engng and Remote Sensing*, **5**, 94–9.

Estes, J. E. and Simonett, D. S. (1975) Fundamentals of image interpretation. In *Manual of Remote Sensing*, American Society of Photogrammetry, Washington DC, 869–1076.

Eule, H. W. (1959) Tree crown measurements and relationships between crown size, stem diameter and growth for beech in north-west German thinning trials. *Allg. Forst.-u.Jagdstg.*, **30**, 185–201.

FAO (1981) *Towards the Year 2000*, Food and Agriculture Organization of the United Nations (FAO), Rome.

FAO (1985) Forest Resources 1980. International Year of the Forest 1985, Food and Agriculture Organization of the United Nations (FAO), Rome.

FAO (1986) (based on work of J. W. van Roessel) Guidelines for forestry information processing. Forestry Paper 74.

Feree, M. J. (1953) Estimating timber volumes aerial photographs. Technical Publication NY State College of Forestry 75, New York State.

Ferns, D. C., Zara, S. J. and Barber, J. (1984) Application of high resolution spectroradiometry of vegetation. *Photogramm. Engng and Remote Sensing*, **50**, 1275–35.

Fischer, W. (1963) An application of radar to geological interpretation. *Proc. 1st Symp. on Remote Sensing of Environment, University of Michigan.*

Fitzpatrick-Lins, K. (1981) Comparison of sampling procedures and data analysis for a land-use map. *Photogramm. Engng and Remote Sensing*, **47**, 343–51.

Flores-Rodas, M. A. (1985) Forestry beyond 2000 – a world perspective. In *Report of the 10th UN/FAO International Training Course on Applications of Remote Sensing to Monitoring Forest Lands*, Food and Agriculture Organization of the United Nations (FAO), Rome, 1–11.

Forstreuter, W. (1988) Inventory of tropical rainforest in the southern provinces of the Republic of Guinea based on Landsat TM data and GIS processing. *Proc. IUFRO Division 4 Meeting, Finland.*

Francis, D. A. (1957) Use of aerial survey methods in Scandinavian forestry. *Emp. For. Rev.*, **38**, 266–76.

Fritchen, L. J., Balick, L. K. and Smith, J. A. (1982) Interpretation of night-time imagery of a forested canopy. *J. Appl. Meteorol.*, **21**, 730–34.

Fritz, N. L. (1967) Optimum methods of using infrared sensitive colour film. *Photogramm. Engng and Remote Sensing*, **33**, 1128–38.

Fuchs and Tanner (1986) Surface temperature measurements of bare soils. *J. Appl. Meteor.*, **7**, 303–5.

Gates, D. M. (1980) *Biophysical Ecology*, Springer-Verlag, New York.

Gates, D. M. and Tantraporn, W. (1952) The reflectivity of deciduous trees and herbaceous plants in the infrared to 25 microns. *Science*, **115**, 613–16.

Gausman, H. W. (1974) Leaf reflectance of near infrared. *Photogramm. Engng and Remote Sensing*, **40**, 183–92.

Gaussen, H. (1948) *Carte de la végétation de la France*, Service de la Carte de la Végétation de la France, Toulouse.

Gering, L. R. and May, M. (1990) The use of aerial photographs and angle gauge sampling of tree crown diameter for forest inventory, in *State of the art methodology of forest inventory* (eds V. J. Bauer and T. Cunia)(Symposuim Proceedings), USDA Forest Service Technical Report PNW GTR 263, Portland, pp. 263–8.

Gregory, S. (1978) *Statistical Methods and the Geographer*, Longmans, London.

Gut, D. and Hohle, J. (1977) High altitude photography: aspects and results. *Photogramm. Engng and Remote Sensing*, **43**, 1245–55.

Gwynne, M. D. (1982) Global environment monitoring system (GEMS) of UNEP, *Environmental Conservation*, **9**, 35–41.

Hannibal, L. W. (1952) Aerial photo-interpretation in Indonesia, Second Asia-Pacific Forestry Commission (FAO), 1–23.

Haralick, R. M. and Fu, King-Sun (1983) Pattern recognition and classification. In *Manual of Remote Sensing* (ed. R. N. Colwell), American Society of Photogrammetry, Washington DC, chapter 18.

Heinsdijk, D. (1957) The upper storey of tropical forest. *Trop. Woods*, **107**, 68–84.

Heinsdijk, D. (1958) The upper storey of tropical forests. *Trop. Woods*, **108**, 31–45.

Hellden, U. and Olsson, K. (1982) The potential of Landsat MSS data for wood resources monitoring. Report 52, Lunds University, Sweden.

Heller, R. C. (technical co-ordinator) (1975) Evaluation of ERTS-1 data for forest and rangeland survey, USDA FS Research Paper PSW-112, Pac. SW. For. and Range Exp. Sta., Berkeley, California.

Heller, R. C. (1978) Case applications of remote sensing for vegetation damage assessments. *Photogramm. Engng and Remote Sensing*, **44**, 1159–66.

Heller, R. C., Aldrich, R. C. and Bailey, W. F. (1963) Identification of tree species on large scale panchromatic and colour photographs. 29th Meeting, American Society of Photogrammetry, Washington DC.

Heller, R. C. and Ulliman, J. J. (1983) Forest resources assessments. In *Manual of Remote Sensing* (ed. R. N. Colwell), American Society of Photogrammetry, Washington DC, chapter 34.

Henniger, J. and Hildebrandt, G. (1980) *Bibliography of publications on damage assessment in forestry and agriculture by remote sensing techniques*, University of Freiburg, FRG.

Hepting, G. N. (1963) Climate and forest disease. *Ann. Rev. Phytopath.*, **1**, 31–50.

Herwitz, R., Peterson, D. L. and Eastman, J. R. (1990) Thematic Mapper detection of changes in leaf area of closed canopy pine plantations in central Massachusetts. *Remote Sens. Environ.*, **29**, 129–40.

Heygi, F. (1986) Remote sensing applied to monitoring forest lands. *Report of the 10th UN/FAO International Training Course on Applications of Remote Sensing*, RSC Series 40, FAO, Rome, 157–66.

Hielkema, J. U. and Howard, J. A. (1976) Application of remote sensing techniques for improving desert locust survey and control. *Final Report AGD(RS) 4/76*, FAO, Rome.

Hielkema, J. U., Tucker, C. J., Howard, J. A. and van Ingen-Schenau, H. A. (1986) The FAO/NASA/NLR Artemis System: An integrated concept for environmental monitoring by satellite in support of food/feed security and desert locust surveillance. *20th International Symposium on Remote Sensing of Environment*, Nairobi, Kenya (ERIM Proceedings, Ann Arbor, Michigan).

Higgins, G. M. and Kassam, A. H. (1981) FAO's agro-ecological approach to determination of land potential. *Pedologie*, **31**, 147–68.

Hildebrandt, G. (1960) Ein Zeitvergleich von Wald-taxationen mit und ohne Luftbildern, *9th International Congress of Photogrammetry*, London.

Hildebrandt, G. and Kenneweg, H. (1969) Information über die waldvegetation aus farbigen Luftbilderen, Erfahrungen und Erwartungen, Bildmessungund Luftbildwesen, 165–70.

Hildebrandt, G. and Kenneweg, H. (1970) The truth about false colour film: a German view. *Photogramm. Rec.*, **6**, 448–51.

Hildebrandt, G. and Kadro, A. (1984) Aspects of countrywide inventory and monitoring of actual forest damage in Germany. *Bildmessung und Luftbildwesen*, **52**, 201–16.

Hirai, M., Watnabe, K., Tsuru, H., Miyaki, M. and Kimura, M. (1975) Development of geostationary meteorological satellite (GMS) of Japan. *Proc. 11th International Symposium on Space Technology and Science, Tokyo*, 461–65.

Hirsch, S. N. (1968) Project fire scan summary of 5 years progress in airborne infrared fire detection. *Proc. 5th Symposium on Remote Sensing of Environment, Ann Arbor, Michigan.*

Hock, J. C. (1984) Monitoring environmental resources through NOAA's orbiting satellites. *ITC Journal*, 26368.

Hoffer, R. M., Flemming, M. D., Bartolucci, S. M. *et al.* (1979) Digital processing of Landsat MSS and topographic data to improve capabilities for computerized mapping of forest cover, Technical report, LARS, Purdue University, Indiana.

Hoffer, R. M. *et al.* (1975) Computer aided analysis of Skylab multispectral data in mountainous terrain for land use, forestry, water resources and geologic applications. LARS Information Note 121275, Purdue University, Indiana.

Hoffer, R. M., Dean, M. E., Knowlton, D. J. and Latty, R. S. (1982) Evaluation of SLAR and Simulated Thematic Mapper Data for Forest Cover Mapping using computer-aided analysis techniques. LARS Report 083182, Purdue University, Indiana.

Holdridge, L. R. (1966) The life zone system. *Adsonia*, **6**, 199–203.

Horne, A. I. D. (1984) Forest cover monitoring by remote sensing in Great Britain. *Proceedings EARSEL/ESA Symposium on integrated approaches in remote sensing*, ESA, Paris, 100–7.

Horne, A. I. and Rothnie, B. (1984) Use of optical SAR 580 data for forest and non-woodland tree survey. *Proc. ESA SAR Final Workshop, Ispra, Italy.*

Howard, J. A. (1958) The land system approach to the survey, mapping and gazetting of the Kigwe-Rubugwa-Uyui Forest Reserve. Report, Forest Department, Morogoro, Tanzania.

Howard, J. A. (1959) The classification of woodland in western Tanganyika by type-mapping from aerial photographs. *Empire Forestry Review*, **38**, 348–64.

Howard, J. A. (1961) Brechfa Forest Working Plan, Part I, UK Forestry Commission, Cardiff.

Howard, J. A. (1965) Small scale photographs and land resources in Nyamweziland, East Africa. *Photogramm. Engng and Remote Sensing*, **32**, 287–93.

Howard, J. A. (1966) Spectral energy relations of isobilateral leaves. *Aust. J. Biol. Sci.*, **19**, 757–66.

Howard, J. A. (1967a) Unpublished data, School of Forestry, University of Melbourne.

Howard, J. A. (1967b) Ecological analysis in photo-interpretation. *Proceedings 2nd Symposium on Photo-Interpretation*, Commission VII, ISP, Paris.

Howard, J. A. (1969) Aerial photointerpretative and reflective properties of eucalypts. PhD thesis, University of Melbourne, Melbourne.

Howard, J. A. (1970a) *Aerial Photo-Ecology*, Faber and Faber, London.

Howard, J. A. (1970b) Multi-band concepts of forested land units. *International Symposium of Photo-Interpretation*, Vol. 1, International Archives of Photogrammetry, Dresden.

Howard, J. A. (1970c) Stereoscopic profiling of land units from aerial photographs. *Australian Geographer*, 11, 259–68.

Howard, J. A. (1970d) Stereoscopic profiling and photogrammetric description of woody vegetation. *Australian Geographer*, 11, 359–72.

Howard, J. A. (1971) Reflective foliaceous properties of tree species. In *Application of Remote Sensors in Forestry* (ed. G. Hildebrandt), Druckhaus Rombach, Freiburg, FRG.

Howard, J. A. (1972) Unpublished data, School of Forestry, University of Melbourne, Australia.

Howard, J. A. (1974) Aerial albedos of natural vegetation in south-eastern Australia. *ERIM Symposium*, Ann Arbor, Michigan, 1301–7.

Howard, J. A. (1975) FAO Statement, UN Committee on the Peaceful Uses of Outer Space, United Nations, New York.

Howard, J. A. (1979) Satellite remote sensing for land-use and forest mapping in developing countries. *ECE Proceedings*, RSC Series, FAO, Rome.

Howard, J. A. (1980) *Aerial albedos of natural surfaces in south-eastern Australia, ERIM Symposium*, Ann Arbor, Michigan, 1301–7.

Howard, J. A. (1983) Capacity and mechanisms of international cooperation: remote sensing applications. RSC Series, FAO, Rome.

Howard, J. A. (1984a) International cooperation in remote sensing applications. *Earth-Orient Applic. Space Techno.*, 4, 205–9.

Howard, J. A. (1984b) International cooperation in remote sensing applications. *Space Technol.*, 4, 295–309.

Howard, J. A. (1990) Remote sensing application in inventory and monitoring: a critique. Proc. Global Natural Resource Assessment and Monitoring. *Amer. Soc. Photogramm. and Remote Sensing*, 592–601.

Howard, J. A. and Kosmer, H. (1967) Monocolour mapping by microfilm. *Photogramm. Engng and Remote Sensing*, 33, 1299–1302.

Howard, J. A., Watson, R. D. and Hessin, T. D. (1971) Spectral reflective properties of *Pinus ponderosa* in relation to the copper content of the soil. *7th Intern. Symp. Remote Sensing of Environment*, ERIM, Michigan.

Howard, J. A. and Barton, I. J. (1973) Instrumentation for mapping solar radiation from light aircraft. *Am. J. Appl. Op.*

Howard, J. A. and de Kock, R. B. (1974) Additive viewing as an interpretative technique for satellite imagery studies. *14th International Technical Scientific Meeting on Space, Rome, Italy*, AGD(RS) 4/74, FAO, Rome.

Howard, J. A. and Gittens, W. J. (1974) Multiband photographic sensing with miniature cameras. *Revue Photointerpretation*, Paris, 71, 8–16.

Howard, J. A. and Schwaar, D. C. (1978) Role and application of high altitude photography in the humid tropics with special reference to Sierra Leone. *Proceedings, International Symposium on Observation and Inventory of Earth Resources and Endangered Environment*, International Society of Photogrammetry (ISP)/ International Union of Forest Research Organizations (IUFRO), Freiburg, FRG, 1403–30.

Howard, J. A., Barrett, E. C. and Hielkema, J. U. (1979) *The Application of Satellite Remote Sensing to Monitoring of Agricultural Disasters*, Pergamon Press, Oxford, 231–40.

Howard, J. A. and Mitchell, C. W. (1980) Phyto-geomorphic classification of the landscape. *Geoforum*, **11**, 85–106.

Howard, J. A. and Schade, J. (1982) *Towards a Standardized Hierarchical Classification of Vegetation for Remote Sensing*. Bulletin 2, Remote Sensing Series, FAO, Rome.

Howard, J. A., Lorenzini, M. and Zandonella, A. (1984) *A procedure for the extraction of vegetation information from NOAA AVHRR data, ERIM Symposium*, Paris.

Howard, J. A. and Travaglia, C. (1985) Short course training of remote sensing users from developing countries. *Proceedings, International Conference on Training for Remote Sensing Users*, GDTA, Toulouse, 193–209.

Howard, J. A., Kalensky, Z. D. and Blasco, F. (1985) Concepts for global mapping of woody vegetation using remote sensing data. RSC Series, **32**, FAO, Rome, Italy.

Howard, J. A. and Mitchell, C. W. (1985) *Phytogeomorphology*, Wiley, New York.

Howard, J. A. and Odenyo, V. A. O. (1986) FAO remote sensing activities relevant to agriculture and food security in Africa. *20th International Symposium on Remote Sensing of Environment, Nairobi, Kenya* (ERIM, Ann Arbor, Michigan).

Howard, J. A. and Reichert, P. (1986) Advances in satellite remote sensing for forestry in developing countries. Invited paper, IUFRO Congress, Yugoslavia.

Howard, J. A. and Lantieri, D. (1987) Vegetation classification, land systems and mapping using SPOT Multispectral Data. *Proc. SPOT 1 First in-Flight Results*, CNES, Toulouse, 137–50.

Hudson, W. D., Amsterburg, R. J. and Meyers, W. L. (1976) Identifying and mapping forest resources from small scale colour infrared air-photos, Research Report 304, Agricultural Experimental Station, Michigan State University.

Husch, B. (1963) *Forest Mensuration and Statistics*, Ronald Press, New York.

Husch, B., Miller, C. I. and Beers, T. W. (1972) *Forest Mensuration*, Ronald Press, New York.

Idso, S. B. and De Witt, C. T. (1970) Light relations in plant canopies. *Appl. Optics*, **9**, 177–84.

Ilvessalo, Y. (1950) On the correlation between crown diameter and the stem of trees. *Comm. Inst. Forestalis Fenn.*, **38**, 5–32.

International Commission on Illumination (ICI) (1957) *International Lighting Vocabulary*, Vol. 1, ICI, Paris, section 20.

Isaev, A. S. (1986) The application of remote sensing for the study of Siberian forest. *Biology International*, 9–11.

ITC (1985) Guideline specifications for vertical black-and-white aerial photography. *ITC Journal*, 53–6.

Itten, K. I., Nanayakkara, S. D., Humbel, R., Bichsel, M. and Sommer, M. (1985) Inventorying and monitoring of Sri Lankan forests using remote sensing techniques. *Proceedings IUFRO Conference, Zurich*, 93–8.

Jaakola, S. and Hagner, O. (1988) Multi-sensor remote sensing for forest monitoring. *Proceedings IUFRO Meeting*, University of Helsinki, 146–58.

Jacobs, M. R. (1932) *Ideas for the use of aerial photography in forestry, especially in underdeveloped countries* (in German). Doctoral thesis, Tharandt School of Forestry, Germany. (Available from the Forestry and Timber Bureau Library, Canberra, Australia.)

Johnson, E. W. (1957) The limit of parallax perception. *Photogramm. Engng and Remote Sensing*, **23**, 933–7.

Johnson, P. L. (ed.) (1969) *Remote Sensing in Ecology*, University of Georgia Press, Athens.

Joyce, A. T. and Sader, S. A. (1986) The use of remotely sensed data for the monitoring of forest change in tropical areas. *20th Int. Symposium on Remote Sensing of Environment, Nairobi, Kenya*, 363–77.

Justice, C. O., Townsend, J. R., Holben, B. N. and Tucker, G. J. (1985) Analysis of the phenology of global vegetation using meteorological satellite data. *Int. J. Remote Sensing*, **6**, 1271–8.

Kadro, A. (1989) Use of Landsat-TM data for forest damage inventory. *Proceedings, Workshop on Earthnet Pilot Project on Landsat TM Applications*, ESA SP-1102, Frascati, Italy.

Kalensky, Z. D. (1974) Personal communication.

Kalensky, Z. D. (1986) Forest mapping based on computer analysis of remote sensing data. *Report of the 10th UN/FAO International Training Course: Applications of Remote Sensing to Monitoring Forest Lands*, RSC Series **40**, 73–80.

Kalensky, Z. D. (1989) Personal communication.

Kalensky, Z. D. and Scherk, L. R. (1975) Accuracy of forest mapping from Landsat computer compatible tapes. *Proc. 19th International Symposium on Remote Sensing of Environment, Ann Arbor, Michigan*, 1159–67.

Kalensky, Z. D. and Nielsen, U. (1978) Resource mapping based on stereo-orthophotomaps, Canadian Forest Service, Ottawa.

Kalensky, Z. D., Kardono, D., Potts, T. F. and Michino, T. (1978) Thematic Map of Lombok Island from Landsat computer compatible tapes. *Proc. 12th International Symposium on Remote Sensing of Environment*, ERIM, Michigan, 1349–65.

Kandler, G. (1985) Die epidemiehypothese als Erklärung des Waldsterbensr. In *Was Wir Uber des Waldsterben*, Verlag, Cologne, 117–25.

Karteris, M. A. (1985) Mapping of forest resources from a Landsat diazo colour composite. *Int. J. Remote Sensing*, **6**, 1792–1811.

Kelsh, H. T. (1940) The slotted templet method for controlling maps made from aerial photographs. *US Dept. Agric. Misc. Publ. 404*.

Kenneweg, H. (1973) Interpretationen von Luftaufnahmen für die Erforschung und Gestaltung von Vegetationsbestanden in Westdeutschen Ballungsräumen. *Proceedings IUFRO Symposium* (Remote Sensing including Aerial Photography), University of Freiburg, FRG, 87–116.

Kharin, N. G. (1975) Spectral reflectance characteristics of the USSR main tree species. Academy of Sciences, Ashkhabad, Turkmen SSR.

Kilmeyer, A. and Epp, H. (1983) Use of small format aerial photography for land use mapping and resources monitoring. *ITC Journal*, 285–7.

Kimes, D. S. (1983) Dynamics of directional reflectance factor distributions for vegetation canopies. *Appl. Opt.*, **22**, 1364–72.

Kittredge, J. (1944) Estimation of the amount of foliage of trees and stands. *J. For.*, **42**, 393–404.

Klein, W. H. (1982) Estimating bark beetle-killed lodgepole pine with high altitude panoramic photography. *Photogramm. Engng and Remote Sensing*, **48**, 733–37.

Kneppeck, I. D. and Ahern, F. J. (1989) A comparison of images from a push-broom scanner with normal colour aerial photographs for detecting scattered recent conifer mortality. *Photogramm. Engng and Remote Sensing*, **55**, 333–7.

Knipling, E. B. (1969) Physical and physiological basis for the reflectance of visible and near infrared radiation from vegetation. *Remote Sens. Environ.*, **1**, 155–59.

Konecny, G. (1985) The photogrammetric camera experiment on Space-lab-1. *International Seminar: Surveying by Satellite for the Developing World*, Trinity College, Dublin, 8–15.

Krinov, E. L. (1947) Spectral reflectance properties of natural formations, USSR Academy of Sciences (translation: G. Belkov, 1953, Natural Resources Council of Canada, T-439).

Kückler, A. W. (1967) *Vegetation Mapping*, Ronald Press, New York.

Kummer, R. H. (1964) A use for small scale photography in forest management. *Photogramm. Engng and Remote Sensing*, **30**, 527–32.

LaBau, V. J. and Winterberger, K. C. (1988) Use of four-phase sampling design in Alaska Multiresource vegetation inventories. *Proc. IUFRO Division 4 Meeting, Finland.*

Ladouceur, G., Trotter, P. and Allard, R. (1982) Zeiss Stereotop modified into an analytical stereoplotter. *Photogram. Engng and Remote Sensing*, **48**, 1577–80.

Langdale-Brown, I. (1967) Personal communication.

Langley, P. G., Aldrich, R. C. and Heller, R. C. (1969) Multistage sampling of forest resources by using space photography. 2nd NASA Earth Resources Aircraft Program Status Review, Houston, Texas, 19-2/-21.

Landgrebe, D. A., Hoffer, R. M. and Goodrick, F. E. (1972) An early analysis of ERTS-1 data. *Proceedings of the NASA Symposium on Preliminary Results of ERTS Data, Greenbelt, Maryland*, 21–38.

Lantieri, D. (1986) Evaluation of SPOT simulation data for forest mapping and monitoring semi-arid areas (Kenya), 10th UN/FAO International Training Course, Rome, 143–50.

Lauer, D. T. and Benson, A. S. (1973) Classification of forest lands with ultra-high altitude, small scale, false colour infrared photography. *Symposium on Remote Sensing in Forestry*, International Union of Forest Research Organizations (IUFRO), Freiburg, FRG.

Lillesand, T. M. and Keifer, R. W. (1979) *Remote Sensing and Image Interpretation*, Wiley, New York.

Loetsch, F. G. (1957) A forest inventory in Thailand. *Unisylva*, **11**, 174–90.

Loetsch, F. and Haller, K. E. (1964) *Forest Inventory*, BLV Verlagsgesellschaft, Munich.

Losee, S. T. (1951) Photographic tone in forest interpretation. *Photogramm. Engng and Remote Sensing*, **17**, 785–99.

Lueder, D. R. (1960) *Aerial Photographic Interpretation*, MacGraw-Hill, New York.

Lyons, E. H. (1966) Fixed airbase 70 mm photography, a new tool for forest sampling. *For. Chron.*, **42**, 420–37.

Maclean, C. D. (1981) Timber volume stratification on small-scale aerial photographs. *J. For.*, 739–40.

McNamara, P. J. (1959) Air photo mapping of Western Australian forests, ANZAAS Conference, Perth, Australia.

MacSiurtain, M. P., Joyce, P. M., Muirgheasa, N. O., Hoey, M. J. and McCormick, N. (1989) *Use of second generation earth observation satellite data in the implementation of the forest management models within the less favoured areas in the Republic of Ireland*, Department of Forestry, University College, Dublin.

Mallingreau, J. P. and Tucker, C. J. (1987) The contribution of AVHRR data to measuring and understanding global processes: large scale deforestation in the Amazon basin. *IGARSS Proceedings*, Ann Arbor, Michigan, 484–9.

Marble, D. F. and Pequet, D. J. (1983) Geographic information systems and remote sensing. In *Manual of Remote Sensing* (ed. R. N. Colwell), American Society of Photogrammetry, Washington DC, 923–58.

Maruyasu, T. and Nishio, M. (1962) Experimental studies on colour aerial photographs in Japan. *Photogrammetria*, **18**, 87–106.

May, J. R. (1960) A stand aerial volume table for *Eucalyptus obliqua*. School of Forestry, University of Melbourne, Australia.

Mead, R. and Meyer, M. (1977) Landsat digital data application to forest vegetation and land use classification in Minnesota, Research Report 77-6, University of Minnesota.

Measuronics Corp (1980) Flying video cameras for data collection. MC, Great Falls, Montana, USA.

Meisner, D. E. (1986) Fundamentals of airborne video remote sensing. *Remote Sensing of Environment*, **19**, 63–79.

Meyer, D. (1961) A reflecting projector you can build. *Photogramm. Engng and Remote Sensing*, **27**, 76–8.

Meyer, M. P. (1977) Personal communication.

Meyer, H. A. and Worley, D. P. (1957) Volume determination from aerial stand volume tables and their accuracy. *J. For.*, **55**, 368–72.

Meyer, M. P., Meisener, D. E., Johnson, W. L. and Lindstrom, O. M. (1983) Small aircraft 35 mm photography and video imagery applications to developing country surveys. *National Conference on Resource Applications, San Francisco, August*.

Mikhailov, V. Y. (1961) The use of colour sensitive films in aerial photography in the USSR. *Photogrammetria*, **17**.

Milne, G. (1936) *A Provisional Soil Map of East Africa*. Amani Memoirs, Amani, Tanzania.

Mitchell, C. W. (1973) *Terrain Evaluation*, Longmans, London.

Mitchell, C. W. and Howard, J. A. (1978) *Final summary report on the application of Landsat imagery to soil degradation mapping at 1 : 5 000 000*. AGL Bulletin 1/78, FAO, Rome.

Moessner, K. E. (1954) A simple test for stereoscopic perception. *Paper 144. Central States Forest Experimental Station, US Forest Service*.

Moessner, K. E. (1963) A test of aerial photo classifications for use in forest management. *US Forest Service Research Paper Intermontain Forest Range Experimental Station*.

Monsi, M. and Saeki, T. (1953) Über den Lichfaktor in den Pflanzengesellschaften un seine Bedentung für die Staffproduktion. *Jap. J. Bot.*, **14**, 22–52.

Moore, C. S. (1968) Photography as a method of measuring apple trees. Report E. Malling Res. Stn, 111–15.

Morain, S. and Simonett, D. (1967) K-band radar in vegetation mapping. *Photogramm. Engng and Remote Sensing*, **33**, 730–38.

Morgan, J. (1978) Introduction to the Meteosat System, MDMD/Met/JM-bd/833, European Space Operations Centre, Darmstadt, FRG.

Murtha, P. A. (1972) A guide to aerial photographic interpretation of forest damage in Canada. Canada Forest Service Publication 1292.

Murtha, P. A. (1983) Some air-photo scale effects on Douglas fir damage type interpretation. *Photogramm. Engng and Remote Sensing*, **49**, 327–35.

Myers, B. J. (1976) Tree species identification on aerial photographs: the state of the art. *Australian Forestry*, **39**, 180–92.

Myers, B. J. (1978) Separation of eucalypt species on the basis of crown colour on large-scale colour aerial photographs. *Aust. For. Res.*, **8**, 139–51.

Myers, N. (1980) *Conservation of Tropical Moist Forests*, National Academy of Sciences, Washington DC.

Myers, W. L., Evans, B. M., Baumer, G. M. and Heivly, D. (1989) Synergism between human interpretation and digital pattern recognition in preparation of thematic maps. *Photogrammetria* (to be published).

Nelson, R. F. (1983) Detecting forest canopy change due to insect activity using Landsat MSS. *Photogramm. Engng and Remote Sensing*, **49**, 1303–14.

Nelson, R. (1989) Regression and ratio estimators to integrate AVHRR and MSS data. *Remote Sens. Environ.*, **39**, 201–16.

Nielsen, U., Aldred, A. H. and Macleod, D. A. (1979) A forest inventory in the Yukon using large scale photo sampling techniques. For. Man. Inst. Inf. Report FMR-X-121, Canadian Forestry Service, Ottawa.

Norwine, J. and Gregor, D. H. (1983) Vegetation classification based on advanced very high resolution radiometer (AVHRR) satellite imagery. *Remote Sensing of Environment*, **13**, 69–87.

Nyyssonen, A. (1955) On the estimate of growing stock from aerial photographs. *Comm. Inst. Forestalis Fenn.*, **46**(1), 57.

Nyyssonen, A., Poso, S. and Keil, E. M. (1966) The use of aerial photographs in the estimation of some forest characteristics. Institute of Forest Mensuration and Management, University of Helsinki.

Oesberg, R. P. (1987) The Meteor satellite system. United Nations/FAO International Remote Sensing Training Course, Rome, pp. 80–89.

Olson, C. E. (1972) Remote sensing of changes in morphology and physiology of trees under stress. Final Report, NASA, Washington DC.

Ovington, J. D. (1957) Dry matter production by *P. sylvestris*. *Ann. Bot.*, **21**, 287–314.

Pacheco dos Santos, A. and de Morais Novo, E. M. (1978) Deforestation planning for cattle grazing in Amazon Basin using Landsat data. Report INPE-1225-PE/126, Instituto de Pesquisas Espacias, Brazil.

Paelinck, P. (1958) Note sur l'estimation du volume des peuplements à limba à l'aide des photos aériennes. *Bull. Agric. Congo Belge*, **49**, 1045–54.

Paivinen, R. and Witt, R. (1988) Application of NOAA/AVHRR data for tropical forest cover mapping in Ghana. *Proc. IUFRO Division 4 Meeting, Helsinki, Finland.*

Petrie, G. (1970) Some considerations regarding mapping the earth from satellites. *Photogramm. Record*, **6**, 590–624.

Poore, M. E. (1963) *Advances in Ecological Research* (ed. J. B. Cragg), Academic Press, London.

Raunkiaer, C. (1934) *The Life Forms of Plants and Statistical Plant Geography*, Oxford University Press, Oxford.

Richards, P. W. (1952) *The Tropical Rainforest*, Cambridge University Press, Cambridge.

Richards, J. A., Woodgate, P. W. and Skidmore, A. K. (1987) An explanation of enhanced radar backscattering from flooded forests. *Int. J. Remote Sensing*, **8**, 1093–100.

Rogers, E. J. (1961) Application of aerial photographs and regression technique for surveying Caspian forests of Iran. *Photogramm. Engng and Remote Sensing*, **29**, 811–16.

Rosenfield, G. H., Fitzpatrick-Lins, K. and Ling, H. S. (1982) Sampling for thematic map accuracy testing. *Photogramm. Engng and Remote Sensing*, **48**, 131–37.

Sabins, F. F. (1978) *Remote Sensing – Principles and Interpretation*, W. H. Freeman, San Francisco.

Sader, S. A. (1986) Analysis of effective radiant temperatures in a Pacific Northwest forest using thermal infrared multispectral scanner data. *Remote Sensing of Env.*, **19**, 105–15.

Sader, S. A. (1987) Forest biomass, canopy structure and species composition relationships with multipolarization L-band synthetic aperture radar data. *Photogramm. Engng and Remote Sensing*, **53**, 193–202.

Sader, S. A. (1988) Remote sensing investigations of forest biomass and change detection in tropical regions. *Proc. IUFRO Division 4 Meeting, Helsinki, Finland.*

Sader, S. A., Waide, R. B., Lawrence, W. T. and Joyce, A. T. (1989) Tropical forest biomass and successional age class relationships to a vegetation index derived from Landsat TM data. *Remote Sens. Environ.*, **28**, 143–56.

Sala, R. M. (1981) *The State of the World Population*, UN Fund for Population Activities, New York.

Salisbury, J. W. (1986) Preliminary measurements of leaf spectral reflectance in the 8–14 μm region. *Int. J. Remote Sensing*, **7**, 1879–86.

Satoo, T., Kunugi, R. and Kumekawa, A. (1956) Amount of leaves and production of wood in aspen second growth in Hokkaido. Bull. For. 32, Tokyo University.

Sayn-Wittgenstein, L. (1960) Recognition of tree species on air photographs by crown characteristics. Technical Note 95, Canadian Forest Research Division, Ottawa.

Sayn-Wittgenstein, L. and Aldred, A. H. (1967) Tree volumes from large scale aerial photographs. *Photogramm. Engng and Remote Sensing*, **33**, 69–73.

Sayn-Wittgenstein, L., de Milde, R. and Inglis, C. J. (1978) Identification of tropical trees on aerial photographs, Canadian Forest Service, Ottawa.

Schade, J. (1985) Assessment of forest cover of Angola based on the visual interpretation of imagery, RSC Series, FAO, Rome.

Schade, J. (1987) Video Remote Sensing – new ways for the Philippine national forest resources inventory, *FRI Report No. 12, Philippine–German Forest Resources Inventory (FRI) Project*, GTZ, Bonn, Germany.

Schimper, A. F. (1903) *Plant Geography on a Physiological Basis*, Clarendon Press, Oxford.

Schmid, P. (1985) Monitoring tropical forest cover in Sri Lanka using digital satellite image analysis. *Proc. IUFRO Conference on Inventorying and Monitoring Endangered Forests*, 99–103.

Schnurr, J. (1988) Forest and land use inventory based on satellite data (Landsat-MSS, Landsat-TM, SPOT) for a subtropical region in Mexico. *Proceedings IUFRO Meeting, Finland*.

Schulte, O. W. (1951) The use of panchromatic, infrared and colour aerial photography in the study of plant distribution, Quebec, Canada. *Photogramm. Engng and Remote Sensing*, **17**, 688–74.

Schwarr, D. (1975) Personal communication.

Schwaar, D. C. (1978) Review and selection of remote sensing imagery for land resources surveys. *Proc. First National Remote Sensing Seminar*, Freetown, Sierra Leone.

Seely, H. E. (1942) Determination of tree heights from shadows in aerial photography. *Photogramm. Engng and Remote Sensing*, **8**, 100–1.

Seely, H. E. (1960) Aerial photogrammetry in forest surveys. *5th World Forestry Conference, Canada*, GP/12/1/B.

Sellers, P. J. (1985) Canopy reflectance, photosynthesis and transpiration. *Int. J. Remote Sensing*, **6**, 1335–72.

Sicco Smith, G. (1971) Applicación de las imagenes de radar en la foto-interpretación de bosques humidos tropicales, Centro Interpretación Aerial Fotografía (CIAF), Bogota.

Sicco Smith, G. (1978) SLAR for forest type classification in a semi-deciduous tropical region. *ITC Journal*, **3**, 385–400.

Singh, K. D. (1990) The conceptual framework of FAO forest resources assessment 1990. Proc. International Conference Global Natural Resource Monitoring and Assessments – Preparing for the 21st Century. *Amer. Soc. Photogramm. and Remote Sensing*, **1495**, 527–33.

Smith, J. W. (1958) Effect of the earth's curvature and reflection on the mensuration of vertical photographs. *Photogramm. Engng and Remote Sensing*, **24**, 751–6.

Smith, J. H. (1965) Biological principles to guide estimation of stand volumes. *Photogramm. Engng and Remote Sensing*, **31**, 87–91.

Smith, J. H., Lew, Y. and Dobie, J. (1960) Intensive assessment of factors influencing photo cruising. *Photogramm. Engng and Remote Sensing*, **26**, 463–9.

Smith, J. A., Lin, T. L. and Ranson, K. J. (1980) The Lambertian assumption and Landsat data. *Photogramm. Engng and Remote Sensing*, **46**, 1183–4.

Snedecor, G. and Cochran, W. G. (1978) *Statistical Methods*, Iowa State University Press, Ames.

Snyder, J. P. (1981) *Space oblique mercator projection: mathematical development*, USGS Bulletin 1518.

Soerkerno, (1976) Personal communication.

Spencer, R. (1974) Supplementary aerial photography with 70 mm and 35 mm cameras. *Aust. For.*, **37**, 115–25.

Spencer, J. D. (1979) Fixed-base large-scale photographs for forest sampling. *Photogrammetria*, **35**, 117–40.

Spencer, R. D. (1985) Large-scale aerial photo comparison for detecting pine dieback. *Aust. For.*, **48**, 102–8.

Spencer, R. D. and Howard, J. A. (1971) Fixed airbase helicopter photography in forest plantation studies. *Rev. Photo-Interpretation*, **71**(4/5), Editions Technip, Paris.

Spencer, R. D. and Hall, R. J. (1988) Canadian large-scale aerial photographic systems (LSP). *Photogramm. Engng and Remote Sensing*, **54**, 475–82.

Spurr, S. H. (1948) *Aerial Photographs in Forestry*, Ronald Press, New York.

Spurr, S. H. (1960) *Photogrammetry and Photo-Interpretation*, Ronald Press, New York.

Stellingwerf, D. (1963) Volume determination on aerial photographs. *ITC Series B*, **18**, Enschede, Netherlands.

Stellingwerf, D. A. (1966) Interpretation of tree species and mixtures on aerial photographs. *Proceedings*, Commission VII, ISP, Paris.

Stellingwerf, D. A. (1973) Application of aerial volume tables and aspects of their construction. *Proc. IUFRO Conference on Remote Sensing*, Freiburg, Germany, 211–49.

Stellingwerf, D. A. and Lwin, S. (1985) The accuracy and relative efficiency of Landsat data and orthophotos for determining area and volume of spruce. *Netherlands J. Agric. Sci.*, **33**, 141–50.

Stone, K. H. (1951) Geographic photo-interpretation. *Photogramm. Engng and Remote Sensing*, **17**, 754–71.

Stone, T. A. and Woodwell, G. M. (1985) Analysis of deforestation in Amazonia using shuttle imaging radar. *Int. Geoscience and Remote Sensing Symposium Digest*, IEEE Cat. no. SCH2162-6.

Strahler, A. N. (1964) Quantitative geomorphology of drainage basins and channel networks. In *Handbook of Applied Hydrology*, McGraw-Hill, New York.

Subbaramaya, I., Bhanukumar, O. and Babu, S. V. (1982) The summer monsoon rains over south-east Asia. WMO report 578, Geneva, Switzerland.

Swellengrebel, E. J. (1959) On the value of large scale aerial photographs in British Guiana. *Emp. For. Rev.*, **38**, 56–64.

Tageeva, S. V. and Brandt, A. B. (1961) Study of optical properties of leaves depending on angle of incidence. In *Progress in Photo-Biology*, Elsevier Press, Amsterdam, 163–69.

Taniguchi, S. (1961) Forest inventory by aerial photographs. *Res. Bull. Hokkaido Exp. For. Stat.*, **21**.

Thallon, K. P. and Horne, A. I. (1984) The use of photogrammetry and remote sensing for forestry in Great Britain. *Photogramm. Record*, **11**, 359–70.

Tiwari, K. P. (1975) Tree species identification on large scale aerial photographs at New Forest. *Indian Forester*, **101**, 791–807.

Todd, W. J., George, A. J. and Bryant, N. A. (1979) Satellite evaluation of population exposure to air pollution. *Environ. Sci. and Tech.*, **13**.

Townshend, J. R. (ed.) (1981) *Terrain Analysis and Remote Sensing*, Allen and Unwin, London.

Townshend, J. R. and Justice, C. (1981) Information extraction from remotely sensed data. *Int. J. Remote Sensing*, **2**, 313–29.

Trenchard, M. C. and Artley, J. A. (1981) An application of pan-evaporation estimates to indicate moisture conditions. Agronomy Abstracts, Annual Meeting American Society of Agronomy, Atlanta, Georgia.

Trevett, J. W. (1986) *Imaging Radar for Resources Surveys*, Chapman and Hall, London.

Tucker, C. J., Townshend, J. R. and Goff, T. E. (1985) African land-cover classification using satellite data. *Science*, 227, 369–75.

UNESCO (1973) *International Classification and Mapping of Vegetation*, UNESCO, Paris.

UNESCO (1976) *Computer Handling of Geographic Data* (ed. R. F. Tomlinson), UNESCO Press, Paris.

US National Research Council (1970) *Remote Sensing with Special Reference to Agriculture and Forestry*, USNRC, Washington DC.

van Roessel, J. W. (1986) Guidelines for forestry information processing. Forestry Paper 74, Forest Department, FAO, Rome.

Viksne, A., Liston, T. C. and Sapp, C. D. (1970) SLAR reconnaissance of Panama. *Photogramm. Engng and Remote Sensing*, 36, 253–59.

Vink, A. P. (1964) Practical soil surveys and their interpretation for agricultural purposes. *ITC Series B*, no. 16.

Vlcek, J. (1983) Videography: some remote sensing applications. American Society of Photogrammetry Annual Convention.

Vogel, F. A. (1986) Applications of remote sensing. National Agricultural Statistics Service, USDA, Washington DC.

Watson, R. D. (1971) Spectral and bidirectional reflectance of pressed vs unpressed fiberfax. *Appl. Opt.*, 10, 1685.

Wear, J. F. and Bongberg, J. W. (1951) Use of aerial photographs in the control of insects. *J. For.*, 49, 623–33.

Weber, F. P. (1965) Aerial volume table for estimating cubic foot losses of white spruce and balsam fir in Minnesota. *J. For.*, 63, 25–9.

Weber, F. P. (1971) The use of airborne spectrometers and multispectral scanners for previsual detection of ponderosa pine trees under stress from insects and disease. In *Monitoring Forest Lands from High Altitudes and from Space*, NASA, Washington DC.

Welander, E. (1952) Estimation of mixture of species, tree height, density and volume with aerial photographs. *Norrlands SkogsvFörb. Tidskr.*, 2, 209–41.

Welch, R., Jordan, T. R. and Ehlers, M. (1985) Comparative evaluations of the geodetic accuracy and cartographic potential of Landsat-4 and Landsat-5 thematic mapper image data. *Photogramm. Engng and Remote Sensing*, 51, 124–62.

Wilson, E. (1920) Use of aircraft in forestry and logging. *Canadian Forestry Mag.*, 16, 439–44.

Wong, C. L. and Blevin, W. R. (1967) Infrared reflectance of plant leaves. *Aust. J. Biol. Sci.*, 20, 501.

Xu, G. H., Tang, S. Z., Zhang, D. H. and Li, Z. Q. (1985) The experiment on the application of Landsat MSS data to forest inventory in Linjiang, Chinese Academy of Sciences, Beijing.

Young, H. E., Call, F. M. and Tryon, T. C. (1963) Multimillion acre forest inventories based on air photos. *Photogram. Engng and Remote Sensing*, 29, 641–44.

Zieger, E. (1928) Ermittlung von Bestandmassen aus Flugbilden mit Hilfe des Hugershoff-Heydeschen Autokartographen. *Versuchsanstalt zu Tharandt*, **3**, 97–127.

Zsilinszky, V. G. and Palabekiroglu, S. (1975) Volume estimates of deciduous forests by large-scale photo sampling. *Proceedings of Workshop on Canadian Forest Inventory Methods*, University of Toronto Press, Toronto, 201–13.

Index

Page numbers in italics refer to Figures and Tables, those in bold refer to the definitions of remote sensing terms